Molecular Aspects of
Gene Expression in Plants

EXPERIMENTAL BOTANY

An International Series of Monographs

CONSULTING EDITOR

J. F. Sutcliffe

School of Biological Sciences, University of Sussex, England

Forthcoming Title

Molecular Aspects of
Gene Expression in Plants

Edited by

J. A. BRYANT

Department of Botany,
University College, Cardiff, Wales

1976
ACADEMIC PRESS
London   New York   San Francisco

A Subsidiary of Harcourt Brace Jovanovich, Publishers

ACADEMIC PRESS INC. (LONDON) LTD.
24/28 Oval Road,
London NW1

United States Edition published by
ACADEMIC PRESS INC.
111 Fifth Avenue
New York, New York 10003

Library of Congress Catalog Card Number: 76 1071
ISBN: 0 12 138150 1

PRINTED IN GREAT BRITAIN BY PAGE BROS (NORWICH) LIMITED

Contributors

C. M. BRAY, *Department of Biochemistry, University of Manchester, Manchester M13 9PL, England.*

J. A. BRYANT, *Department of Botany, University College, P.O. Box 78, Cardiff CF1 1XL, Wales.*

D. GRIERSON, *Department of Physiology and Environmental Studies, Nottingham University School of Agriculture, Sutton Bonington, Loughborough LE12 5RD, England.*

A. J. TREWAVAS, *Department of Botany, University of Edinburgh, The King's Buildings, Mayfield Road, Edinburgh EH9 3JH, Scotland.*

Preface

Research on the biochemistry of plants has tended to lag behind research on bacteria and animals. This seems particularly true of the topic dealt with in this book, namely genes and their function as seen at the molecular level. The reasons why progress has been slower with plants than with other organisms are difficult to define, but two are frequently suggested. First, plants are regarded as more difficult to work with than other organisms, and secondly, in the minds of many biochemists, plants are often thought to be less important and less interesting than, for example, mammals. Whilst there may be some element of truth in the first of these suggestions, the second is certainly without foundation. The ability of plants to carry out photosynthesis is one very important and interesting feature of their metabolism which clearly sets them apart from animals. Another important feature of plants is their plasticity in development, whereby both form and function may be drastically modified by environmental and other factors. This is an extremely interesting characteristic, to which the regulation of gene expression is directly relevant. Further, with a growing awareness throughout the world of the need to produce more food from plants, plant biochemistry must be given more attention. During the next decade we shall undoubtedly see a good deal of effort directed at improving yield and quality of plant proteins. A number of different approaches will be used, including selective breeding programmes, direct attempts to modify the genetic content of plant cells and attempts to regulate gene expression artificially. An understanding of plant biochemistry in general, and of the biochemistry of plant genes and their expression in particular, is clearly of fundamental importance in supporting these efforts.

Since gene structure, function and expression in plants is clearly both interesting and important, it is particularly unfortunate that there is little available literature which gives the topic a good overall coverage. Indeed, the decision to write this book was taken after hearing numerous *cris de coeur*, from both students and teachers of plant science, that they could not find a suitable book to cover courses on gene expression; it has thus been written in order to fill a very obvious gap. It is hoped that the book will be particularly helpful to advanced undergraduate students, post-graduate students and other research workers, and to lecturers (especially those who have to lecture on gene expression, but whose research interests lie elsewhere).

In writing this book I have been fortunate enough to be able to call upon the services of three very able co-authors. All three are actively engaged in research in the fields on which they have written and have been able to make many positive and helpful suggestions about the contents of the book. Dr Donald Grierson in particular gave up many hours of his time to discuss the book with me in the early stages of its preparation.

In addition to my co-authors, there are many other people to whom I wish to express my gratitude. Professor James Sutcliffe, the consulting editor for the series, made many helpful suggestions during the writing of the book. The staff of Academic Press have always been ready to give help and advice when needed. I am indebted to the numerous publishers and authors who have so readily granted permission to reproduce figures and tables from their publications. I am especially grateful to those authors who allowed access to their work prior to publication: Professor Lawrence Bogorad, Professor John Ellis, Dr Jonathan Gressel, Dr Martin Hartley, Dr Christopher Leaver and Dr Peter Payne. Many people have had a hand in typing the various drafts of the manuscript, and although it is perhaps not entirely fair to single out some individuals rather than others, I particularly wish to thank Mary Mathieson, Diana Parsons and Sally Stafford. Above all, it is my privilege and pleasure to thank my wife, Marjorie, who has been a constant source of support, encouragement and help throughout the whole period during which the book was being prepared.

July, 1976   J.A.B.

Contents

1. Nuclear DNA

J. A. BRYANT

I. INTRODUCTION

It is perhaps the most widely known tenet of modern biology that DNA (deoxyribonucleic acid) carries genetic information. Despite the now universal acceptance of the identification of DNA as the chemical basis of heredity, the evidence leading to this conclusion has been obtained only relatively recently. By 1940, it had been well established that inherited characteristics are con-

1

trolled by genes which are arranged in a linear fashion along the chromosome. Chromosomes were known to consist of DNA and proteins, and it was widely assumed that the specificity of individual genes resided in the protein rather than the DNA. The reasons for this assumption were two-fold. Firstly, proteins were known to occur in a very large number of highly specific forms. Secondly, the structure of DNA seemed too simple to be capable of much variety, since it had been shown to consist only of four deoxyribonucleoside monophosphates (deoxyadenosine, deoxycytidine, deoxyguanosine and deoxythymidine monophosphates) each consisting of a heterocyclic, nitrogenous base, a deoxyribose residue and phosphoric acid (Fig. 1.1). The deoxyribonucleoside monophos-

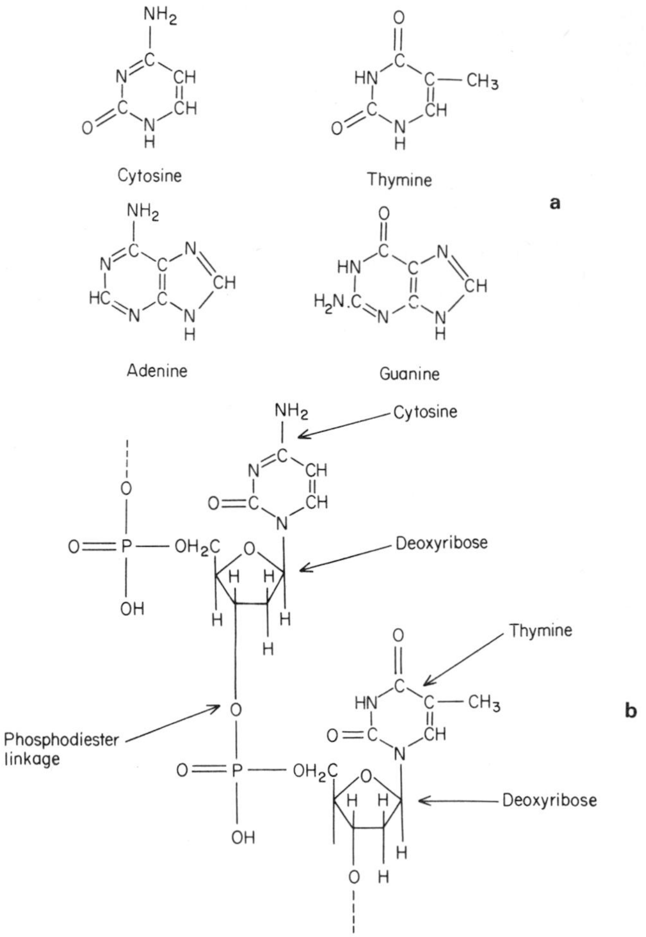

FIG. 1.1. (a) Structures of the four major bases of DNA. (b) Phosphodiester linkage between two deoxyribonucleoside monophosphate residues.

phates were known to be linked together by phosphodiester linkages between carbon 3′ of one deoxyribose and carbon 5′ of the next, and it was thought that DNA was a simple repeating structure, the basic unit of which was a tetra-nucleotide. ·

In the early 1940s the first piece of evidence which pointed to DNA as being the genetic material was obtained. Avery, MacLeod and McCarty showed that cells of a non-virulent, non-encapsulated strain of *Diplococcus pneumoniae* are transformed to give virulent, encapsulated cells following the uptake of purified DNA extracted from virulent, encapsulated cells. Further, the newly acquired characteristics are inherited from generation to generation [14]. Transforma-tion of bacterial cells by DNA from cells of a different strain of the same species has since been demonstrated for a great variety of different characters, includ-ing the ability of cells to utilize particular nutrients, to manufacture particular enzymes and successfully to invade particular hosts. The widely divergent nature of the characters which can be transferred by DNA, and the inheritance of the characters in daughter cells arising from transformed cells by cell division, support the hypothesis put forward by Avery and his associates, namely that DNA carries the information necessary for specifying inherited characteristics.

The identification of DNA as the genetic material in bacteria led to the idea that DNA is also the genetic material in higher organisms (i.e. those organisms possessing complex chromosomes). However, the evidence that specificity of genes in higher organisms resides in DNA was, and indeed still is, largely circumstantial, although nonetheless convincing. Firstly, it was noted that DNA occurs in all organisms that have been examined (with the exception of certain viruses which contain RNA but no DNA). This suggests that DNA has a basic and universal role in all living cells; further, since the role of DNA in bacterial cells is informational, it is extremely likely that this is also its role in cells of higher organisms. Secondly, it was established that in a given species, the amount of DNA per somatic cell nucleus is the same for all types of cell. Further, the DNA content of gamete nuclei is exactly half that of somatic cell nuclei [314, 315, 496]. The DNA content of a nucleus is therefore directly related to the chromosome content of the nucleus (diploid or haploid). Thirdly, Chargaff established that the overall base composition of DNA is constant for all cells of a given species, but varies from species to species [78]. This suggests that the base composition is conserved from one cell generation to the next. These features concerning the DNA of higher organisms were taken as strong evidence that the DNA of such organisms, like that of bacteria, functions as the genetic material. This view clearly is fundamentally different from the view that proteins are responsible for the specificity of genes, and it is therefore remark-able that the "DNA hypothesis" was so widely accepted by 1950, only 6 years after the publication of the work of Avery and his colleagues.

The acceptance of the hypothesis that DNA is the genetic material led to a great deal of interest being generated in the structure of DNA. Any hypothesis concerning the structure of DNA needed to accommodate two interesting features. Firstly, Chargaff made the important observation that in any sample of DNA, the molar percentage of the purine base, adenine (A) is equal to the molar percentage of the purimidine base, thymine (T), and that the molar percentage of the purine base, guanine (G), is equal to the molar percentage of the pyrimidine base, cytosine (C) [78]. This finding was slightly modified by Wyatt, who demonstrated that in higher organisms, and particularly in plants, some of the cytosine (up to 25% in wheat, *Triticum*) is in fact 5-methyl-cytosine (5-MeC) [530]. The data of Chargaff and Wyatt may be summed up as

$$(A) = (T), (G) = (C) + (5\text{-Me-C}).$$

This summary is generally known as Chargaff's rule.

Secondly, the X-ray crystallographic studies carried out by Wilkins, Franklin and their colleagues [141, 516] had shown that the DNA molecule is a fibrous structure 2·0 nm in diameter with a regularly repeated unit every 3·4 nm along the long axis of the fibre. In addition to the periodicity of 3·4 nm, there is also a periodicity of 0·34 nm. The most straightforward interpretation of these data is that the DNA molecule is a helix of pitch 3·4 nm, with the nitrogenous bases set perpendicular to the long axis of the molecule, and spaced at intervals of 0·34 nm. Further, the intensity of the X-ray diffraction pattern suggested that the helical structure is a double helix rather than a single helix.

The model for DNA structure which best fits the known physico-chemical features of DNA is that proposed by Watson and Crick in 1953 [503]. Watson and Crick were fortunate to have access to the data from Wilkins' laboratory prior to publication, and by a process of trial and error, intuitive insight and model-building, eventually arrived at a hypothesis for the structure of DNA which is consistent both with the X-ray diffraction data and with the known stereochemical features of the deoxyribonucleoside monophosphates. The model proposed by Watson and Crick embodies the following postulates:

(i) that the DNA molecule contains two polynucleotide chains, each chain being a right-handed helix, with ten base pairs to each turn of the helix;

(ii) that the two chains are interlocked and twisted about the same axis (i.e. are coiled plectonemically);

(iii) that the phosphate groups are on the outside of the helix, thus forming a double sugar-phosphate backbone, with the bases on the inside of the helix, set with their planes at 90° to the long axis of the helix;

(iv) that the two polynucleotide chains are antiparallel (i.e. run in opposite "chemical directions": see Fig. 1.17);

(v) that the two chains are held together by hydrogen bonding, and further

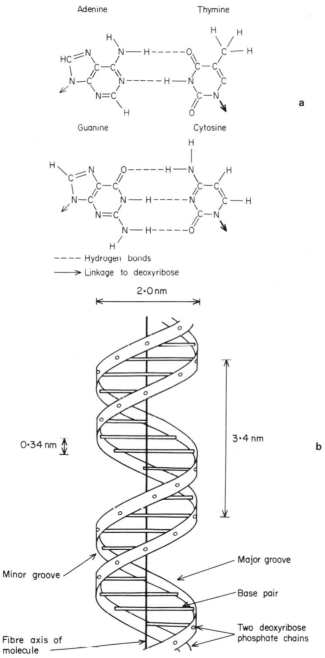

FIG. 1.2. (a) Structures of base pairs. (b) Diagrammatic representation of the double helical structure of DNA.

that the hydrogen bonding is base-specific, only occurring between adenine and thymine, or between guanine and cytosine. Postulate (v) is clearly consistent with Chargaff's rule.

There have been some slight modifications to the model since 1953, but the main features have been confirmed by subsequent experimentation. The structures of the base pairs and of the double helix are shown diagrammatically in Fig. 1.2.

On the basis of their model, Watson and Crick made two suggestions concerning the biological function of DNA [504]. Firstly, the molecule involves specific base pairing. A given base sequence in one polynucleotide chain automatically governs the base sequence in the other. Thus, DNA replication could occur by separation of the polynucleotide chains, followed by use of the separated chains as templates for the synthesis of two new complementary chains, giving rise to two double helices in place of one. Secondly, the model places no restriction on the order of bases in the polynucleotide chain, and it was therefore proposed that the precise sequence of bases is the code specifying the genetic information. In 1953, there was no experimental evidence to support either of these suggestions. However, both suggestions have been widely confirmed by subsequent experimental work, such that the terms "semi-conservative replication" and "triplet code of bases" have become part of the dogma of molecular biology.

Despite the very rapid advances in our knowledge of the structure and function of DNA since 1944, much detail remains to be elucidated. This is particularly true of eukaryotic organisms in general, and even more particularly true of plants. The remainder of this chapter is a consideration of what is known about the structure, function and metabolism of nuclear DNA in plants. The DNAs of chloroplasts and mitochondria, discovered in the early 1960s, are dealt with in Chapter 4.

II. PROPERTIES OF NUCLEAR DNA IN PLANTS

A. Extraction

In order to characterize DNA, it is necessary to extract it from the plant and purify it. Procedures for preparing purified native DNA from plant cells generally employ the following techniques:

(i) The cells are broken open by means suitable for the material under investigation. For higher plants, this usually means homogenizing the material in a weak salt solution (sometimes buffered) containing a surface-active detergent such as sodium lauryl sulphate (also known as sodium dodecylsulphate) or tri-*iso*-propylnaphthalene sulphonic acid.

(ii) The homogenate is lysed with detergent and then deproteinized by denaturing the protein with phenol or chloroform.

(iii) Nucleic acids are precipitated from the deproteinized lysate by the addition of ethanol.

(iv) The nucleic acids are dissolved in a buffered salt solution, treated with ribonuclease and then deproteinized again.

(v) DNA is precipitated with ethanol.

The precipitated DNA may be further purified by repeated cycles of dissolving in buffer followed by reprecipitation with ethanol or by dissolving in buffer and then dialysing extensively. The DNA thus extracted retains many, if not all, of its chemical properties and some of its biological activity. In one respect, however, the DNA is not in its native state. A DNA molecule is very long in relation to its width and is also somewhat brittle because of the restraints imposed by hydrogen bonding and by base-stacking forces (electrostatic attractions between adjacent bases in a polynucleotide chain). The molecule is therefore vulnerable to hydrodynamic shear, and breaks easily during the extraction procedure. Extracted DNA therefore has a lower molecular weight than native DNA. DNA extracted by the type of procedure outlined above generally has a molecular weight of around 1×10^7 daltons. The molecular weight of native nuclear DNA in the plant cell nucleus may be of the order of 1×10^{12} daltons (see section IIIA).

For most purposes it is not necessary to separate nuclear DNA from the DNA of mitochondria and plastids. The DNA in the mitochondria makes up 1% or less of the total DNA in a plant cell, and, if non-green cells are used for extraction, the plastid DNA contributes between 1% and 5% of the total (see Chapter 4). These levels of contamination are regarded as acceptable in many types of experiment. However, if nuclear DNA is required for detailed analytical studies, it is extracted from either whole nuclei or from chromatin (see section IIIB) previously isolated from the plant tissue.

B. Base composition

For determination of overall base composition of DNA, the DNA must first be hydrolysed. Treatment with 98% formic acid for 30 min at 175°C or with 12 N perchloric acid for 1 h at 100°C are two commonly employed procedures for hydrolysis of DNA. Both procedures break down DNA to its constituent bases, although treatment with perchloric acid causes extensive demethylation of 5-methyl-cytosine [26]. The bases may be separated by paper chromatography, eluted from the chromatogram and estimated spectrophotometrically [26]. Base compositions of some higher plant and algal DNAs are given in Table 1.I. Several features are apparent from Table 1.I. Firstly, the DNA of plants obeys

Chargaff's rule (section I). Secondly, in higher plants a significant proportion of the cytosine is methylated, as first reported by Wyatt for *Triticum* DNA. Thirdly, the G–C content of higher plant DNA varies from 34 to 46% in different species, although most higher plant DNAs fall in the range 40–46% G–C. Algae show a greater interspecific variation in the G–C content of their DNA.

TABLE 1.I. Base compositions of plant DNAs

Plant	Adenine	Thymine	Guanine	Cytosine	5-Me-Cytosine
			Mol %		
Daucus carota	26·7	26·8	23·2	17·3	6·0
Cucurbita pepo	30·2	29·0	21·0	16·1	3·7
Phaseolus vulgaris	29·7	29·6	20·6	14·9	5·2
Gossypium hirsutum	32·8	32·9	16·9	12·7	4·6
Triticum sp.	27·3	27·1	22·7	16·8	6·0
Zea mays	26·8	27·2	22·8	17·0	6·2
Allium cepa	31·8	31·3	18·4	12·8	5·4
Scenedesmus accuminatus	18·7	17·5	32·9	30·9	
Rhabdonema adriaticum	31·4	31·7	18·6	18·3	

Because of the large amount of DNA in a plant cell nucleus (section III), it is impossible to determine the precise base sequence of the DNA. Even if the base sequence could be analysed at the improbable rate of one base pair per second, it would take between 30 and 3000 years (depending on the plant species involved: see Table 1.IV) to complete the analysis! However, it is possible to determine the frequency with which the 16 possible dinucleotide sequences occur in the DNA, using a technique known as nearest neighbour frequency analysis. In this procedure, DNA is extracted and purified, and then used as a template for the synthesis of DNA *in vitro* by exogenous DNA polymerase. The incubation system contains the four deoxyribonucleoside 5′-triphosphates, one of which is radioactively labelled with ^{32}P. The deoxyribonucleoside phosphates are incorporated into DNA as monophosphates, a pyrophosphate group being split off for each phosphodiester linkage formed. After synthesis, the labelled DNA is hydrolysed with deoxyribonuclease II and phosphodiesterase to yield deoxyribonucleoside 3′-monophosphates. The ^{32}P-phosphate is therefore now attached to the deoxyribonucleoside adjacent to the one to which it was originally linked (Fig. 1.3). The deoxyribonucleoside 3,-monophosphates are separated by chromatography. Estimation of the radioactivity in each of the four deoxyribonucleoside 3′-monophosphates gives the frequency with which each occurs next in the DNA molecule to the deoxyribonucleoside phosphate originally supplied in a radioactive form (Fig. 1.3). The whole procedure is repeated a further three times, so that the nearest-neighbour frequencies for all four deoxyribonucleoside phosphates can be worked out. This procedure has been carried out with DNA from a number of higher plants and algae. In each case, the frequencies of the 16 possible dinucleotides (nearest-neighbour pairs) depart significantly from random (Table 1.II). This is, of course, consistent with the view that the specific sequence of bases carries the genetic information.

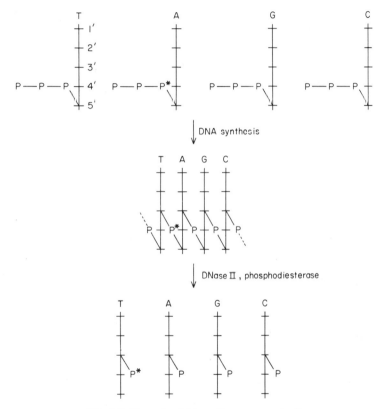

FIG. 1.3. Nearest-neighbour frequency analysis.

Nearest-neighbour frequency analysis can also be used to demonstrate that the two strands of the DNA molecule are anti-parallel. This can best be illustrated by consideration of a dinucleotide sequence in the DNA molecule, for example G5′p3′A (usually written GpA). In this nomenclature, the "p" represents the phosphodiester linkage between adjacent deoxyribonucleotides. In the shortened form (e.g. GpA), the deoxyribonucleoside which is attached to

J. A. Bryant

TABLE 1.II. Nearest neighbour frequency analysis of DNA from *Chlamydomonas reinhardi.*[a]

Dinucleotide sequence	Observed frequency (%)	"Expected" frequency (%)[b]
ApA	6·0	(5·3)
ApC	6·0	(6·2)
ApG	6·0	(6·2)
ApT	5·4	(5·3)
CpA	7·7	(6·2)
CpC	7·4	(7·2)
CpG	6·3	(7·2)
CpT	5·7	(6·2)
GpA	4·4	(6·2)
GpC	9·2	(7·2)
GpG	7·1	(7·2)
GpT	5·5	(6·2)
TpA	5·3	(5·3)
TpC	4·6	(6·2)
TpG	7·3	(6·2)
TpT	5·9	(5·3)

[a] From data presented in ref. 461.
[b] Expected frequency if nucleotide sequence is random (within the constraints imposed by the overall base composition).

the phosphate group via the 5′ carbon of the deoxyribose is always written first. If the two strands of the DNA molecule are antiparallel then the complementary sequence to G5′p3′A (GpA) will be C3′p5′T (TpC), whereas if the two strands are parallel, the complementary sequence will be C5′p3′T (CpT). Reference to Table 1.II shows that, within the limits of experimental error, the frequencies of GpA and TpC are the same, whereas CpT occurs with the same frequency as ApG. These results can only be interpreted in terms of the two strands being anti-parallel.

C. Thermal denaturation

The hydrogen bonds which hold the two strands of a DNA double helix together, and the electrostatic attractions ("base-stacking" forces) which maintain the position of the bases in each strand are heat labile. Heating a solution of

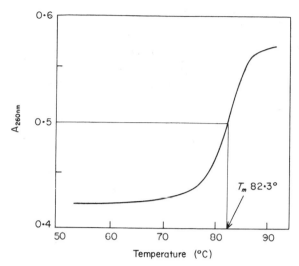

FIG 1.4. Thermal denaturation profile of DNA from *Pisum sativum* (garden pea). Denaturation was carried out in 0·1M NaCl (from ref. 62).

DNA therefore results in separation of the two strands in the double helix. The regular structure of a native double helix, governed by hydrogen bonding and base-stacking forces, imposes some limitations on the resonance behaviour of the bases. Separation of the two strands releases the bases from these limitations, resulting in an increased resonance, and hence an increased absorbance in the ultraviolet region of the spectrum. This is known as hyperchromicity. Thermal denaturation can therefore be detected by monitoring the DNA at its absorbance peak (260 nm) in a spectrophotometer. Figure 1.4 shows the thermal denaturation profile (melting curve) of nuclear DNA from *Pisum sativum* (garden pea). Two features are clear from this profile. Firstly, the transition from double-stranded to single-stranded DNA is very sharp, and takes place over a small temperature range. Secondly, the absorbance at 260 nm after denaturation is 1·35 times as high as the absorbance at 260 nm of the native DNA (i.e. hyperchromicity = 35%). The temperature corresponding to half the maximum hyperchromicity is known as the melting temperature or T_m. The melting temperature of DNA is affected by the strength of the counterion (usually sodium) present in the DNA solution. High sodium ion concentration confers stability on the double helix; increasing the concentration of sodium ions in the DNA solution therefore causes an increase in the value of T_m. The T_m of DNA is also affected by the G–C content of the DNA. The G–C base pair is more stable than the A–T base pair since it is held together by three hydrogen bonds rather than two (Fig. 1.2). Further, intra-chain electrostatic interactions are stronger in G–C-rich molecules than in A–T-rich molecules. G–C-rich

DNA therefore has a higher T_m than A–T-rich DNA. The relationship between G–C content and T_m can be used to calculate the G–C content of a DNA sample from a T_m value, using the equation

$$G\text{–}C = 2 \cdot 44 \, (T_m - 81 \cdot 5 - (16 \cdot 6 \log M))$$

where M is the concentration of the counterion. For example, the G–C content of *Triticum* (wheat) DNA calculated from T_m is 46% [204]. This agrees very well with results derived from chemical determinations of base composition (46–47% G–C).

The relationship between G–C content and T_m means that if a given molecule of DNA contains regions within it which differ markedly in G–C content from the average G–C content of the whole molecule, then those regions will have a markedly different T_m from the rest of the molecule. This results in heterogeneous denaturation behaviour. Such heterogeneity is rarely visible in the denaturation profile, unless extremely sensitive apparatus is used for recording the profile; the heterogeneity can be detected, however, by plotting the "derived" melting curve, where the ratio of hyperchromicity over a small rise in temperature (0·5°C) to total hyperchromicity is plotted against temperature. Figure 1.5 shows the "derived" melting curve for *Pisum sativum* DNA. It is very obviously heterogeneous, containing regions which are denatured at temperatures below the T_m and regions which are denatured above the T_m. This feature substantiates the implication arising from nearest-neighbour frequency analysis, namely that the base sequence of a DNA molecule is ordered and non-random.

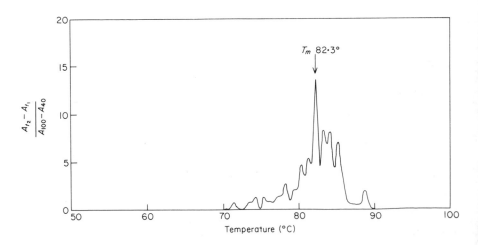

FIG. 1.5. Derived melting curve of DNA from *Pisum sativum* (garden pea). Denaturation was carried out in 0·1M NaCl (from ref. 61).

D. Buoyant density

When dense solutions of caesium chloride are centrifuged at high speed $(25\,000$ rev min^{-1} or higher) for long periods (at least 48 h) a linear concentration gradient of caesium chloride is formed in the centrifuge tube. If DNA is also present in the caesium chloride solution, the DNA takes up a position in the gradient which corresponds to its buoyant density. At this position, the centrifugal force acting on the DNA molecule is counterbalanced by the density of the caesium chloride, and the system is therefore in equilibrium. The buoyant density of DNA in caesium chloride is strongly influenced by the base composition of the DNA. DNA molecules which are G–C-rich have a greater buoyant density than A–T-rich DNA molecules. G–C-rich DNA therefore forms a buoyant band nearer the bottom of the centrifuge tube than A–T-rich DNA. For DNAs which contain no 5-methyl-cytosine, the relationship between buoyant density and G–C content is given by the equation

$$\text{Buoyant density} = 1{\cdot}660 + 0{\cdot}098\ (\text{G–C})\ \text{g.cm}^{-3},$$

where (G–C) is the molar fraction of G + C. Accurate determination of buoyant density in an analytical ultracentrifuge therefore gives a method for determination of overall base composition. The DNA under investigation is dissolved in caesium chloride, together with "marker" DNA species of known buoyant densities. A small volume of the solution $(0{\cdot}3–0{\cdot}4$ ml) is placed in the appropriate cell of the ultracentrifuge and centrifuged to equilibrium. The ultracentrifuge cells have quartz glass sides through which their contents may be photographed while the centrifuge is running. When equilibrium is reached, the cell contents are photographed in ultraviolet light. The photograph is scanned with a micro-densitometer, giving a trace with peaks corresponding to the bands of DNA in the ultracentrifuge cell (Fig. 1.6). The buoyant density of the DNA under investigation can then be calculated relative to the marker DNAs. It is also possible to calculate the buoyant density of DNA from observations of its sedimentation and equilibrium behaviour in the ultracentrifuge by more fundamental mathematical methods, without using marker DNAs.

DNAs which contain 5-methyl-cytosine exhibit a departure from the linear relationship between G–C content and buoyant density [245]. The presence of 5-methyl-cytosine leads to a decrease in buoyant density, the magnitude of the effect being proportional to the amount of 5-methyl-cytosine in the DNA. Since up to 25% of the cytosine residues in plant DNA may be methylated, use of buoyant density determinations to estimate G–C content of plant DNAs is subject to error (Table 1.III).

As well as giving an estimate of the buoyant density of DNA, ultracentrifugation of nuclear DNA from eukaryotic organisms, including plants, has

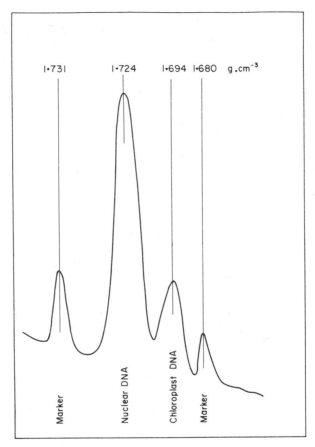

FIG. 1.6. Microdensitometer trace of u.v. photograph obtained after analytical caesium chloride density gradient centrifugation of DNA from cells of *Chlamydomonas*, together with two marker DNAs (from ref. 397).

TABLE 1.III. Relationship between buoyant density and G–C content (i.e. mol % (G + C + 5Me–C)) for plant DNAs

Species	Buoyant density of DNA g cm⁻³	G–C content calculated from buoyant density	Actual G–C content
Phaseolus vulgaris	1·697	38	41
Triticum sp.	1·702	43	46
Zea mays	1·701	42	46
Allium cepa	1·691	32	37

shown the presence of satellite DNAs in a wide variety of species. A satellite DNA is a DNA which is different enough from the bulk of the DNA ("main band DNA") in base composition to take up a different equilibrium position from the main band DNA. The satellite is revealed in the micro-densitometer trace either as a "shoulder" of the main peak or as a completely separate peak (Fig. 1.7). Satellite DNAs have been detected in many plants, covering a very wide taxonomic range [209]. Most plant satellite DNAs are denser than the corresponding main band DNA, implying that satellite DNA has a higher G–C content than main band DNA. The best documented exception to this is the satellite DNA from *Linum* (flax) which is less dense than *Linum* main band DNA (Fig. 1.7).

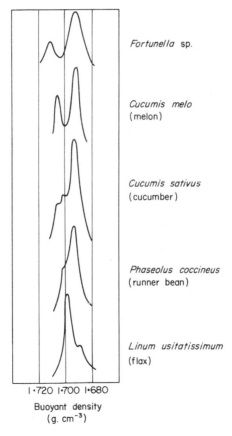

Fig. 1.7. Microdensitometer traces of u.v. photographs obtained after analytical caesium chloride density gradient centrifugation of DNA from a number of higher plants. The traces clearly show the presence of satellite DNA (from ref. 209).

J. A. Bryant

The origin of some plant satellite DNAs has been questioned. It has been shown that certain of the satellites detected in DNA extracted from *Raphanus sativus* (radish), *Vicia faba* (broad bean) and from various members of the Cucurbitaceae are not of plant origin but are in fact bacterial DNA species [354, 400]. Nevertheless, it is clear that the presence of native satellite DNA is a genuine feature of DNA extracted from many plants. Ingle and his colleagues have carried out an extensive study of the satellite DNAs in a wide range of higher plants [209]. Satellite DNAs are particularly common in the Cucurbitaceae, some species possessing more than one satellite DNA.

By use of the preparative ultracentrifuge, it is possible to centrifuge large quantities of DNA (several hundred micrograms) in gradients of caesium chloride. After centrifugation, the gradients are fractionated. A number of methods are available for gradient fractionation; one of the most widely used is illustrated diagrammatically in Fig. 1.8. A fine capillary tube, connected to a length of silicon rubber tubing, is inserted to the bottom of the ultracentrifuge tube. Flow is induced in the tubing by a peristaltic pump, thereby sucking out the contents of the ultracentrifuge tube, starting with the bottom of the gradient and finishing with the top. As the contents of the ultracentrifuge tube pass out of the silicon rubber tubing, an equal number of drops is counted into each of up to 100 small test tubes (Fig. 1.8). The fractions thus obtained are diluted with a small volume of buffered, dilute sodium chloride solution, and assayed for absorbance at 260 nm. The fractions containing DNA may then be retained for further analysis. Using this preparative procedure, good yields of satellite DNA have been obtained from a number of plants.

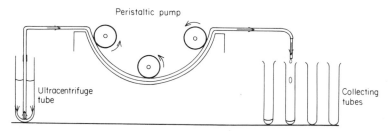

FIG. 1.8. Diagram of a system used for fractionation of caesium chloride density gradients formed in a preparative ultracentrifuge.

The satellite DNAs from plants investigated so far prove to be of two types. Firstly, the satellite DNA may correspond to the very simple, highly reiterated DNA sequences detected by renaturation kinetics (section IIE, below). Secondly, most plant satellite DNAs, although apparently consisting of a single component, as deduced from their behaviour in gradients of caesium chloride, are in fact heterogeneous. For example, the satellite DNA of *Cucumis melo* (melon) contains two components as detected by heterogeneity in the melting curve and in the renaturation kinetics of the satellite [27, 428]. One of the components is a simple, highly reiterated DNA; the other is a much more complex DNA which is less highly reiterated. The difference between the melting temperatures of the two components of melon satellite DNA suggests that the two have different base compositions. It is therefore surprising that they form a single satellite band in caesium chloride gradients. The probable reason for this is that the two types of DNA are covalently linked, perhaps in some type of alternating arrangement, where long sequences of the less complex DNA are interspersed with shorter sequences of the more complex DNA [428].

E. Renaturation of denatured DNA

DNA which has been denatured by heat, renatures again when the temperature is lowered slowly. Two types of renaturation are recognized. Type I renaturation is a rapid reversal of thermal denaturation. It is caused by renaturation of parts of the DNA molecule which are incompletely denatured by heat treatment and which therefore remain "in register". Since the G–C base pair is more stable than the A–T base pair, it is believed that Type I renaturation (or "snap-back") is caused by G–C-rich regions ("nucleation sites") in the DNA. Further, if the G–C-rich regions occur with any degree of frequency, then the nucleation sites can bring about Type I renaturation of a large proportion of the DNA, as has been observed for the DNA of *Zea mays* (maize) (Fig. 1.9). Type II renaturation is a renaturation of strands which are completely separated by denaturation. This type of renaturation, generally referred to as "reannealing" or "reassociation", is dependent on collisions between complementary strands. The rate of reannealing therefore depends on the concentrations of the two strands of DNA, and the reaction thus exhibits second order kinetics.

In addition to being denatured by heating, DNA is also denatured by treatment with alkali, since an increase in pH alters the charge distribution in the bases and hence disrupts the hydrogen bonding and electrostatic interactions. Because G–C-rich and A–T-rich regions of the DNA molecule are equally vulnerable to denaturation by alkali, alkali-denatured DNA does not exhibit Type I renaturation (snap-back). Therefore, if alkali-denatured DNA is neutralized and placed in a spectro-photometer cell, the reassociation of completely separated strands may be followed by monitoring the decrease in absorbance at 260 nm. It has already been mentioned that the process of reannealing follows second-order kinetics, being dependent on the concentration of DNA. This means that at a given DNA concentration, the rate of reannealing is inversely related to the overall size (complexity) of the genome represented by the DNA. (The larger the genome, the fewer the copies repre-

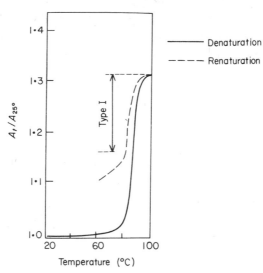

FIG. 1.9. Type I renaturation ("snap-back") of DNA from *Zea mays* (maize). Experiments were carried out in the presence of 0·17M Na$^+$ (from ref. 58).

sented by a given amount of DNA.) This is illustrated in Fig. 1.10, which shows the reannealing kinetics of DNAs from several sources where the total genome size (i.e. amount of DNA per cell or per virus particle) is known. The genomes of eukaryotic cells in general, and of higher plants and animals in particular, are

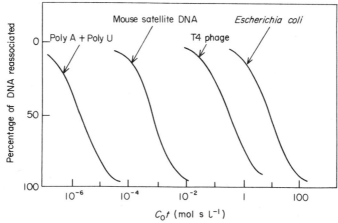

FIG. 1.10. Reassociation kinetics of various types of DNA. The term "Cot" is the concentration of DNA (in moles of nucleotide per litre) multiplied by time (in seconds) elapsed from the start of reassociation (from ref. 56, copyright 1968 by the American Association for the Advancement of Science).

very large and complex, containing large amounts of DNA. It would therefore be expected that DNA from higher organisms should reanneal very slowly. However, Britten and his colleagues made the surprising discovery that, provided the DNA is sheared to give small fragments (of the order of $1-10 \times 10^5$ daltons), DNA from higher organisms, including plants, exhibits multiphasic reannealing kinetics [56]. The slowest rate of reannealing is consistent with the large size of the eukaryotic genome. However, a large proportion of eukaryotic DNA (up to 30% in some plants and animals) reanneals at least as rapidly as, and in some instances much more rapidly than, bacteriophage DNA. In addition to the fast and slow phases of reannealing exhibited by eukaryotic DNA, there is also a less well defined intermediate phase, involving up to 50% of the total DNA.

In many organisms, the DNA which exhibits rapid reannealing is different enough in buoyant density from the remainder of the DNA to form a satellite band when the DNA is centrifuged in caesium chloride (section IID, above). With such organisms, it is relatively easy to purify the rapidly-reannealing-DNA in order to study its reannealing kinetics in detail. However, a large number of animals and plants do not possess satellite DNA. Nevertheless, they do possess rapidly-reannealing-DNA [28, 56]. Isolation of rapidly-reannealing-DNA from such organisms is achieved by use of columns of hydroxyapatite [56]. Hydroxyapatite selectively binds double-stranded DNA. Passage of denatured, neutralized DNA through a hydroxyapatite column after allowing reannealing to proceed for a short time therefore results in the binding of only the rapidly-reannealing fraction. The temperature of the eluting buffer is then increased. This causes the bound DNA to become thermally denatured, leading to its elution from the column as single-stranded DNA.

The rapidly-reannealing-DNAs which have received most attention are the satellite DNA of mouse (*Mus musculus*), a light A–T-rich satellite and the satellite of guinea-pig (*Cavia porcellus*), a heavy, G–C-rich satellite. The difference in buoyant density between these satellite DNAs and their respective main-band DNAs has facilitated isolation of the satellites in relatively large quantities. The characterization of mouse and guinea-pig satellite DNAs is relevant to our understanding of all rapidly-reannealing-DNAs, and will therefore be described in some detail.

Mouse satellite DNA reanneals extremely rapidly (Fig. 1.10) and is clearly less complex than the DNA of the simplest bacteriophage. The rate of reannealing is in fact indicative of a basic unit consisting of approximately 300 nucleotide pairs. Knowing that this DNA makes up 10% of the total DNA, and knowing the total amount of DNA per mouse cell, Britten and Kohne calculated that satellite DNA must be repeated about one million times in the mouse genome. For this reason, the satellite DNA is also termed "reiterated" DNA. Further, since the reiterated DNA is detectable in unsheared DNA, with a fragment size of 10^5 nucleotide pairs ($6 \cdot 5$–7×10^7 daltons), it is likely that many of the repeats are in tandem with each other, giving tracts of around 330 (i.e. $10^5/300$) copies of reiterated DNA in parts of the mouse genome [56].

The size assigned to the units of reiterated DNA in mammals has been questioned. The two strands of mouse satellite DNA and of guinea-pig satellite are so different in average base composition that they may be separated by centrifugation of the satellite in gradients of alkaline caesium chloride [137, 438]. Southern has used this technique to make preparations of the individual strands, then subjecting the two strands to partial sequence analysis, using the depurination technique [438]. The DNA is treated with dilute formic acid plus diphenylamine. This causes a breakdown of the purine bases. The pyrimidine tracts which remain are then analysed by hydrolysis and chromatography. On the basis of this sequence analysis, Southern has suggested that the basic unit in guinea-pig satellite DNA is only six nucleotide pairs long. The structure proposed by Southern for this basic unit is as follows:

5′CCCTAA3′
3′GGGATT5′

Southern suggests that the individual copies of this basic unit have diverged from each other by point mutation, giving a large number of variations of the basic sequence. It is further suggested that this sequence divergence causes the rate of reannealing to be decreased by up to two orders of magnitude. Similar results have been obtained for mouse satellite DNA, although in mouse satellite, the basic unit is A–T-rich, and is said to be 20 nucleotide pairs long [499]. If this interpretation is correct, then the mouse genome contains about 15 million imperfectly matched copies of the basic unit of reiterated DNA, and further, some of the tracts of tandem repeats must contain about 4500 copies. However, on the basis of more recent experiments [41], Britten and his colleagues have suggested that the results obtained by sequence analysis do not necessarily support the view that the basic unit in rodent satellite DNA is 6–20 nucleotide pairs long, and further, that the degree of sequence divergence proposed by Southern could not in any case reduce the reannealing rate by more than 20%. Britten therefore still holds the view that the basic unit in mouse satellite DNA is at least 250 nucleotide pairs long.

Of the reiterated DNAs which are known from plants, that making up part of the satellite DNA of *Cucumis melo* (melon) has received most attention. The reannealing rate of reiterated DNA from melon has been measured by Bendich and Anderson [27]. They followed the reannealing reaction by continuous monitoring of absorbance at 260 nm in a spectrophotometer, and estimated the basic unit of reiterated DNA in melon to be about 400 nucleotide pairs in length. This would mean that the melon genome contains around one million copies of the reiterated DNA. Little or no mismatching occurred during reannealing (as indicated by the sharp thermal denaturation profile of the reannealed DNA). Reiterated DNAs from other plants vary in complexity from being similar to melon reiterated DNA (e.g. *Linum usitatissimum*, flax) to being up to 25 times more complex than melon reiterated DNA (e.g. *Phaseolus coccineus*, runner bean). It is thus clear that the reiterated DNAs of higher plants are in general more complex than those of rodents [209].

The function of reiterated DNA in both plants and animals is unknown. However, there is good evidence that it does not have a coding function. The base sequence of guinea-pig satellite DNA, for example, is such that any RNA transcribed from it would contain a very high proportion of chain termination triplets (see Chapter 3). Further, the technique of DNA–RNA hybridization, even when employed under the most sensitive conditions, fails to detect any RNA in mouse cells which is complementary to mouse satellite DNA [139].

The location of reiterated DNA within the chromosome provides further evidence that reiterated DNA does not have a coding function. The use of Giemsa stain, which under appropriate conditions selectively stains reiterated DNA, indicates that a large proportion of the reiterated DNA in both plants and animals is located in the constitutive heterochromatin [151, 466,]. Heterochromatin is the name given to the regions of the chromosomes which remain permanently condensed throughout the cell cycle, which do not uncoil in telophase, and to which no genes map. The location of reiterated DNA in heterochromatin has been confirmed in the mouse by the technique of *in situ* molecular hybridization [227, 342]. In this technique, DNA is denatured *in situ* in a histological preparation on a microscope slide. The slide is then immersed in a solution containing either radioactive denatured satellite DNA or radioactive RNA synthesized *in vitro* on a satellite DNA template. The hybridization of the radioactive nucleic acid with the regions of the chromosome corresponding to heterochromatin is then revealed by autoradiography.

Although the bulk of the reiterated DNA is clearly associated with the heterochromatin, there is good evidence that reiterated DNA sequences also occur throughout the length of the chromosome, since extensive shearing of the main-band DNA of mouse causes the release of satellite DNA sequences, suggesting that some satellite DNA sequences occur as integral parts of molecules which are otherwise of main-band composition [139]. This observation is supported by the results of Stern and Hotta [446], who supplied ^{3}H-thymidine to developing pollen grains at different stages of meiosis in *Lilium*. They showed that during the repair-synthesis of DNA which takes place after crossing over, the DNA which becomes labelled is of a different average base composition from main-band DNA (although not different enough to appear as a satellite DNA), and further, reanneals very rapidly. The most straightforward interpretation of these data is that the strand breakage which is necessary for crossing over occurs in sequences of spacer DNA which lie between the genes, and that this spacer DNA is in fact reiterated DNA. The question of the function of the bulk of the reiterated DNA (i.e. that located in the heterochromatin), however, still remains unanswered. Several ill-defined roles for this reiterated DNA have been proposed including an involvement in chromosome pairing, in control of gene expression, in control of the cell cycle, or in the maintenance of chromosome structure.

The existence of a class of DNA which reanneals faster than expected from the size of the eukaryotic genome, but more slowly than the reiterated DNAs, has already been mentioned. This "intermediate" DNA often exhibits heterogeneous reannealing kinetics, and is clearly less well defined than reiterated DNA. Indeed, the intermediate DNA of a given organism almost certainly represents a number of different types of DNA. One type of DNA which may be defined as "intermediate" DNA in the context of reannealing kinetics is the DNA coding for ribosomal RNA. The genes coding for ribosomal RNA are highly amplified, some plant species containing as many as 20 000 copies per diploid genome [4 11] (see Chapter 2). (The term "redundancy" which is used by many authors to describe a state of permanent gene amplification will not be used here, since in the opinion of the present author the implications inherent in the term are highly misleading.) By virtue of the high number of copies, the DNA coding for ribosomal RNA reanneals faster than DNA corresponding to genes which exist as single copies ("unique sequences"). Similarly, the genes coding for transfer RNA are also highly amplified [476] and therefore make up a part of the "intermediate" DNA. The more complex sequences interspersed with the highly reiterated sequences in satellite DNA (section IID) may also be classed as "intermediate" DNA. Other than these examples, the nature of "intermediate" DNA is largely unknown, but in general terms we can state that it probably represents families of similar (but not identical) sequences of low complexity plus families of identical or almost identical sequences of greater complexity (e.g. amplified genes).

The rapidly reannealing and "intermediate" DNAs make up 50–70% (or even more) of the DNA of eukaryotic cells (as judged from reannealing kinetics). The remainder of the DNA renatures extremely slowly, and is widely assumed to represent unique sequences. Whether or not these unique sequences can be considered to be equivalent to genes is a matter of controversy, since the amount of DNA represented by the fraction which reanneals very slowly is much bigger than the amount of DNA needed to account for all the genes required by a eukaryotic cell (see section III).

III. PLANT CHROMOSOMES

A. The amount of DNA per nucleus

The dense and relatively specific staining of DNA with the Feulgen stain has provided a means by which the amount of DNA in a cell nucleus may be accurately estimated. Tissue sections are mounted on slides, stained with Feulgen stain, and then scanned with a microdensitometer. This gives a measure of the intensity of the stain in the nucleus. Stain intensity may then be related to DNA concentration by reference to a series of standards, after correction for the appropriate controls. Table 1.IV shows the amounts of DNA per nucleus for several higher plants and for one unicellular alga. Two very interesting features are clear from the data presented in the table. Firstly, plant cell nuclei contain very large quantities of DNA, compared with bacterial cells. Most bacteria contain $0 \cdot 01-0 \cdot 05 \times 10^{-12}$ g of DNA per haploid cell, whereas diploid plant cells contain $1 \cdot 0-75 \times 10^{-12}$ g per nucleus, the actual amount being dependent on the species. The lengths of DNA double helix represented by such amounts of DNA are almost unbelievable. For example, $1 \cdot 0 \times 10^{-12}$ g of DNA represents 300 mm of double helix. The total nuclear DNA complement of *Vicia faba* (broad bean) therefore represents about 9 m of double helix, whilst each chromosome of *V. faba* contains, on average, 750 mm of double helix. Analysis of the amount of DNA per nucleus in terms of information potential is also very interesting. One $\times 10^{-12}$ g of DNA represents 1×10^9 nucleotide pairs. This is enough DNA to code for $8 \cdot 5 \times 10^5$ proteins each of molecular weight 50 000. In these terms, the nuclear DNA content of *Lilium longiflorum* represents sixty million genes. It is obvious from these figures that the amount of DNA present in a plant cell nucleus is hundreds or even thousands of times greater than is needed to code for proteins, assuming that the gene coding for each protein is present as a single copy.

TABLE 1.IV. Nuclear DNA content and chromosome number in plants

Species	Amount of DNA per diploid nucleus (10^{-12} g)	Diploid chromosome number
Anemone virginiana	21	16
Anemone fasciculata	52	14
Aquilegia sp.	1·2	
Vicia faba	29	12
Vicia sativa	4	12
Allium cepa	34	16
Allium fistulosum	26	16
Allium sibiricum	15	
Lilium longiflorum	72	
Euglena gracilis	3 (per haploid nucleus)	

B

The second interesting feature apparent from Table 1.IV is the very great interspecific variation in DNA content shown by higher plants. This contrasts markedly with the much greater uniformity shown by mammals (nearly all of which contain 4.0–6.0×10^{-12} g of DNA per diploid nucleus). The interspecific variation does not appear to be related to genetic diversity, since very closely related plant species, with the same chromosome number, may have widely differing nuclear DNA contents. Nor does the variation seem related to the complexity of the organism, since some higher plant species contain less DNA per nucleus than does the unicellular alga *Euglena gracilis*.

The very large amounts of DNA in the nuclei of eukaryotic cells has led many cell biologists to propose that eukaryotic chromosomes may be multi-stranded (i.e. may be composed of several or even many DNA molecules lying side by side). However, the bulk of the relevant experimental evidence suggests that despite the enormous amount of DNA in a eukaryotic chromosome, a chromosome is in fact one very long DNA double helix [515]. Of this evidence, the following points are particularly worthy of mention:

(i) The classical experiments carried out by Taylor, Woods and Hughes [470], which are summarized diagrammatically in Fig. 1.11, showed that the

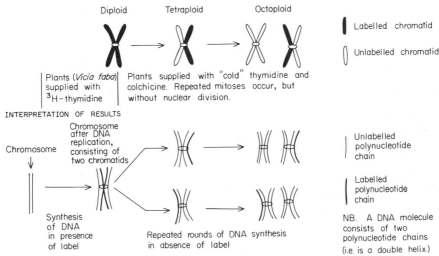

FIG. 1.11. Semi-conservative replication of chromosomes in meristematic cells in roots of *Vicia faba* (broad bean). The plants were supplied with radioactive thymidine for a short period, in order to label the replicating DNA. The plants were transferred to a non-radioactive thymidine solution which also contained colchicine (colchicine inhibits spindle formation and thus prevents the completion of nuclear division). Successive rounds of DNA replication took place, causing the accumulation of diploid, then tetraploid and then octoploid sets of replicated chromosomes. The location of radioactivity in the chromosomes was detected by autoradiography (redrawn from ref. 470).

chromosomes of *Vicia faba* replicate semi-conservatively. This is exactly what is expected if each chromosome is one DNA double helix, since the DNA also replicates semi-conservatively (see section IV).

(ii) It has proved possible to "visualize" in the electron microscope the DNA of the lamp-brush chromosomes of amphibians by using the technique developed by Kleinschmidt [251]. The chromosomes are treated with protease, and the DNA released is floated on a solution of protein. The DNA spreads at the air–water interface, and may then be deposited on an electron microscope grid. The dimensions of the DNA fibres released from these giant chromosomes are consistent with the chromosomes containing one DNA double helix [311].

(iii) Attempts to measure the molecular weight of chromosomal DNA have largely been frustrated because of the susceptibility of DNA molecules to hydrodynamic shear. However, Petes and Fangman [361, 362] developed a method whereby DNA may be released from yeast (*Saccharomyces cerevisiae*) nuclei by osmotic lysis without shear. At the same time, the DNA becomes dissociated from the chromosomal proteins. The molecular weight of the DNA may then be estimated by sedimentation velocity measurements in the ultracentrifuge or by measurement of its length in electron micrographs. The average molecular weight of yeast DNA measured by these procedures corresponds well with the average molecular weight predicted if each chromosome is one DNA double helix.

(iv) Experiments carried out by Rees and his colleagues strongly suggest that the differences in DNA content between closely related species, possessing the same number of chromosomes, may be explained in terms of differences in chromosome length, and not differences in the "strandedness" of the chromosomes [380]. *Allium cepa* and *Allium fistulosum* differ markedly in nuclear DNA content. The two species are very closely related, and have the same number of chromosomes ($2n = 16$). Hybrids between the two species are fertile. Examination of the chromosomes of hybrid plants during the meiotic pairing of homologous chromosomes shows that the chromosomes of *A. cepa* are longer than their homologues in *A. fistulosum*. This results in the formation of bivalents with unpaired loops and overlaps (shown diagrammatically in Fig. 1.12). The most straightforward interpretation of these results is that the larger amount of DNA in *A. cepa* has arisen by additions along the length of the chromosomes, rather than by increasing the number of DNA double helices in the chromosome. This is again consistent with the chromosome being equivalent to one DNA double helix.

(v) Two facets of the genetic behaviour of chromosomes, namely mutation and crossing over, involve specific events at specific points in the chromosome. It is extremely difficult, although admittedly not impossible, to envisage mechanisms whereby either of these events could take place in a multi-stranded chromosome. This consideration alone has led geneticists to incline strongly to

FIG. 1.12. Diagram illustrating some of the configurations observed during meiotic pairing of homologous chromosomes in hybrids between two species which have the same chromosome number but different amounts of DNA per nucleus.

the hypothesis that a chromosome is one DNA double helix [515], although it is clear that there is also a good deal of experimental evidence to support the hypothesis.

The reason for the very large amount of DNA in a plant cell remains a matter for speculation. One hypothesis which gained much support in the 1960s was that the large amount of DNA in eukaryotic chromosomes represented a large-scale amplification of genes. The "master and slaves" model of Callan [70] was one popular version of the amplification hypothesis. However, the only well documented instances of permanent gene amplification involve the genes coding for ribosomal RNA, transfer RNA, and in animal cells for the histones. (The amplification of simple sequences to form the rapidly reannealing, reiterated DNA, is not gene amplification, since this DNA does not have an informational function.) Further, work with specialized animal cells which manufacture one or two specific proteins in large quantities strongly suggests that the genes coding for those proteins are not amplified, but are in fact "unique sequences". The gene amplification hypothesis, although valid in a small number of instances, cannot therefore explain the large size of the eukaryotic genome.

We have already noted that the highly reiterated DNA which exhibits rapid reannealing kinetics has no informational function, and it is also likely that the bulk of the "intermediate" DNA (with the exception of the amplified genes) also has no coding function. In fact, it seems probable that a large proportion of the DNA in a plant or animal cell has a function other than coding. Two pieces of experimental evidence are relevant here. Firstly, Nagl has shown that where two closely related species have widely differing amounts of DNA, the species with the higher amount of DNA often has proportionally more heterochromatin (i.e. reiterated DNA) than the species with the lower amount of DNA [323]. This raises the possibility that the interspecific variation in plants with respect to nuclear DNA content arises from variation in the amount of DNA which has no informational role. The second line of evidence concerns the genus *Vicia*. Maher and Fox have studied several species of *Vicia*, all of which have 12 or 14 chromosomes in a diploid set, but which show great interspecific variation in nuclear DNA content [292]. Using the technique of RNA–DNA hybridization (Chapter 2), Maher and Fox measured the number of genes coding for ribosomal RNA, and compared the numbers with the nuclear DNA content in each species (Table 1.V). There is some correlation between the amount of DNA per nucleus and the number of genes per nucleus which code for ribosomal RNA. However, the interspecific variation in the number of genes coding for ribosomal RNA is much less than the interspecific varuation in nuclear DNA content. For example, *V. faba* has 7 times as much nuclear DNA as *V. sativa*, but only 2·6 times as many genes coding for ribosomal RNA. It is tempting to deduce from these data that increases in nuclear DNA content have arisen largely by increases in DNA which has no coding function, rather than by increases in the number of gene copies. It would be very interesting, were it possible, to select a gene which is present as a unique sequence (i.e. two copies per diploid nucleus) in *V. sativa*, and to estimate how many copies of that gene are present in each nucleus of *V. faba*. Unfortunately, we are not able to carry out this experiment with current methodology.

TABLE 1.V. Nuclear DNA content, chromosome number and reiteration frequency (level of amplication) of genes coding for ribosomal RNA in the genus *Vicia*.[a]

Species	Diploid chromosome number	Amount of DNA per diploid nucleus (10^{-12} g)	rRNA genes per diploid nucleus
Vicia faba	12	29	9500
Vicia narbonensis	14	14	6250
Vicia sativa	12	4	3750

[a] From reference 292.

Although it is likely that a large proportion of the DNA in the nucleus of a eukaryotic cell does not have an informational role, it should not be regarded as having no function. Indeed, Crick has proposed a model for chromosome structure in which sequences of DNA which have an informational role (i.e. genes) are interspersed with much longer sequences whose main function is one of control [94]. Crick envisages that the control of gene transcription may be exerted at the chromosome level by changes in the configuration of the non-informational DNA. The model is obviously a rather generalized one, and, as Crick himself has pointed out, is very speculative. However, the hypothesis has received a good deal of support, and, perhaps more importantly, has stimulated a good deal of research on the role of DNA configuration in the control of gene action.

B. Chromatin

The large amount of DNA in the nucleus of a eukaryotic cell imposes the necessity for supercoiling of the DNA in order to pack it into the nucleus. The physico-chemical properties of DNA make it extremely unlikely that it should spontaneously assume a stable supercoiled configuration. It is therefore not surprising that the DNA in a eukaryotic nucleus occurs as a complex with proteins. The complex is called chromatin, and the individual aliquots of chromatin (which retain their discrete identities throughout the life of the cell) become visible as chromosomes during mitosis and meiosis. There are two classes of protein in chromatin. These are the histones, which are basic proteins, rich in the basic amino acids lysine and arginine, and the acidic, or non-histone chromatin proteins.

The histones are easily the more abundant of the two classes of protein, at least in terms of mass, and further they govern many of the more obvious properties of chromatin. For these reasons, some authors use the term "chromatin" to denote a DNA–histone complex. (This usage will be avoided here; the term "nucleohistone" will be used to denote a DNA–histone complex, whilst the term "chromatin" will imply the presence of the acidic proteins in addition to DNA and histones.) Analysis of the histones from a very wide variety of organisms has revealed three very interesting features. Firstly, in any chromatin sample, the mass of histones is approximately equal to the mass of DNA [40, 195, 427]. Secondly, as a class, the histones show only a very limited heterogeneity [40, 195, 427]. There are five main classes of histones:

f1,   lysine-rich
f2b, slightly lysine-rich
$f2a_2$, slightly arginine-rich
$f2a_1$, arginine-rich
f3,   arginine-rich

A number of different forms of nomenclature have been used in classifying histones. In most instances, the terms used bear little relation to histone structure, but may be related to procedures for fractionating histones. The nomenclature used here is currently the most popular. These five histones are present in all chromatin samples so far examined. Indeed, in plants no other types of histone are known, although additional, apparently unique histones, have been detected in some species of birds and fish. Thirdly, the structure of each of these classes of histones shows remarkably little interspecific variation [40, 195]. The most extreme example of this concerns histone $f2a_1$: the amino acid sequence of histone $f2a_1$ in calf thymus differs in only two amino acid residues from the sequence of histone $f2a_1$ in the pea, *Pisum sativum* [103]. Sequence conservation, although not as marked as in histone $f2a_1$, is also a feature of the other classes of histones [195]. The close similarity of the sequences of histones in widely differing species implies that histones play a very basic and general role, and further that a very large part of the amino acid sequence is necessary for that role.

The binding of histones to DNA is very much reduced if the DNA is denatured [195]. This implies that a feature of the structure of native, double-stranded DNA is necessary for histone binding. The dimensions of a polypeptide chain are such that a histone molecule could fit into the major groove of the DNA double helix (see Figs 1.2 and 1.13). Experiments involving the use of antibiotics and dyes which bind specifically in the minor groove of DNA have indicated that the binding of these compounds is not affected by the presence of histones, which lends support to the view that histones bind in the major groove of the double helix [427]. However, there is in fact no *direct* evidence that the histones bind in the major groove. Indeed, if the currently popular model for chromatin structure is correct, then the histones are not associated specifically with either of the grooves of the DNA helix (see below). The nature of the binding between histones and DNA is an electrostatic attraction between the negatively charged phosphate groups of DNA and the positively charged basic amino acids (lysine and arginine) in the histones. Indeed, since in chromatin there is an approximate equality between the number of phosphates in DNA and the number of lysine plus arginine residues in the histones, many investigators took the view that nucleohistone was an electrostatically neutral complex, with all the DNA phosphates bound to basic amino acids. However, the work of Itzhaki [212] has shown that in native chromatin from rat thymus, about 40% of the DNA phosphates will bind positively charged molecules such as Toluidene Blue or poly-L-lysine, and therefore cannot be bound to histones. Similar results have been obtained for nucleohistone [349]. It has also been shown that about 25% of the lysine residues in the histones of native chromatin can be acetylated by treatment with acetic anhydride, and must therefore be free [427]. Further, the distribution of lysine and arginine residues is asymmetric, with a greater abundance of these basic amino acids towards the ends of the molecules [427]. On the basis of these data, it has been suggested that in native chromatin or in nucleohistone, DNA is bound to histones for only about 60% of its length [212, 427]. The remainder of the DNA is thought to be closely associated with histones although not bound to the histones; the parts of the histone molecules which are not bound to DNA are probably held in place simply by virtue of the fact that they are covalently linked to parts of the histone molecules which are bound to DNA. Estimates of the molecular weight of the fragments of DNA which are released from chromatin by controlled shearing (and which therefore are presumably not bound to histones) suggests that the tracts of "free" DNA may be of the order of 4×10^3 nucleotide pairs long [494].

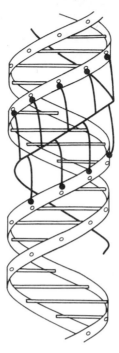

FIG. 1.13. Diagram illustrating the possible interaction between a group of positively charged amino acids in a protein and the negatively charged phosphate groups in the backbone of the DNA double helix.

However, the validity of this result has been placed in doubt by data recently obtained by Kornberg and his colleagues. Using a specific nuclease from *Staphylococcus*, they have been able to break down chromatin (extracted from nuclei by gentle osmotic lysis) into subunits. Each subunit contains 185–200 base pairs of DNA, and two molecules each of histones f2b, $f2a_2$, f2a and f3 [255a]. Kornberg proposes that the DNA is wound round the outside of the histone octomer, and that the histones are arranged in such a way that the basic amino acids are available for binding with the DNA phosphates. This subunit of chromatin is referred to as the "nucleosome". Histone f1 is not involved in the structure of the nucleosome, but is envisaged as being associated with it. Kornberg suggests that this subunit structure pertains to all the chromatin (i.e. both that being transcribed and that not being transcribed) and further suggests that the stretches of the free DNA between the nucleosomes (which are sites attacked by the *Staphylococcus* nuclease) are only a few base pairs long [135a]. Some of the DNA in the nucleosome is less tightly attached than the remainder to the histones, since more extensive nuclease treatment degrades up to 46% of the DNA of the nucleosome. This is clearly consistent with the suggestion made earlier that only about 60% of the total length of DNA is actually bound to histones. The DNA which is degraded by nuclease digestion of nucleosomes varies between 25 and 140 base pairs in length [258a]. Since each nucleosome contains about 200 base pairs of DNA helix, this suggests that the binding of DNA to histones in the nucleosomes is variable. These data clearly contrast with earlier findings that DNA sequences up to 4×10^3 base pairs long are "free" of histones (p. 29).

The binding of histones to DNA has three major consequences. Firstly, the double helical structure of DNA is very much stabilized against thermal denaturation. Nucleohistone therefore has a much higher T_m than pure DNA (Fig. 1.14). Secondly, the presence of histones bound to the DNA causes supercoiling, as detected by electron microscopy, X-ray analysis and spectroscopic analysis of nucleohistone and of chromatin [349, 427, 517]. The supercoiling probably arises by interactions (hydrogen bonding, electrostatic repulsion and attraction, etc.) between amino acids in different regions of the histone molecules, causing the DNA–histone complex to become tightly coiled [94, 427]. This is somewhat analogous to the imposition of a tertiary structure on a protein molecule by interactions between amino acids in different parts of the molecule. The DNA of a eukaryotic cell nucleus is therefore packed into the available space and stabilized by virtue of its binding with the histones. Indeed, of the proteins in chromatin, the histones alone seem responsible for chromatin structure, since nucleohistone has the same overall structure and configuration as native chromatin.

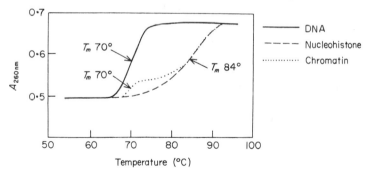

FIG. 1.14. Thermal denaturation profiles of DNA, nucleohistone and chromatin of *Pisum sativum* (garden pea). The experiments were carried out in the presence of 0·016M Na⁺ (from ref. 40).

The third consequence of the binding of histones to DNA is a functional one. In the early 1960s, Bonner and his colleagues [40] carried out a series of experiments which in some respects were well ahead of their time. They showed that nucleohistone from pea (*Pisum sativum*) supports *in vitro* RNA synthesis, using added RNA polymerase (extracted either from pea or from *Escherichia coli*) at only 5% of the rate achieved with purified pea DNA (Fig. 1.15). Similar results have since been obtained with nucleohistone and DNA from a wide range of organisms. The availability of DNA as a template for RNA synthesis is therefore very much reduced by the presence of histones, or, in Bonner's words "the histones are agents of gene repression" [40]. The extent of the repression (95% or greater) is perhaps surprising in view of the finding mentioned previously that about 40% of the DNA phosphates in nucleohistone or in chromatin are unbound. This supports the suggestion that even the regions of the DNA molecule which are not bound to the histones are nevertheless associated with the histones.

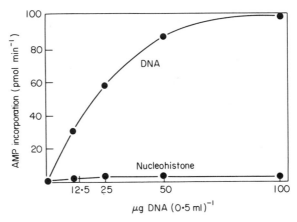

FIG. 1.15. The template activities of DNA and nucleohistone from *Pisum sativum* (garden pea) DNA and nucleohistone were used as templates in a reaction mixture containing radioactive ATP and non-radioactive CTP, GTP and UTP, together with essential co-factors, plus RNA polymerase. The incorporation of radioactivity into an acid-insoluble product was taken as a measure of the rate of RNA synthesis (from ref. 40).

The second group of chromatin proteins, known either as the non-histone proteins or as the acidic chromatin proteins, exhibits much more variety than the histones [195, 427]. There are many more acidic proteins than histones (20–30 as opposed to five), and unlike the histones, the acidic proteins show a good deal of interspecific variation in terms of molecular weight [195]. The ratio of acidic proteins to histones is very variable, the mass of acidic proteins comprising 5–25% of the total chromatin mass. In general, tissues or organs which exhibit a high metabolic activity are characterized by a high content of acidic chromatin proteins, whereas metabolically inactive tissues or organs contain lower amounts of acidic chromatin proteins [195, 427].

Several of the enzymes involved in nucleic acid metabolism, including RNA polymerase, one of the DNA polymerases and at least one of the deoxyribonuc-leases, are tightly bound to the chromatin *in vivo* and it seems likely that these enzymes are amongst the acidic chromatin proteins. However, none of these enzymes has been identified positively in a total acidic chromatin protein preparation, although each has been isolated from chromatin by the usual procedures of enzyme solubilization and purification.

Of the acidic chromatin proteins which are not enzymes, none has been identified as having a specific, unique function. However, it is becoming clear that as a group, these proteins have marked effects on the interaction between histones and DNA. Much of the evidence for this again comes from the work of Bonner and associates. Figure 1.14 shows, in addition to the thermal denaturation profiles of DNA and nucleohistone, the thermal denaturation profile of native chromatin. It is clear that the DNA in chromatin is thermally denatured in a biphasic manner. The first step in the hyperchromicity, repre-senting the denaturation of 20–25% of the DNA, has the same T_m as pure DNA, whereas the second step has the same T_m as nucleohistone. The obvious inference from this is that the presence of the acidic chromatin proteins modifies the interaction between histones and DNA, releasing part of the DNA from stabilization against thermal denaturation.

The other feature discovered by Bonner and his colleagues is that chromatin has a much higher template activity than nucleohistone. Thus, chromatin from the apical buds of pea supports *in vitro* RNA synthesis at 20–25% of the rate obtained with purified DNA (cf. nucleohistone, 5%). Further, there is a correlation between the proportion of the DNA in chromatin which exhibits the same T_m as pure DNA, and the efficiency of chromatin as a template. The chromatin from pea cotyledons supports *in vitro* RNA synthesis at only about 10% of the rate obtained with pure DNA; the proportion of DNA in cotyledon chromatin which is thermally denatured at the T_m of pure DNA represents about 5% of the total DNA.

From these data, the inferences may be drawn that the acidic chromatin proteins derepress part of the DNA (i.e. make it available as a template), that this derepression probably represents a modification of the interaction between DNA and histones, and that the extent of the derepression is tissue- or organ-specific. (Bonner also obtained evidence that the DNA sequences involved in the derepression are tissue- or organ-specific: see Chapter 6.) These inferences are supported by the observations mentioned earlier, that chromatin from metabolically active tissues contains a greater proportion of acidic proteins than chromatin from metabolically inactive tissues.

So far, the evidence cited supporting the idea that the acidic proteins modify the repressive activity of histones has been largely circumstantial. We have noted the differences in behaviour between nucleohistone and native chromatin, and attributed the differences to the acidic proteins. It is perhaps worth mentioning that neither Bonner nor other investigators working in this area in the 1960s in fact suggested that the differences in behaviour between nucleo-histone and chromatin might be caused by the acidic proteins. Evidence that the acidic proteins do modify the interaction between DNA and histones comes from the work of Paul and More [349]. They worked with chromatin, nucleo-histone and DNA from calf thymus (Table 1.VI). As in pea, the chromatin has a much higher template activity for *in vitro* RNA synthesis than nucleohistone. However, the template activity of nucleohistone becomes the same as that of chromatin if the acidic proteins are added back to the nucleohistone prepara-tion. The modification of the repressive action of histones by the acidic chromatin proteins almost certainly involves those sequences of the DNA which are not actually bound to the histones, since Paul and More observed, as previous authors have done, that the proportion of unbound DNA phosphate in nucleohistone is the same as in native or reconstituted chromatin. Similar, although less unequivocal, results have recently been obtained with plant cells [302].

TABLE 1.VI. Properties of chromatin, nucleohistone and DNA from calf thymus.[a]

	Template activity with exogenous RNA polymerase	Percentage of "free" phosphate
DNA	100	100
Native chromatin	22	47
Nucleohistone	3	49
Reconstituted chromatin	24	46

[a] From ref. 349.

For an understanding of the relationship between chromatin structure and gene action, it is important to know whether the acidic chromatin proteins exhibit any specificity, both in terms of the occurrence of different proteins in different types of cell, and in terms of the ability of proteins to recognize particular DNA sequences. It is also important to know the relationship between the non-histone proteins and nucleosomes, since it has been shown that actively transcribed chromatin has the same general nucleosome structure as non-transcribed chromatin [135a]. Research aimed at answering these ques-tions is being carried out in many laboratories but as yet the findings are inconclusive. We can at present only speculate that the non-histone proteins make the DNA available for transcription by interacting in some way with the nucleosomes. It is clear that a large number of problems remain to be solved before we reach an understanding of the process of transcription. Some of these problems are dealt with in later chapters. The major problem to be dealt with here is the problem of quantity. Even 30% of the DNA in the nucleus of a eukaryotic cell is far in excess of the cell's apparent information requirements (see section IIIA). Despite this, there is evidence from some animal cells that newly synthesized RNA, extracted from the nucleus, may be complementary to up to 30% of the nuclear DNA. Not only do eukaryotic cells possess much more DNA than they apparently need for informational purposes, but they actually transcribe more than they apparently need. The "extra" DNA in chromatin is clearly not redundant (in the strict sense of the term), and elucidation of its function remains one of the most interesting problems in molecular biology.

IV. DNA SYNTHESIS

A. Replication of DNA

The idea that DNA replicates semi-conservatively (i.e. that each of the strands in a DNA molecule acts as a template for the synthesis of a new complementary strand, giving two molecules each consisting of one old and one new strand) was first proposed by Watson and Crick in 1953 (see section I). Evidence that the DNA of the bacterium *Escherichia coli* replicates semi-conservatively was published by Meselson and Stahl in 1958 [306]. They grew synchronously dividing cultures of *E. coli* for several generations in a medium containing heavy nitrogen (^{15}N), causing all the nitrogenous compounds, including DNA, to become more dense. The cells were then washed and transferred to a medium containing normal nitrogen (^{14}N). At the time of the transfer, and subsequently at intervals corresponding to the generation time, samples of cells were taken and DNA was extracted from them. The DNA was centrifuged to equilibrium in gradients of caesium chloride (Fig. 1.16). At the time of the density transfer, all the DNA was dense, as expected. After one round of DNA synthesis, all the DNA was of a density intermediate between that of heavy ^{15}N-DNA and normal ^{14}N-DNA. This is exactly consistent with the view that DNA synthesis results in the formation of molecules with one new and one old strand. After two rounds of DNA synthesis in normal ^{14}N medium, half the DNA was of normal density and half was of intermediate density. This again is exactly what would be expected if DNA replicates semi-conservatively (see Fig. 1.16).

This density transfer technique has since been used to show that the nuclear DNA in synchronously dividing cells of the unicellular algae *Chlamydomonas* and *Euglena* [81, 293], and in synchronously dividing cultured mammalian cells, replicates semi-conservatively. Until recently, it has proved very difficult to obtain synchronously dividing cultures of higher plants (see Chapter 5), so

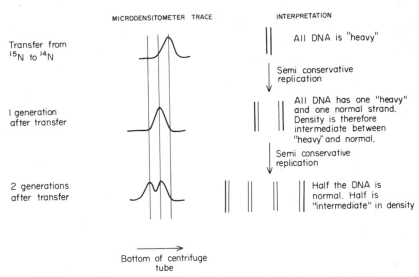

FIG. 1.16. Diagrammatic summary of the results obtained by Meselson and Stahl demonstrating that the DNA of *Escherichia coli* is replicated semi-conservatively (redrawn from ref. 306).

semi-conservative replication of DNA has not been conclusively demonstrated in higher plants (although semi-conservative replication of higher plant chromosomes has been shown: section III). However, in view of the demonstration of semi-conservative replication in organisms as different as bacteria, algae and mammals, there is no reason to suppose that DNA replication in the nuclei of higher plants is not semi-conservative.

At about the same time as Meselson and Stahl first demonstrated semi-conservative replication in *E. coli*, Kornberg isolated from the same bacterium an enzyme which brings about the semi-conservative synthesis of DNA. The enzyme is now known as "DNA polymerase I". Subsequent work by Kornberg and his colleagues [255] has led to an understanding of the mechanism of the enzyme's action, although some aspects of the mechanism are still not resolved. The first stage of replication is the formation of a single-stranded break ("nick") in the DNA by the action of DNA endonuclease (an enzyme which is able to break internal phosphodiester linkages in a DNA molecule). The breakage of one strand enables the two strands of the DNA molecule to unwind and separate. The polymerase enzyme then proceeds with the synthesis of the two new strands. The synthesis of DNA by the enzyme is completely dependent on the presence of magnesium ions, all four deoxyribonucleoside triphosphates (deoxyadenosine, deoxycytidine, deoxyguanosine and thymidine triphosphates) and DNA. The newly synthesized DNA exhibits identical properties to the DNA which is used as a template; the new DNA is therefore a faithful copy of the template (as predicted by the hypothesis of semi-conservative replication). The synthesis of DNA takes place by the addition of a deoxyribonucleoside-5′- monophosphate (Fig. 1.17). Since the substrates for polymerization are

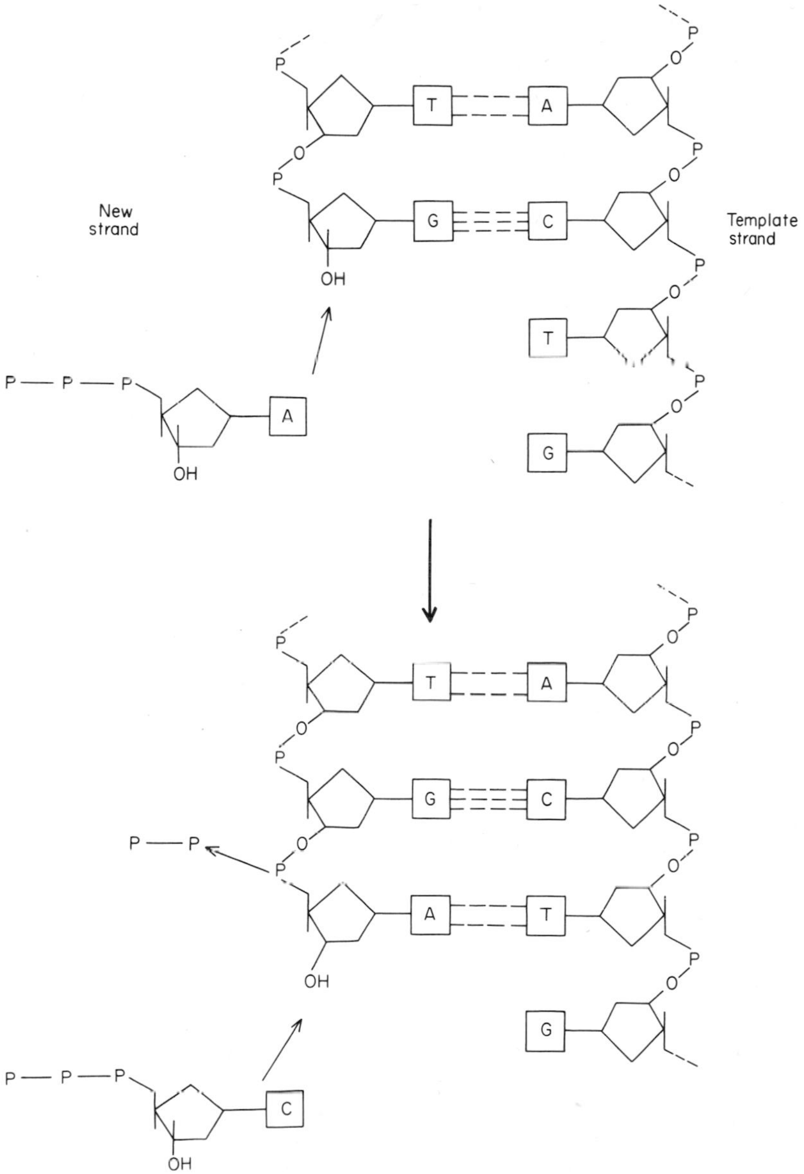

FIG. 1.17. DNA synthesis: the incorporation of a deoxyribonucleoside monophosphate at the 3′ end of a growing poly-deoxyribonucleotide chain.

deoxyribonucleoside-5'-triphosphates, it is obvious that a pyrophosphate must be eliminated for every phospho-diester linkage formed.

The synthesis of DNA in one chemical direction (see Fig. 1.17) means that the enzyme must work in two physical directions, since the two strands of a DNA molecule are anti-parallel. This means that one of the new strands must be synthesized discontinuously (see Fig. 1.18). In fact, Okazaki [399] has claimed that synthesis of both new strands is discontinuous, although the mechanism of synthesis outlined here only requires discontinuous synthesis of one new strand. The fragments of newly synthesized DNA ("Okazaki pieces") are then joined by a DNA ligase.

Since Kornberg's discovery and characterization of DNA polymerase I, two more DNA polymerases, DNA polymerase II and DNA polymerase III, have been discovered in *E. coli* cells. These enzymes are also able to synthesize DNA in a semi-conservative manner, and the precise role of the three polymerases in DNA replication and repair is still a matter for discussion.

The amount of DNA in a single chromosome of a eukaryotic cell vastly exceeds that in a bacterial "chromosome". A chromosome of broad bean (*Vicia faba*) contains about 600 times as much DNA as a cell of *Escherichia coli*. Despite this, DNA replication in *Vicia faba* is completed in 8 h at 25°C (compared with 40 min at 37°C in *E. coli*). Allowing for the difference in temperature, *Vicia faba* apparently replicates its DNA about 100 times as fast as *E. coli* [71]. This high rate of replication is possible because DNA synthesis is initiated at many points (possibly several hundred) along the length of the chromosome [71]. Replication of DNA proceeds bidirectionally from each initiation point until the replication forks merge with the replication forks of the neighbouring initiation points (Fig. 1.19). DNA synthesis is thus discontinuous in both new strands.

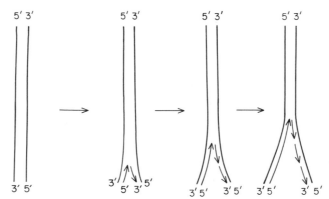

FIG. 1.18. DNA synthesis: discontinuous synthesis of one new poly-deoxyribonucleotide chain and continuous synthesis of the other.

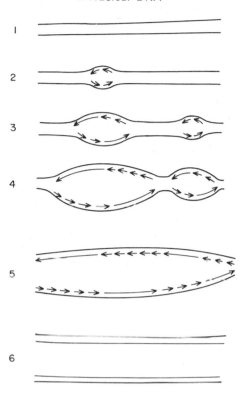

Fig. 1.19. DNA synthesis: initiation points and replication forks in a eukaryotic chromosome, with discontinuous synthesis in both strands.

This again necessitates the existence of a ligase system to join the fragments together. The existence of many initiation points strongly suggests that nicking of one strand of DNA at each initiation point is necessary for the untwisting and separation of the two strands of the DNA helix. It is likely, therefore, that DNA endonuclease is involved in DNA synthesis in eukaryotes as it is in bacteria. The last phase of DNA synthesis in eukaryotic cells is a modification of the newly synthesized DNA. The DNA of higher eukaryotes in general, and of higher plants in particular, contains 5-methyl-cytosine. The 5-methyl-cytosine arises by methylation of cytosine by DNA methylase, following synthesis of the new DNA strands.

The four enzymes involved in the process of DNA synthesis, deoxyribonuclease, DNA polymerase, DNA ligase and DNA methylase, have all been detected in plant cells. However, none of the enzymes has been extensively studied in plants, and our ideas about their operation in plant cells are largely based on results obtained with animal cells.

Deoxyribonuclease has been detected in a number of higher plants and algae. In synchronous cultures of *Chlorella*, deoxyribonuclease activity shows a peak which coincides with the period of net DNA synthesis. A specific inhibitor of enzyme activity is associated with the decrease in deoxyribonuclease activity which occurs after net DNA synthesis is completed [405]. These data provide circumstantial evidence for a role for deoxyribonuclease in DNA synthesis. Unfortunately, no attempts have been made to characterize the *Chlorella* deoxyribonuclease either with respect to mode of action (endo- or exonuclease) or with respect to location in the cell. In higher plants, deoxyribonuclease occurs both in the cytoplasm and as a component of chromatin [444, 447]. The location of the enzyme in chromatin is again suggestive of a role in DNA synthesis, but unfortunately, nothing is known about the mode of action of the chromatin-bound enzyme. In animal cells, the majority of the chromatin-bound deoxyribonuclease is DNA endonuclease and could therefore be involved in the nicking of native DNA for DNA synthesis [488]. There is also some exonucleolytic deoxyribonuclease in the chromatin of animal cells. This is probably concerned with DNA repair (see section IVC).

The available evidence concerning DNA polymerase in eukaryotic cells is somewhat confusing [235]. In animal cells, there are two major DNA polymerases. One is tightly bound to the chromatin and has a molecular weight of about 60 000. The other is cytoplasmic (or "soluble") and has a molecular weight of 100 000–200 000. Traces of the smaller chromatin-bound enzyme may also be detected in the cytoplasm. The activity of the soluble enzyme is high in proliferating cells and low in non-dividing cells; activity of the chromatin-bound enzyme remains more or less constant [76, 77]. This has led to the hypothesis that the soluble DNA polymerase is the enzyme involved in DNA replication, whereas the chromatin-bound polymerase is involved in DNA repair [235]. However, if the soluble enzyme is the replicative enzyme it is difficult to explain why it is located in the cytoplasm, whereas DNA replication occurs in the nucleus.

A number of investigators have recently obtained evidence that the soluble enzyme may be dissociated into two or more sub-units [76, 77, 183]. One of the sub-units is similar, perhaps even identical, to the chromatin-bound enzyme. The dissociation is freely reversible. These findings have led to an alternative hypothesis, namely that the soluble enzyme represents a pool of available sub-units which migrate to the nucleus and become associated with the chromatin when needed for DNA replication [183]. However, the amount of enzyme which migrates into the nucleus must be very small in comparison both with the total pool of undissociated soluble enzyme and with the "basal" level of chromatin-bound enzyme, since it is the soluble enzyme activity which fluctuates in relation to DNA synthesis and not the chromatin-bound activity. The situation is complicated still further by the finding that the cytoplasmic enzyme may form aggregates consisting of varying numbers of sub-units. Whether these aggregates are artefacts of extraction, or whether they have a real physiological role has yet to be determined [235].

Both chromatin-bound and soluble DNA polymerases have been detected and assayed in plants. The enzymes are dependent on the presence of DNA (although the chromatin-bound enzyme must be "solubilized" from the chromatin in order to demonstrate this dependence, since chromatin contains DNA!). The enzymes also require the presence of magnesium ions and all four deoxyribonucleoside-5′-triphosphates. However, there is some activity in a reaction mixture from which up to three of the deoxyribonucleoside phosphates are absent. For example, the chromatin-bound and soluble enzymes from *Pisum sativum* (pea) and the soluble enzyme from *Zea mays* (maize) maintain 20–50% of their maximum activity in the absence of dGTP [390, 450]. This suggests that the enzymes have deoxynucleotidyl terminal transferase activity (i.e. can add deoxyribonucleoside monophosphate residues on to the ends of existing deoxyribonucleotide chains) in addition to DNA polymerase activity. The product of the polymerase reaction is DNA; for the soluble enzymes of *Triticum* and *Zea* and for the chromatin-bound enzyme of *Pisum*, the product has been shown to have identical properties to the DNA used as a template [321, 447, 450].

Purification has been achieved only for the soluble DNA polymerase of *Zea mays* [450], which in the final purification step, may be eluted from a column of diethylaminoethyl-cellulose (DEAE cellulose) as a single peak (Fig. 1.20). Partial purification has been achieved for the soluble polymerases of *Euglena gracilis* [304] and *Pisum sativum* [447], and for the chromatin-bound polymerases of *Beta vulgaris* and *Pisum sativum* [109, 447]. Two of these enzymes apparently exist in more than one form, since they are eluted from ion-exchange columns (such as DEAE cellulose) in more than one peak (Fig. 1.20): the soluble polymerase from *Euglena* is eluted in three peaks and the chromatin-bound polymerase from *Beta* is eluted in four peaks. However, in view of the evidence from animal cells, it seems likely that the multiplicity of forms of these enzymes may be due to the formation of aggregates.

As in mammalian cells, the chromatin-bound and soluble DNA polymerases of plants exhibit different properties from each other. For example, in *Pisum sativum* (pea), the chromatin-bound enzyme has a pH optimum of 7·25 and a magnesium concentration optimum of 5m M, whilst the soluble enzyme has a pH optimum of 8·1 and a magnesium concentration optimum of 15mM [390, 447]. Further characterization, including determination of molecular weights, of the two DNA polymerases from pea and of DNA polymerases from other higher plants is currently in progress. From the limited information available, it is thought that DNA polymerases in higher plants resemble those in mammals, rather than those in lower eukaryotes [61, 447].

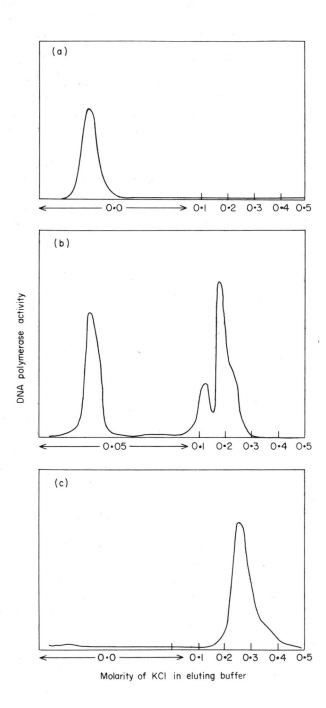

Molarity of KCl in eluting buffer

Information concerning the temporal relationship of the activities of the two enzymes to DNA synthesis in plants is almost non-existent. Most investigators working on this topic have either assayed "total" DNA polymerase, thereby failing to distinguish between the two enzymes, or have assayed one whilst ignoring the other. We are therefore not able to establish any correlation between the relative activities of the enzymes and net DNA synthesis. However, the following data are relevant. In seedlings of pea (*Pisum sativum*) grown at 22°C the initiation of net DNA synthesis occurs 29–30 h after the onset of germination. Prior to the initiation of net DNA synthesis, there is a marked increase in the activities of both the chromatin-bound polymerase and the soluble polymerase. The maximum extractable activity of the chromatin-bound polymerase is only just able to account for the observed rate of net DNA synthesis, whereas the maximum extractable activity of the soluble polymerase is capable of supporting a rate of net DNA synthesis well in excess of the observed rate (Table 1.VII). This is clearly similar to the situation in animals, where the soluble polymerase activity very much exceeds the chromatin-bound polymerase activity in cells which are undergoing net DNA synthesis.

TABLE 1.VII. DNA synthesis and DNA polymerase activities in the embryonic axes of germinating seedlings of *Pisum sativum*.[a]

| | nmol dTMP incorporated per 10 axes per hour at 22°C | | |
| | Actual rate, calculated from rate of net | DNA polymerase activity | |
Age of seedling	DNA synthesis	chromatin-bound	soluble
6 h	0	3·0	165
18 h	0	5·9	180
23 h	0	8·6	405
31 h	8·2	6·8	685
47 h	8·2	8·8	730

[a] From ref. 390.

FIG. 1.20. Ion-exchange chromatography of soluble DNA polymerases on columns of DEAE-cellulose. The enzyme preparations were applied to the columns in a buffer solution containing either KCl at a very low concentration (b) or no KCl (a, c). The columns were washed with the same buffer solution. After the wash, the columns were eluted with linear gradients of KCl in buffer solution. During the wash and during the gradient elution, fractions of column eluate were collected and assayed for DNA polymerase activity. (a) Soluble DNA polymerase from *Zea mays* (drawn from data in ref. 450). (b) Soluble DNA polymerase from cells of *Euglena gracilis* which contained no plastids (drawn from data in ref. 304). (c) Soluble DNA polymerase from *Pisum sativum* (from ref. 447).

DNA ligase, which is necessary for joining the short pieces of newly synthesized DNA, and for repairing the nicks inserted in the DNA at the initiation points, has been isolated from nuclei of *Pisum sativum* [238]. The enzyme shows an absolute requirement for magnesium ions and for ATP and in these respects resembles the ligase of mammalian cells. It is therefore likely that the reaction mechanism is similar to that of the mammalian ligase (Fig. 1.21). DNA methylase has also been isolated from *Pisum* nuclei [230]. The enzyme uses S-adenosyl-L-methionine as a methyl donor for methylation of cytosine in newly synthesized DNA.

B. Control of DNA Synthesis

It is more likely that DNA synthesis is regulated at the biochemical level by control of the availability of deoxyribonucleoside triphosphates rather than by control of their polymerization into DNA, since synthesis of deoxyribonucleoside triphosphates which are not required for DNA synthesis may be regarded as a thermodynamically wasteful process. There are two pathways for the synthesis of deoxyribonucleoside phosphates (Fig. 1.22). In the "major" pathway, deoxyribonucleoside phosphates are derived from ribonucleoside phosphates by reduction. In animals and bacteria, the reduction occurs at the diphosphate level of phosphorylation, and is mediated by ribonucleotide reductase. In the "minor" or "scavenging" pathway, deoxyribonucleosides, particularly thymidine, are phosphorylated by kinases and then enter the major pathway.

Since net DNA synthesis is a periodic event, whereas RNA synthesis is continuous, the most likely control point is the ribonucleotide reductase step (i.e. the first point at which synthesis of deoxyribonucleoside phosphates diverges from the synthesis of ribonucleoside phosphates). The first phosphorylation step in the scavenging pathway, thymidine kinase, is also a possible control point. Indeed, there is evidence from animal cells that both ribonucleotide reductase and thymidine kinase do play a controlling role in DNA synthesis [100]. Both enzymes exhibit very low activities in cells which are not undergoing net DNA synthesis. In the instance of ribonucleotide reductase, the low activity may be caused by the presence of an inhibitor of enzyme activity [262]. In cells which are undergoing net DNA synthesis, activities of both enzymes are high. Further, both enzymes are subject to feedback inhibition by end products. For ribonucleotide reductase, the inhibition has been shown to be mediated by allosteric modification of the enzyme [263].

For plant systems there is little evidence concerning ribonucleotide reductase. Ribonucleotide reductase has been detected in *Euglena* [152]. Like the mammalian enzyme, the enzyme is subject to feedback inhibition. The reaction differs from that in mammalian cells, in that the enzyme in *Euglena* reduces

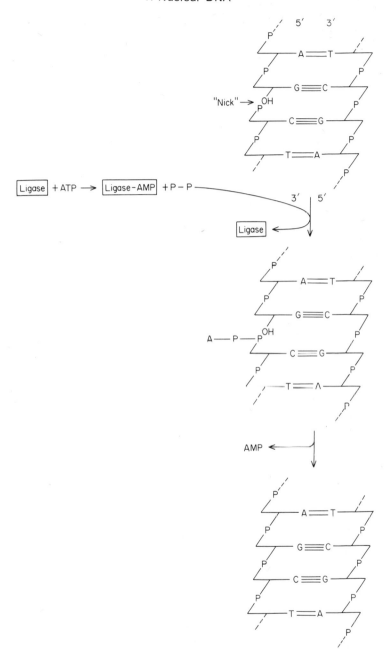

F<small>IG</small>. 1.21. Mode of action of DNA ligase.

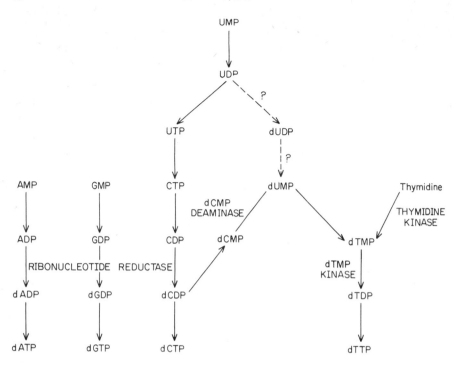

FIG. 1.22. Pathways of deoxyribonucleoside triphosphate synthesis.

nucleoside triphosphates, rather than diphosphates. Ribonucleotide reductase in higher plants appears to be very unstable, and there are therefore no data relating to its possible role in regulating DNA synthesis in higher plant cells. However, other enzymes involved in the synthesis of deoxyribonucleoside triphosphates have been studied. Thymidine kinase has been the most widely investigated, whilst deoxycytidine monophosphate deaminase and thymidine monophosphate kinase have also received some attention. In populations of synchronously dividing plant cells, all these enzymes show a marked periodicity in activity [175, 200, 201, 404, 419]. Their activities increase prior to the onset of net DNA synthesis and then decrease again. This has led to suggestions that each of these enzymes may be involved in the control of DNA synthesis [200, 419]. However, two points need to be raised here. Firstly, whilst it is obvious that the presence or absence of a particular enzyme can exert a coarse control over DNA synthesis, there is no real evidence (e.g. from studies of feedback inhibition or of allosteric effects) that these enzymes have any of the properties normally associated with fine control. Secondly, in some instances, the levels of enzyme activity do not bear any relation to the rate of DNA synthesis. For example, in explants of artichoke (*Helianthus tuberosus*) maximal thymidine kinase activity coincides with the maximum rate of DNA synthesis. However, the activity of the kinase is only 20% of that required to supply all the thymidine phosphates required to sustain the observed rate of DNA synthesis [175]. Therefore, until more is known about the relative importance of the "scavenging" pathway (i.e. thymidine kinase) as compared with the "normal" pathway for synthesis of thymidine phosphates, it is difficult to ascertain whether changes in thymidine kinase activity could regulate the rate of DNA synthesis. It is clear then that a good deal more work is required before any conclusions may be reached concerning control of DNA synthesis in plants.

C. DNA repair

The DNA in living cells is vulnerable to various types of damage, including chemical modifications induced by mutagens or by radiations of various types. In bacterial cells, one of the most common chemical modifications is the formation of chemical bonds between two adjacent pyrimidine bases (in particular between adjacent thymines) in the DNA molecule (Fig. 1.23). Large

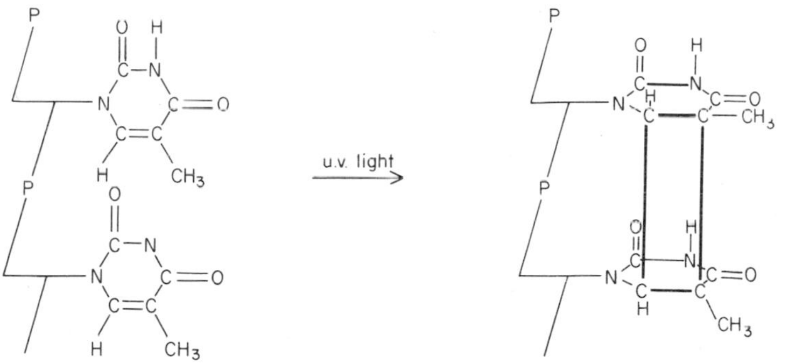

FIG. 1.23. Formation of a thymine dimer.

doses of ultraviolet light are particularly effective in inducing the formation of these pyrimidine dimers. The dimers block the action of DNA polymerase, and hence prevent DNA replication. However, bacteria possess a repair system, in which the dimers may be excised and replaced. The repair process is as follows:

(i) The distorted region of the DNA helix is recognized by an endonuclease, which nicks the strand on the 5′ side of the dimer.

(ii) The tract of the DNA strand containing the dimer is then excised by an exonuclease, starting at the nick and working towards the 3′ end of the strand.

(iii) The gap is then filled by DNA polymerase.

In *Escherichia coli*, DNA polymerase I also exhibits exonuclease activity, which means that stages (ii) and (iii) may be carried out by the same enzyme [255].

(iv) The nick at the 3′ end of the newly inserted sequence of DNA is joined by DNA ligase.

Similar repair mechanisms have been shown to exist in mammalian cells and it seems likely that DNA repair is also a feature of plant cells. However, despite many attempts to detect DNA repair in plants, it has as yet been detected only in protoplasts of carrot, *Daucus carota* [203].

V. DNA TURNOVER

Acceptance of the genetic role of DNA has led to the widely held assumption that DNA is a very stable substance, exhibiting no turnover or metabolic activity once it has been synthesized. There is in fact a good deal of evidence to support the general truth of this assumption, but there is also evidence that under some circumstances, both turnover (i.e. continued synthesis and degradation) of DNA and variation in the amount of DNA per nucleus may occur.

The suggestion that not all the DNA in higher plant cells is metabolically inert was first published in 1959 by Pelc and LaCour [357]. They observed by autoradiographic techniques that newly differentiated, non-dividing root cells of *Vicia faba* (broad bean) incorporate radioactivity from ³H-thymidine into their nuclear DNA. Since no increase in nuclear DNA content occurs in these cells, this observation suggests that the DNA turns over. Following this report, a number of investigators claimed to have isolated "metabolic" or "metabolically labile" DNA from plant cells, having separated it from the bulk of the DNA by ion-exchange chromatography or density gradient centrifugation, after supplying the plants with radioactively labelled precursors of DNA. However, it is now known that the "metabolically labile DNA" was in some instances RNA and in other instances bacterial DNA [62, 63].

More recent investigators have taken precautions to prevent bacterial contamination and to obviate the possibility of RNA becoming labelled in their experiments. Even under these conditions, it is in fact possible to detect DNA turnover both by autoradiography and by extraction of the DNA followed by estimation of the radioactivity in a scintillation counter, after supplying the plants with radioisotope [9, 63, 291]. The turnover has been observed in differentiating and differentiated root cells of *Vicia faba* (broad bean) and of *Pisum sativum* (pea), in differentiated cells in *Spinacea oleracea* (spinach) stems and in cells of *Acer pseudoplatanus* (sycamore) in suspension culture.

The significance of the phenomenon of DNA turnover remains completely unknown. The most obvious interpretation is that DNA turnover represents DNA repair. DNA repair would be expected to occur in both dividing and non-dividing cells, and is a process of degradation and synthesis. However, there are two difficulties with this interpretation. Firstly, the process of DNA repair has only been detected in one plant species (section IVC). Secondly, the magnitude of the turnover (as estimated by the amount of radioisotope incorporated into the DNA which turns over in comparison with the amount of radioisotope incorporated into the stable DNA) seems too great to represent DNA repair alone. It has also been suggested that DNA turnover represents transient gene amplification [356]. According to this hypothesis, each gene is amplified (i.e. many copies of it are made) at the particular stage of cell growth when the particular gene product is required. The copies are then degraded when no longer required. Unfortunately, there are no well documented instances of transient gene amplification in plants. In animals, only the genes coding for ribosomal RNA have been conclusively shown to undergo transient amplification. In oocyte development in amphibia, the number of copies of the polycistronic gene coding for ribosomal RNA increases from the basal level of about 200 to about 1×10^6, and then falls back to the basal level again. Of the other genes investigated in animal systems, the available evidence suggests that even in highly specialized cells, making large quantities of one or two proteins, no transient amplification of the genes coding for those proteins occurs during cell differentiation. Therefore, until more is known about the identity of the DNA which turns over, and how much of the total DNA is involved in the turnover, the significance of DNA turnover remains a matter for conjecture.

A phenomenon which may be associated with DNA turnover is the phenomenon of plasticity in the amount of DNA per nucleus (excluding those changes caused by "polyploidy": see Chapter 5). One of the most interesting examples of plasticity occurs in *Linum* (flax). There is a plastic genotroph (*Pl*) of *Linum*, in which changes in the amount of DNA may be induced by changes in the environment [111]. If *Pl* plants are grown under conditions of high soil nitrogen, the plants become very large (large genotroph, *L*), and there is an increase in the amount of DNA per nucleus. Plants grown under conditions of high soil phosphorus remain small (small genotroph, *S*) and there is a decrease in the amount of DNA per nucleus. The changes in the amounts of nuclear DNA are such that individuals of the *L* genotroph contain 16% more DNA per nucleus than individuals of the *S* genotroph. The characteristics of each genotroph are stable, and can be inherited over many generations (although certain environmental conditions, such as low temperatures, cause both the *L* and *S* genotrophs to revert to the *Pl* genotroph). The difference in nuclear DNA content between the *L* and *S* genotrophs is partly due to a difference in the amount of "intermediate" DNA, as detected by reannealing kinetics [96,

478]. The difference in the amount of intermediate DNA includes a difference in the level of amplification of the genes coding for ribosomal RNA [477]. However, other types of DNA must also be involved, since the change in the number of genes coding for ribosomal RNA could not give more than a 1% difference between the nuclear DNA contents of the L and S genotrophs.

Many differentiated plant cells are "polyploid": instead of containing an amount of DNA equivalent to a diploid chromosome set (i.e. a 2C amount of DNA), they contain 4C or 8C amounts of DNA (and in some instances, even more). This in itself is not evidence for DNA plasticity. It simply means that complete rounds of DNA replication occur in the absence of cell division. However, in some plants, disproportionate replication of parts of the genome may occur. One of the best documented instances concerns the protocorms of the orchid *Cymbidium* [322]. During differentiation and maturation of the protocorm parenchyma cells and of the cells in the roots which grow from the protocorm, repeated cycles of DNA replication occur giving nuclei with 4C, 8C, 16C or 32C amounts of DNA. In addition to these nuclei in which the amounts of DNA indicate the occurrence of complete cycles of DNA replication, there are also nuclei in which the DNA content does not fall into any of the discrete classes listed above. In this latter group of nuclei, disproportionate increases of the heterochromatic (i.e. reiterated) DNA occur, so that the heterochromatin is replicated to a higher level than the euchromatin. In some of these nuclei, the difference in the levels of replication is very marked: 32C for the euchromatin, representing four cycles of DNA replication, and 1024C for the heterochromatin, representing nine cycles of DNA replication. Much of the "extra" replication of the heterochromatin takes place after the replication of euchromatin is complete. Replication of heterochromatic or reiterated DNA can therefore occur in the absence of replication of the rest of the genome. Further, the extent of the extra replication of the heterochromatin, both within an individual nucleus, and in terms of the number of nuclei involved, can be altered by treatment with plant growth substances. It is therefore likely that this extra replication of part of the genome has a physiological role.

A further instance of plasticity in the amount of nuclear DNA also involves only part of the total genome [343]. If explants of the pith of *Nicotiana tabacum* (tobacco) are cultured in nutrient medium, they undergo a period of "de-differentiation" (see Chapter 5) and then start proliferating. This behaviour is in fact exhibited by many mature differentiated plant tissues. During the period of "de-differentiation" of *Nicotiana* explants, nuclear DNA synthesis takes place, as detected by autoradiography. This synthesis is related to the appearance, at 24 h after the initiation of culture, of a heavy satellite DNA. By 48 h after the initiation of culture, the amount of DNA in the satellite band reaches a peak value. From 72 h onwards, the amount of DNA in the satellite band declines and by 120 h, the satellite is once more absent. The appearance and disappearance of the satellite is paralleled by an increase and then a decrease in the amount of DNA per nucleus. Thus, the process of "de-differentiation" is accompanied by a massive, but transient, extra replication of satellite DNA. Unfortunately, the composition of the satellite DNA in *Nicotiana* is not yet known, but by comparison with the satellite DNAs of other higher plants, it may be expected to contain reiterated DNA (see section II).

It is thus clear, both from studies of DNA turnover and from observations of plasticity in the amount of DNA per nucleus, that the nuclear DNA of higher plants is not necessarily completely stable. However, further work is obviously necessary in order to establish the significance of DNA turnover and plasticity.

SUGGESTIONS FOR FURTHER READING

DAVIDSON, J. N. (1972). "The Biochemistry of the Nucleic Acids" (7th edition). Chapman and Hall, London.

FLAMM, W. G. (1972). Highly repetitive sequences of DNA in chromosomes. *Int. Rev. Cytol.* **32**, 2–51.

KEIR, H. M. and CRAIG, R. K. (1973). Regulation of deoxyribonucleic acid synthesis. *Biochem. Soc. Trans.* **1**, 1073–1077.

LEWIN, B. (1974). "Gene Expression", Volume 2: "Eukaryotic Chromosomes". John Wiley, London and New York.

REES, H. and JONES, R. N. (1972). The origin of wide species variation in nuclear DNA content. *Int. Rev. Cytol.* **32**, 53–92.

SIMPSON, R. T. (1973). Structure and function of chromatin. *Adv. Enzymol.* **38**, 41–108.

WATSON, J. D. (1967). "The Double Helix". Weidenfeld and Nicholson, London.

WHITEHOUSE, H. L. K. (1973). "Towards an Understanding of the Mechanism of Heredity" (3rd edition). Edward Arnold, London.

2. RNA Structure and Metabolism

D. GRIERSON

I. INTRODUCTION

Ribonucleic acid molecules are the direct products of genes. They occupy a central role in cell metabolism and a knowledge of the structure and function of the various types of RNA is necessary in order to understand the complex process of protein synthesis. The production of each type of protein molecule is directed by a different RNA species and as the number and type of protein molecules present in a cell determine its nature and function it is highly probable that the regulation of RNA metabolism is of great importance during development and differentiation.

II. GENERAL PROPERTIES OF RNA

RNA molecules are straight chain polymers with molecular weights ranging from about 25 000 to several million daltons. They strongly absorb ultraviolet light because of the presence of heterocyclic bases attached to the pentose sugar, D-ribose, and they have a negative charge due to the phosphate groups which form bridges between the sugar molecules (see Fig. Fig. 2.1). Alkaline hydrolysis of RNA yields a mixture of mononucleotides which may be regarded as the repeating units, each consisting of a sugar, a phosphate group and a base. The four commonly occurring bases are the purines adenine (A) and guanine (G), and the pyrimidines uracil (U) and cytosine (C), (cf. DNA, Chapter 1) although other rare bases are sometimes present, particularly in transfer RNA (see Fig. 2.2). In a polynucleotide chain adjacent sugars are joined by $3' \rightarrow 5'$ phosphodiester bonds and each ribose forms a β-linkage to the

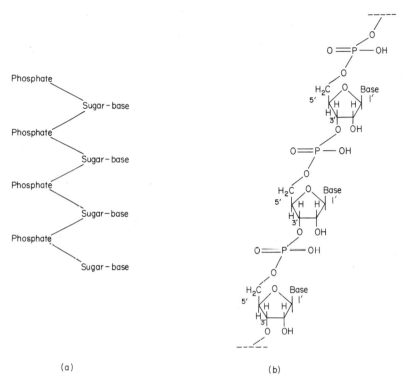

FIG. 2.1. The structure of RNA. (a) Schematic representation. (b) Structural formula of the sugar-phosphate backbone. The "primed" numbers (e.g. 3′) refer to the carbon atoms in the ribose molecule.

N(9) of a purine or N(1) of a pyrimidine. Individual RNA molecules may contain from a few dozen to several thousand nucleotides. Unlike DNA, most RNA molecules are single stranded although they nevertheless have a high degree of secondary structure. This is due firstly to the interaction between successive bases along the polynucleotide chain, a phenomenon known as base stacking, and secondly to the formation of intramolecular hydrogen bonds between complementary sequences of bases. The leads to the presence of short double-stranded segments within the molecule, often referred to as "hairpin loops".

III. RNA EXTRACTION

In order to study the properties of RNA it is necessary to break open the

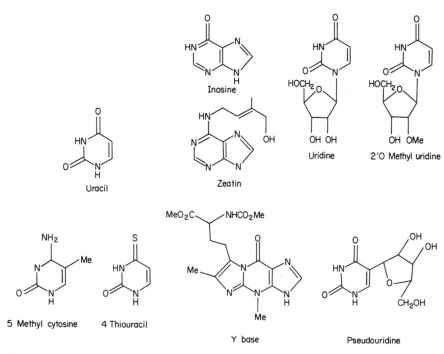

FIG. 2.2. Bases and nucleosides present in RNA. Of the four common bases occurring in RNA, only uracil is shown (adenine, cytosine and guanine are depicted in Fig. 1.1). 5-methyl cytosine is the only methylated base illustrated, but many other base methylations are possible. Inosine is often found in the anti-codon of tRNA and zeatin is sometimes adjacent to the anti-codon. Variations also occur in the ribose molecule, and three nucleosides are shown to illustrate this. Pseudouridine is unusual because it has a C-C linkage between base and ribose. Uridine and 2'-O-methyl uridine are examples of nucleosides without and with a methylated ribose unit.

cells and to extract and separate it from the other types of molecules with which it is associated.

The simplest means of homogenizing plant material is to grind the tissue by hand with a pestle and mortar in a suitable medium. High speed homogenizers which shear or cut the material are often used but these may damage chloroplasts and nuclei and some workers prefer more gentle methods such as chopping with razor blades. In order to minimize damage to cell organelles the cell walls can, in suitable cases, be removed by digestion with enzymes and the resulting protoplasts can be gently lysed and separated into sub-cellular fractions. Once the cells are disrupted, the RNA is very susceptible to degradation by endogenous ribonucleases and precautions must be taken to inhibit the action of these enzymes. If total cell RNA is to be extracted, detergents such as sodium tri-*iso*-propylnaphthalene sulphonate, sodium dodecylsulphate or other nuclease inhibitors may be included in the medium in order to prevent the action of ribonuclease. However, many of the more potent reagents disrupt membranes or denature proteins; this precludes their use during the initial isolation of organelles such as nuclei, chloroplasts and polyribosomes. In such cases the tissue is generally homogenized in the cold in a suitable osmoticum in an attempt to keep the organelles intact. Low temperatures, high pH or high salt concentrations are used to minimize ribonuclease action. Following preparation of the organelles, for example by differential centrifugation, they may be lysed with strong detergents as when extracting total cell RNA.

Most RNA molecules are associated with proteins *in vivo* and it is necessary to remove these during RNA purification. This is most commonly achieved by shaking the homogenized cell contents with aqueous phenol (often containing other deproteinizing agents) to form an emulsion. This treatment denatures the proteins and when the aqueous and phenol layers are separated by a brief centrifugation, the precipitated proteins appear at the interface and insoluble material, such as starch grains and cell walls, sediments to the bottom of the centrifuge tube. The nucleic acids remain in solution and can be precipitated by adding ethanol to a final concentration of about 70% and cooling the solution. After precipitation with ethanol the RNA can be redissolved and reprecipitated to remove traces of phenol and detergents. Alternatively, RNA may be separated from low molecular weight compounds by gel filtration or ultracentrifugation. A treatment with deoxyribonuclease may be incorporated into the purification procedure, or the RNA and DNA may be separated by any of the fractionation methods described later.

Despite its widespread application there are some disadvantages in the use of phenol as a deproteinizing agent. Not only does it selectively precipitate A–T-rich DNA but with some plants, such as *Pinus* and *Dryopteris*, much of the high molecular weight RNA is not recovered. This difficulty may be overcome by the use of chloroform as an alternative deproteinizing reagent. In addition, at least some messenger RNA molecules which contain poly-(adenylic acid) are, under some conditions, precipitated, or alternatively lose the poly-(A)-containing region when treated with phenol. Again the use of chloroform or chloroform–phenol mixtures to remove proteins seems to overcome this problem.

IV. TYPES OF RNA

RNA molecules perform a variety of functions in cell metabolism and different classes are recognized by properties related to their metabolic role, such as molecular size and conformation, nucleotide composition and subcellular location. The three main types of RNA are ribosomal, transfer and messenger

RNA (rRNA, tRNA, mRNA) which are all involved in protein synthesis. These molecules are initially synthesized in the nucleus and are transported to the cytoplasm where protein synthesis takes place. RNA molecules with similar functions, but with slightly different properties, are also synthesized inside chloroplasts and mitochondria, where they are also involved in protein synthesis (see Chapter 4).

A. Ribosomal RNA

Each plant cell generally contains a large number of ribosomes (estimated to be several million) composed of approximately 40% protein and 60% RNA. This rRNA accounts for about 70% of the RNA of the cell and forms an integral part of the structure of the ribosomes. Ribosomal particles are characterized by their sedimentation coefficients which are proportional to the rate of sedimentation in the ultracentrifuge. This technique has been used to show that different types of ribosomes are present in chloroplasts, mitochondria and the cytoplasm. Ribosomes from plant cytoplasm have sedimentation coefficients of approximately 80s and each ribosome consists of two subunits which sediment at 60s and 40s respectively (Fig. 2.3a). When RNA is purified from cytoplasmic ribosomes and analysed, three distinct molecules are normally found to be present in each ribosome. The 40s subunit contains a single RNA molecule with a sedimentation coefficient of 18s and the larger (60s) subunit contains two molecules of 25s and 5s (Fig. 2.3b). The molecular weights of these rRNA molecules, measured in millions, and the approximate number of nucleotides contained in each, are shown in Table 2.I. The corresponding RNAs from the ribosomes of chloroplasts and mitochondria have slightly different properties (see Chapter 4).

B. Messenger RNA

When cells are exposed to radioactive amino acids for short periods of time, newly synthesized proteins are first found associated with ribosomes which are thus shown to be the sites of protein synthesis. It follows that there must be molecules which carry the genetic information for the synthesis of proteins from the DNA in the nucleus to the ribosomes in the cytoplasm. Ribosomal RNA itself is too homogeneous in size and nucleotide composition to be the template for protein synthesis; this function is carried out by mRNA molecules. Ribosomes actively engaged in protein synthesis are often associated in groups called polyribosomes which consist of a string of ribosomes attached to a strand of mRNA (Fig. 2.4). Only about 2% of the total cell RNA is mRNA. It is difficult to purify and because it is present in small amounts it is usually detected by labelling with radioactive tracers. The length and nucleotide

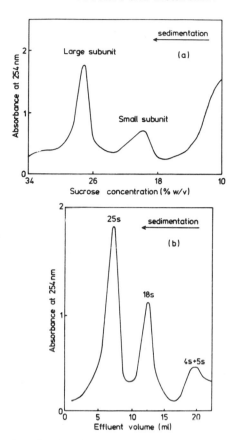

FIG. 2.3. Sucrose gradient fractionation of ribosomal sub-units and ribosomal RNA. (a) Ribosomal subunits from cytoplasmic (80s) ribosomes from *Vicia faba* (broad bean) (from ref. 351). (b) Cytoplasmic RNA from *Brassica rapa* (turnip) (from ref. 266).

sequence of mRNA molecules determine the number and sequence of amino acids in the corresponding proteins. Individual mRNAs coding for particular proteins are homogeneous, but most cells synthesize a large number of different proteins and therefore the mRNA fraction consists of a population of molecules of different sizes and nucleotide sequences. For this reason messenger-like RNA is often referred to as polydisperse RNA.

C. Transfer RNA

The participation of "adaptor molecules" in protein synthesis was postulated by Crick [92] and this suggestion prompted experiments that led to the

D. Grierson

TABLE 2.I The Molecular Weights of Ribosomal RNA

	Large ribosomal subunit	Small ribosomal subunit
(a) 80s CYTOPLASMIC RIBOSOME		
s-value of RNA	25; 18	25 − 26; 5
molecular weight, (daltons)	$1.27 - 1.34 \times 10^6$; 3.8×10^4	$0.69 - 0.7 \times 10^6$
approximate number of nucleotides	3900; 118	2100
(b) 70s CHLOROPLAST RIBOSOME		
s-value of RNA	23; 5	16
molecular weight (daltons)	$1.07 - 1.11 \times 10^6$; 3.9×10^4	0.56×10^6
approximate number of nucleotides	3300; 122	1650
(c) 78s MITOCHONDRIAL RIBOSOME		
s-value of RNA	24; 5	18
molecular weight (daltons)	$1.12 - 1.18 \times 10^6$; 3.8×10^4 (?)	$0.69 - 0.78 \times 10^6$
approximate number of nucleotides	3400; 120 (?)	2200

detection of tRNAs. These are small RNA molecules, with a sedimentation coefficient of about 4s. They have molecular weights of approximately 25 000 and each molecule consists of a single chain of about 75 nucleotides. There are a large number of types of tRNA molecule with different nucleotide sequences but they all have a similar secondary and tertiary structure. During protein synthesis they become charged with specific amino acids which are attached by special enzymes called synthetases. Individual types of tRNA are distinguished by the amino acid they carry and by the anti-codon, a sequence of three bases on the tRNA which are able to form hydrogen bonds with the bases in the codons in mRNA. The base-pairing between the mRNA codons and the anti-codons of the tRNAs ensures that the amino acids are correctly aligned prior to incorporation into nascent polypeptide chains.

V. RNA FRACTIONATION

The three procedures most commonly used for fractionating RNA molecules

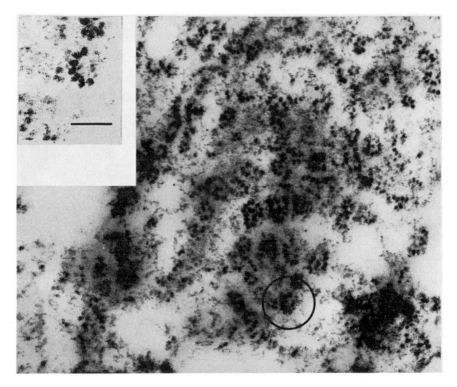

FIG. 2.4. Polyribosomes in a cotyledon cell from *Vicia faba* (broad bean). The inset (upper left) shows polyribosomes (from the ringed area of the large picture) at a higher magnification. The sub-unit structure of the ribosomes is discernible. Scale line — 100 nm. (From ref. 54, copyright 1969 Oxford University Press).

are sucrose gradient centrifugation, chromatography on kieselguhr coated with methylated albumin, and electrophoresis in polyacrylamide gels. Typical results obtained with each of these methods are compared in Fig. 2.5.

A. Sucrose gradient fractionation

This involves the sedimentation of RNA molecules through a gradient of buffered sucrose solution in an ultracentrifuge. The gradient is generally prepared by passing a dense sucrose solution through a simple mixing device into a centrifuge tube while continuously diluting the solution in the mixing chamber with less dense sucrose. A small volume of RNA solution is then layered on top of the sucrose gradient and the contents of the tube are centrifuged at high speed for several hours. Under these conditions RNA

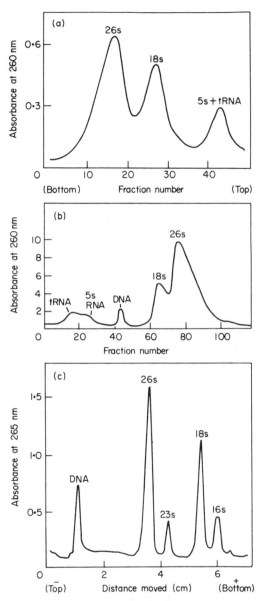

FIG. 2.5. Different methods for fractionating RNA. (a) Sucrose density gradient centrifugation of RNA from soya bean (*Glycine max*). Centrifugation was for 16 h at 23 000 rev min⁻¹ in a 5–20% sucrose gradient (from ref. 208). (b) Fractionation of soya bean nucleic acids by chromatography on columns of methylated albumin-kieselguhr (MAK). Nucleic acids were eluted in a linear gradient of 0·45–1·35 M KCl in potassium phosphate buffer, pH 6·7 (from ref. 436). (c) Electrophoresis of nucleic acids from leaves of tobacco (*Nicotiana tabacum*) in a 2·4% polyacrylamide gel. Electrophoresis was carried out for 3 h (from ref. 161).

molecules enter the gradient and sediment towards the bottom of the tube. The rate of sedimentation is influenced by the size and conformation of the molecules and the gradient of sucrose prevents diffusion of the RNA bands. Generally, the larger the molecule the faster it sediments, although two molecules with similar molecular weights may sediment at different rates if one is more compact than the other.

After centrifugation, fractions are collected from the centrifuge tube and the RNA located by the measuring the ultraviolet absorbance or radioactivity of each fraction (see Chapter 1).

B. Chromatography on Methylated–Albumin–Kieselguhr (MAK) columns

In this method, albumin is first methylated and absorbed to kieselguhr particles. The MAK is then packed into a column and used for chromatography. When nucleic acid molecules are passed through the column they bind to the methylated albumin and can be eluted with solutions of high ionic strength. The degree of binding of nucleic acid molecules to the column is determined by the molecular size, secondary structure and base composition and different classes of RNA are eluted at different salt concentrations. The fractions are normally washed from the column in a gradient of increasing salt concentration and collected and analysed.

C. Polyacrylamide gel electrophoresis

Acrylamide solutions can be polymerized and cross-linked by employing various catalysts to produce a gel of finely controlled pore size. If a solution containing a mixture of nucleic acid molecules is layered on top of the gel and subjected to electrophoresis, the RNAs migrate towards the anode because of their negative charge. The molecules pass through the gel via the pores and larger molecules migrate more slowly than smaller ones because their progress is impeded by the gel. By varying the concentration of acrylamide it is possible to construct gels of different pore sizes suitable for fractionating either large or small molecules (see Fig. 2.31). After electrophoresis the nucleic acid bands can be located by a number of methods, e.g. by scanning the gel with ultraviolet light or by staining. The gel can also be sliced or dried to detect radioactivity by scintillation counting or autoradiography. The fractionation of high molecular weight RNA by gel electrophoresis was first performed by Loening [282]. The method offered much greater resolution than was previously possible and enabled Loening and Ingle to distinguish the rRNAs from chloroplast and cytoplasmic ribosomes, which differ slightly in molecular weight [286]. This is illustrated in Fig. 2.6 in which the RNA from leaves is compared with that from

roots. Loening established that for many RNA molecules the distance moved
during electrophoresis is inversely proportional to the log of the molecular
weight [284]. Although this relationship provides a simple method of estimating the size of many RNA molecules, it is not universally applicable. Molecules
with an unusual base composition or secondary structure may behave anoma-
lously with respect to the standard marker RNAs. For this reason, electrophoresis in the presence of formamide or urea, which unfolds the molecules and
abolishes secondary structure effects, is often a more precise method of
measuring molecular weights [381].

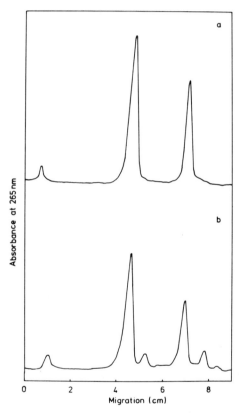

FIG. 2.6. Gel electrophoresis of nucleic acids from roots (a) and leaves (b) of *Phaseolus aureus*
(mung bean). Electrophoresis was carried out for 4 h in 2·4% gels (from ref. 161).

D. The fractionation of low molecular weight RNA

A limited fractionation of nucleic acid samples is achieved by precipitating high molecular weight RNA from solution in the presence of concentrations of sodium chloride or sodium acetate of the order 3–4 M. Low molecular weight RNA and DNA remain soluble under these conditions. Alternatively, DEAE-cellulose chromatography can be used to retard high molecular weight RNA and allow the elution of 5s and 4s RNA separately. This can also be achieved with MAK columns and the 5s and 4s RNA can be completely resolved from each other if shallow salt gradients are used for elution. Molecular sieving on Sephadex G100 also separates high molecular weight RNA from 5s and 4s RNA but achieves very little fractionation of the tRNAs. A more effective method is gel electrophoresis which completely resolves 5s RNA and a broad 4s RNA region. At higher gel concentrations the tRNAs begin to be fractionated on the basis of molecular size. This has only limited success because many different tRNAs are similar in size.

A more subtle approach is to exploit the differences between the tRNAs caused by the various modified bases that they contain, and there are a number of column chromatography systems that achieve this very well [344]. An example of the separation of plant tRNAs by reverse phase chromatography is shown in Fig. 2.7. In this case the tRNA preparations were charged with a radioactive amino acid before fractionation and it can be seen that there are normally several isoaccepting tRNAs that can be charged with the same amino acid. Quantitative changes in the amounts of different isoaccepting tRNAs have been observed to occur during development but the significance of this is not clear. One explanation of this is that it reflects changes in the protein-synthesizing activity of the chloroplasts and mitochondria. Figure 2.8 shows that of the six leucyl tRNAs that can be detected in tobacco leaves, two are present in the mitochondria and two others are confined to the chloroplasts.

VI. DNA–RNA HYBRIDIZATION

A. General principles

Sequence homologies between DNA and RNA can be studied using the technique of DNA–RNA hybridization. This involves the formation *in vitro* of a hybrid double helix with one strand composed of DNA and the other strand of RNA. Such experiments provide information about the genetic origin of RNA molecules.

If a solution of double-stranded DNA molecules is subjected to high temperatures or high pH the two strands separate—a process known as denaturation or melting (see Chapter 1). Upon lowering the temperature or returning the pH to neutrality the two strands can reassociate to form a native double helix. In the presence of a RNA sequence complementary to one of the

D. Grierson

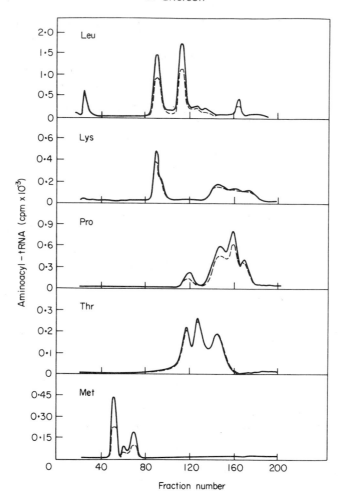

FIG. 2.7. Reverse-phase chromatography of tRNA from roots of pea (*Pisum sativum*) seedlings. Separate samples of total tRNA were charged with one of the following radioactive amino acids: leucine, lysine, proline, threonine, methionine. The samples were then fractionated by reverse-phase chromatography. The charged tRNAs were detected by measuring the amount of attached radioactive amino acid. Note that there are several iso-accepting tRNAs for each of the amino acids used (from ref. 489).

DNA strands, DNA–RNA hybrids are also produced. If the separated DNA strands are attached to an insoluble support, such as nitro-cellulose, this permits the formation of DNA–RNA hybrids in the absence of DNA–DNA reassociation. The formation of such hybrid molecules depends upon the fact that the nucleotide sequence of an RNA molecule is complementary to the

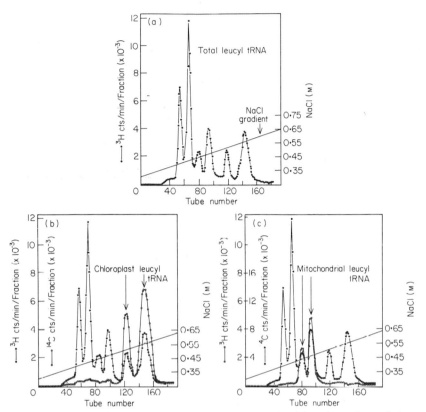

FIG. 2.8. Comparison of leucyl tRNA from chloroplasts, mitochondria and cytoplasm of tobacco (*Nicotiana tabacum*) leaves. (a) Total cellular tRNA charged with ^{3}H-leucine and fractionated by reverse-phase chromatography. (b) Mixture of total cellular leucyl tRNA labelled with ^{3}H-leucine and chromoplast leucyl tRNA labelled with ^{14}C-leucine. (c) Mixture of total cellular leucyl tRNA labelled with ^{3}H-leucine and mitochondrial leucyl tRNA labelled with ^{14}C-leucine. The NaCl gradient used for elution is shown (from ref. 166).

sequence in one of the two strands of the DNA from which it was transcribed. The complementary sequences can therefore form hybrids stabilized by hydrogen bonds. The rate of the hybridization reaction is governed by a number of factors such as the nucleic acid concentration, the reaction time, the salt concentration and the temperature. Highly radioactive RNA is normally used in such experiments so that the hybrid can be detected by the presence of radioactivity. At the end of the reaction, RNA not specifically involved in hybrid formation is selectively digested with low concentrations of ribonuclease. This does not affect the RNA involved in hybrid formation which is more resistant to ribonuclease because it forms a helical structure with a strand of DNA.

B. The genes for rRNA

One common hybridization procedure is to attach single-stranded DNA to nitrocellulose filters and to expose separate DNA samples to increasing concentrations of RNA so that a saturation curve is obtained. A typical result observed when rRNA is hybridized to nuclear DNA is shown in Fig. 2.9a. At low RNA concentrations the reaction does not normally reach completion but as the RNA concentration is increased, a greater proportion of the available sites in the DNA becomes saturated with rRNA. A similar type of curve is obtained if the hybridization is carried out at a single RNA concentration and the reaction time is varied (Fig. 2.9b). The $25s$ and $18s$ rRNAs hybridize independently with DNA to produce different saturation values. The combined numerical value of the two plateaux is equal to that obtained when the two types of RNA are hybridized together. This shows that the $25s$ and $18s$ rRNA sequences are coded for by separate DNA sequences or cistrons. Often the ratio of the two saturation plateaux is approximately $2:1$, the same as the ratio of the molecular weights of the two rRNAs, and this indicates that there are equal numbers of cistrons for each type of rRNA (see Fig. 2.10). In some hybridization experiments deviation from this behaviour has been detected. This will occur if there is some sequence homology between the two types of rRNA or if the RNA samples used for hybridization are not pure.

If the total amount of DNA is known, together with the specific radioactivity of the rRNA used in the reaction, the percentage of DNA that codes for rRNA may be calculated from the saturation plateau. This figure varies with the species studied but, generally speaking, between $0 \cdot 1$ and $1 \cdot 0\%$ of the nuclear DNA of higher plants is complementary to rRNA. This means that there are several thousand genes for rRNA in each diploid nucleus (see Table 2.II). Similar experiments with $5s$ and tRNA show that there are also multiple genes coding for these molecules [476].

The functional significance of such a multiplicity of genes for these RNA molecules is quite clear. A plant cell often contains several million ribosomes, and actively growing cells, perhaps dividing once every 24 h, must synthesize this number of new ribosomes every day. As it takes several minutes to transcribe a single gene for rRNA, particularly at the low temperatures at which plants often grow, it follows that in order to manufacture ribosomes at the required rate there must be multiple sites of synthesis. This is achieved partly by having multiple genetic copies and partly by transcribing each gene at a number of points simultaneously (see Fig. 2.18).

To be sure that hybrid formation reflects a true sequence homology between the DNA and RNA strands it is important to establish the degree of base-pairing which gives an estimate of the specificity of the hybrid. Although ribonuclease is often used to remove mismatched sequences, short stretches of

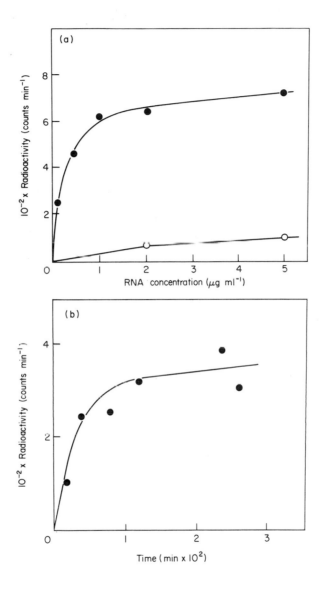

FIG. 2.9. Saturation hybridization of DNA with ribosomal RNA. (a) Purified 25s rRNA from *Phaseolus aureus* (mung bean) was hybridized to 10 μg aliquots of DNA for 3 h at 68°C in the presence of 0·99 M Na⁺ (closed circles), or to DNA minus the genes coding for rRNA, under the same conditions (open circles). The specific activity of the RNA was 30 000 counts min⁻¹ μg⁻¹. See also Fig. 2.12 (from ref. 159). (b) Purified 25s rRNA was hybridized, at a concentration of 1·5 μg ml⁻¹, for different times to 5 μg aliquots of DNA at 65°C in the presence of 0·99M Na⁺ (from ref. 161).

D. Grierson

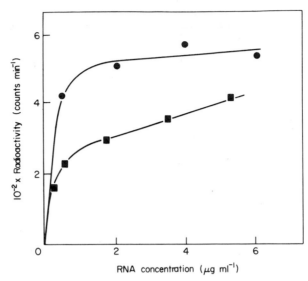

FIG. 2.10. Hybridization of 25s and 18s rRNA to DNA. ^{32}P-labelled 25s or 18s rRNA was hybridized to 5 μg aliquots of DNA for 3 h at 70°C in the presence of 0·33 M Na$^+$. The nucleic acids were from *Phaseolus aureus* (mung bean) (from ref. 161).

TABLE 2.II. Number of ribosomal RNA genes in different plants[a]

Plant	%DNA hybridized	Number of genes per haploid complement
Swiss chard (*Beta vulgaris cicla*)	0·20	1150
Maize (*Zea mays*)	0·18	3100
Wheat (*Triticum vulgare*)	0·092	2100
Cucumber (*Cucumis sativum*)	0·96	4400
Onion (*Allium cepa*)	0·090	6650
Pea (*Pisum sativum*)	0·17	3900
Artichoke (*Helianthus tuberosus*)	0·022	260

[a] From ref. 211

unpaired bases may be unaffected by this treatment. A better test is to study the thermal denaturation of the hybrid. A melting profile with a sharp transition indicates a good hybrid. As with DNA, the T_m is related to the base composition of the nucleic acid. For 25s rRNA from *Pisum* (which has a G–C content of 53·7%) hybridized to DNA, T_m is 88·5°C (Fig. 2.11).

In higher plants the genes for rRNA generally have a higher content of G and C than the bulk of the DNA. In consequence, when DNA is fractionated by

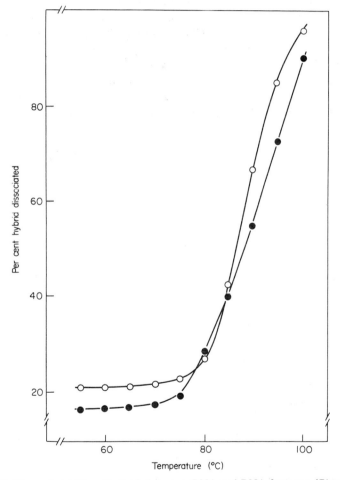

FIG. 2.11. Thermal stability of hybrids between rRNA and DNA from pea (*Pisum sativum*). Hybrids were formed between DNA and either 25s rRNA or 18s rRNA, and then subjected to treatment at different temperatures. The amount of radioactive rRNA released at different temperatures was taken as a measure of the thermal dissociation of the hybrid molecules (from ref. 411).

buoyant density centrifugation in caesium chloride (Chapter 1, section IID) the rRNA genes may separate from the main band DNA [411]. The position of these genes can be shown by hybridizing fractions from the gradient with rRNA as illustrated in Fig. 2.12. Using this approach and exploiting the fact that the number of rRNA genes is greatly amplified in eggs of the toad *Xenopus*, Birnstiel has isolated the genes for rRNA in pure form and studied them in great detail [34]. The results show that the two types of rRNA cistrons alternate along the DNA molecule and that they are separated by sequences that do not code for rRNA (Fig. 2.13). Such an organization is entirely consistent with the polycistronic transcription of both sequences of rRNA as part of a single precursor molecule (see section VII). Although plant rRNA genes have not been studied in such detail, the general principles of their organization seem to be similar to those established for *Xenopus* [481].

Cytogenetic investigations in the fruit-fly *Drosophila* and in *Xenopus*, combined with DNA–RNA hybridization experiments, have established that the genes for rRNA are localized in the nucleolus. This is also true of plants [481].

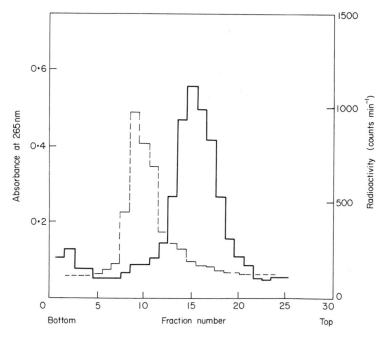

Fig. 2.12. Hybridization of rRNA to DNA fractions from a caesium chloride gradient DNA from mung bean (*Phaseolus aureus*) was centrifuged in caesium chloride for 72 h at 28 000 rev min^{-1} at 25°C. The gradient was fractionated as described in Chapter 1, and each fraction was used for hybridization with 25s rRNA. The solid histogram shows the distribution of DNA; the dashed histogram indicates the hybridization (from ref. 162).

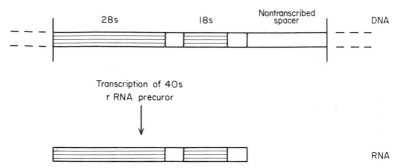

FIG. 2.13. The repeating sequence of DNA which codes for ribosomal RNA in *Xenopus* (clawed toad) (from ref. 510).

A very elegant method of demonstrating this is to hybridize highly radioactive rRNA to a cytological preparation in which the DNA has been denatured but the chromosome morphology has not been destroyed. The site where the rRNA hydridizes can then be detected by autoradiography. When this is done, the results show that the genes coding for rRNA are indeed in the nucleolus [49].

C. The hybridization of mRNA to DNA

Hybridization may also be used to characterize mRNA and the sequences coding for it. Theoretically, it is possible to use this technique to determine what proportion of the genome is transcribed, and how many gene copies code for a particular mRNA, and to compare different populations of mRNA. There are however two major technical difficulties. Firstly, it is difficult to separate mRNA from the other RNA species in the cell, and very difficult to purify individual mRNA species. Secondly, the genes coding for mRNAs are often represented only twice in the diploid genome (i.e. are "unique" sequences). Reaction rates are very slow and hybridization may not reach completion. This problem may be overcome by using very high RNA concentrations and long incubation times, often with both DNA and RNA in solution. An alternative approach is to use only a trace of very highly radioactive RNA in the presence of an excess of DNA. Under these conditions, the rate of hybridization, in comparison to the renaturation rate of the DNA (see Chapter 1) gives an indication of the number of DNA sequences complementary to the RNA under investigation.

VII. SYNTHESIS OF RIBOSOMAL RNA

As mentioned earlier, ribosomal RNA is present in cells in large amounts. Each type of rRNA is homogeneous in molecular weight and nucleotide composition and is therefore easy to identify and purify. Further, the genes that code for rRNA occur in multiple copies and can be separated from the bulk of the cellular DNA. With these operational advantages it is not surprising that our knowledge of the synthesis and properties of rRNA is quite extensive. In addition to providing information about ribosome biogenesis the study of rRNA metabolism serves to illustrate a number of important principles that are probably also applicable to mRNA metabolism.

A. rRNA precursors in blue-green algae

The synthesis and metabolism of RNA molecules is often studied by feeding cells with radioactive precursors of RNA such as ^{32}P-orthophosphate, or a ^{3}H or ^{14}C-labelled base, e.g. uracil, or a nucleoside, such as uridine. Inside the cell these compounds are converted into nucleoside triphosphates, the actual substrates for the RNA synthesizing enzymes, and the radioactivity becomes incorporated into RNA. When cells are labelled for a short time with ^{3}H-uridine or ^{32}P-phosphate and the RNA extracted and fractionated by gel electrophoresis, a number of radioactive RNA peaks can generally be detected superimposed upon a background of polydisperse RNA [164]. Figure 2.14 shows the results of such an experiment with the blue-green alga *Tolypothrix distorta*. After about 15 min labelling with ^{3}H-uridine, the main peaks of radioactivity correspond to RNAs with molecular weights of $1 \cdot 22$ and $0 \cdot 76 \times 10^6$ daltons. These RNA fractions are in fact macromolecular precursors of rRNA. They are difficult to detect by u.v. scanning of the gel because although highly radioactive they are present only in small amounts. The metabolic fate of these molecules may be studied if after a short labelling period the cells are washed, to remove radioactive uridine, and replaced in a medium containing an excess of unlabelled uridine. RNA is then extracted from different batches of cells at intervals, fractionated by gel electrophoresis and the amount of radioactivity in each fraction is determined. For a short time after removal of ^{5}H-uridine from the culture medium, radioactivity already taken up by the cells continues to accumulate in the non-rRNA fractions (Fig. 2.15). However, these RNA molecules are metabolically unstable and the amount of radioactivity in each eventually declines. A third RNA with a molecular weight of $0 \cdot 68 \times 10^6$ daltons becomes transiently labelled but after a time the amount of radioactivity in this fraction also decreases. At about the same time there is a rapid increase in the rate of accumulation of radioactivity in rRNA (Fig. 2.15).

From this type of observation it is concluded that the rRNA sequences are initially synthesized as part of larger precursor molecules which are subsequently processed by special enzymes which remove the non-rRNA and convert the precursors to mature rRNA. In *Tolypothrix distorta* the larger

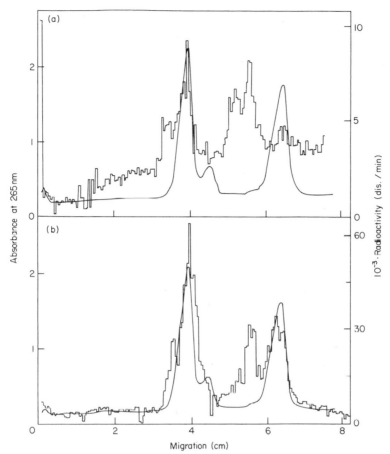

FIG. 2.14. Pulse-labelled RNA from the blue-green alga *Tolypothrix distorta*. (a) RNA from cells labelled for 30 min with ³H-uridine. (b) RNA from cells labelled for 30 min and then incubated in non-radioactive medium for a further 2·5 h. In each instance, DNA was removed by deoxyribonuclease treatment prior to gel electrophoresis (from ref. 164).

rRNA (molecular weight $1·1 \times 10^6$ daltons) is formed by processing of the $1·22 \times 10^6$ daltons precursor. In contrast, the $0·76 \times 10^6$ daltons RNA is first converted into an intermediate with a molecular weight of $0·68 \times 10^6$ daltons and this latter molecule finally gives rise to the smaller rRNA (molecular weight $0·56 \times 10^6$ daltons). The results of similar studies with the blue-green alga *Anacystis nidulans* have also revealed rRNA precursors. In this species, however, the precursor of the larger rRNA is not always observed and there appears to be only a single processing step during the production of the smaller rRNA from its precursor [107, 414, 464]. It has also been shown in *Anacystis*

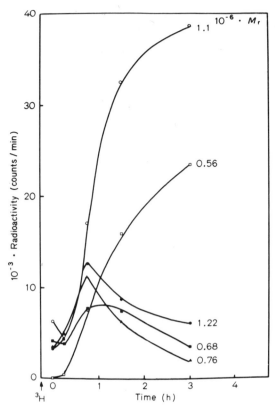

FIG. 2.15. The distribution of radioactivity in different RNAs from *Tolypothrix distorta* during a "pulse-chase" experiment. Cells were incubated for 15 mins in ^{3}H-uridine and then transferred to a non-radioactive medium. Samples of cells were harvested at intervals and the RNA was extracted and fractionated by gel electrophoresis. The radioactivity in each RNA fraction was determined (from ref. 164).

nidulans that the rRNA sequences become further modified at some stage after synthesis by the introduction of methyl groups, and this is probably a general phenomenon during rRNA synthesis.

B. rRNA precursors in higher plants

In contrast to the situation in blue-green algae and bacteria, where the rRNA precursors are only slightly larger than the final rRNA products, in higher plants and animals much larger precursor molecules are synthesized [163, 285, 358]. Figure 2.16 shows the rapidly-labelled RNA from artichoke (*Helianthus tuberosus*) tuber tissue fractionated by gel electrophoresis. The peaks of radioactivity correspond to RNA fractions with molecular weights of 2·3, 1·4 and 0·9 × 10⁶ daltons. These molecules have only a short half-life and do not normally accumulate in large amounts. The mature 25s and 18s rRNAs (molecular weights 1·3 and 0·7 × 10⁶ daltons) are located by their optical density. They have a much lower specific radioactivity than the precursors themselves because only the recently synthesized rRNA molecules are radioactive. As shown in Table 2.III these rapidly labelled RNA components have a very similar base composition to rRNA. This was one of the first indications that they are in fact rRNA precursors. The polydisperse RNA has a much lower content of G and C than has rRNA and is more similar in composition to the bulk of the nuclear DNA. In some plant tissues, including roots of pea (*Pisum sativum*) and tubers of artichoke (*Helianthus tuberosus*), there is a single large precursor molecule with a molecular weight within the range 2·2–2·6 × 10⁶ daltons [391]. In other instances, including carrot (*Daucus carota*) roots and cultured sycamore (*Acer pseudoplatanus*) cells, there is an addition a second, even larger precursor (Fig. 2.17, refs 90, 268).

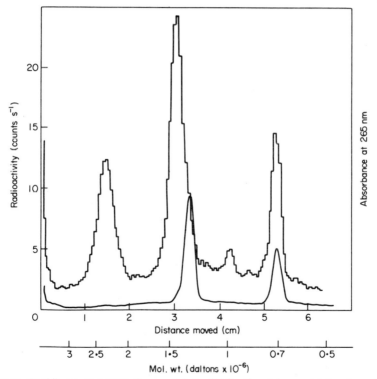

FIG. 2.16. Rapidly-labelled RNA from artichoke (*Helianthus tuberosus*) tuber tissue. Tuber explants were cultured under aseptic conditions for 30 h, then incubated in ³²P-orthophosphate for 15 min, followed by non-radioactive phosphate for 45 min. Nucleic acids were extracted. The DNA was removed with deoxyribonuclease and the RNA was fractionated by electrophoreis in 2·4% polyacrylamide gels (from ref. 391).

D. Grierson

TABLE 2.III. Base composition of artichoke RNA components[a]

Molecular weight of RNA (daltons × 10⁻⁶)	Type of RNA	Mol%				
		C	A	G	U	G+C
>3	Heterogeneous	19·5	32·0	26·9	21·8	46·4
2·3	Precursor	21·1	27·7	30·0	21·3	51·1
1·4	Precursor	21·0	28·8	31·3	18·7	52·3
1·3	Ribosomal	20·1	29·2	31·1	19·4	51·2
0·7	Ribosomal	20·3	26·9	28·3	24·5	48·6
	Long Label					
1·3	Ribosomal	21·6	28·1	31·5	18·9	53·1
0·7	Ribosomal	20·5	27·6	28·1	23·9	48·6

[a] From ref. 391

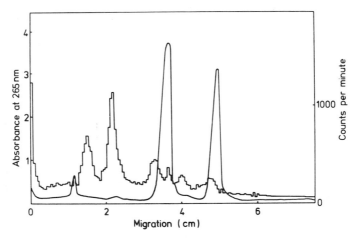

FIG. 2.17. Rapidly-labelled RNA from cells of sycamore (*Acer pseudoplatanus*) grown in suspension culture. Cells were labelled for 40 min with ³H-uridine. The nucleic acids were extracted and fractionated by gel electrophoresis (from ref. 161).

When pulse-labelled cells are separated into nuclear and cytoplasmic fractions prior to nucleic acid extraction, the precursors to rRNA are detected in the nucleus. It is here that they are first transcribed from the multiple genes coding for rRNA that are clustered in the nucleolus. When carefully prepared nucleolar preparations are viewed with the electron microscope the repeating units of rRNA sequences along the DNA can be clearly seen, each with a number of growing chains of rRNA precursor attached (Fig. 2.18). If plant tissues are labelled for a short time and then incubated in non-radioactive medium, the processing of the precursors and the transport of newly synthesized rRNA to the cytoplasm can be followed. As shown in Fig. 2.19 the large precursor molecules never leave the nucleus but are converted to smaller products by special processing enzymes a few minutes after synthesis. After a lag period, during which time the precursor RNAs may undergo further modification, radioactive 18s rRNA begins to accumulate in the cytoplasm, followed after a slightly longer delay by 25s rRNA [268]. Shortly after synthesis of the precursor molecule the rRNA sequences become methylated at specific sites. In eukaryotes, most if not all of the methyl groups are attached to the 2′ carbon atom of the ribose moeity of the rRNA but only about 1% of the residues are modified in this way. From the molecular weight estimates it is clear that the precursor molecule is sufficiently large to contain the combined sequences of the 25s and 18s rRNA plus between 10 and 30% non-rRNA. In hybridization experiments between DNA and radioactive precursor RNA, the formation of hybrids can be inhibited in the presence of an excess of non-radioactive rRNA. Furthermore, both 25s and 18s rRNA compete with the precursor for sites on the DNA [162]. This indicates that the precursor contains sequences of both types of rRNA. Confirmation of this has been obtained by partial sequence analysis. When RNA molecules are digested by ribonuclease enzymes a large number of fragments is produced. The number and properties of the fragments depend upon the source of the RNA and the specificity of the enzymes used. Under controlled conditions a particular RNA sequence always gives rise to the same fragments and these can be separated by two-dimensional electrophoresis to give a characteristic "fingerprint". The fingerprint patterns obtained from 25s and 18s rRNA from rye (Secale cereale) using T1 ribonuclease are shown in Fig. 2.20. It can be seen that there are some particularly large fragments that are peculiar to each type of rRNA. As one would expect, a mixture of both types of rRNA contains all of these fragments. Analysis of the fingerprint pattern obtained from the rRNA precursor shows that it also contains fragments characteristic of both 25s and 18s rRNA, confirming that both types of rRNA sequences are present in the precursor molecule [415]. (The 5s RNA which is present in the larger subunit of the ribosome appears to be synthesized separately.) A generalized scheme of the pathway of processing of the rRNA precursor in plants is shown in Fig. 2.21. It must be stressed that certain points of detail are still in doubt: the precise arrangements of the 18s and 25s rRNA and the non-rRNA sequences within the precursor are not known. This question and the role of the processing intermediates requires investigation using sequencing techniques. Modification of the precursor by methylation and removal of the excess RNA probably occurs in the nucleolus but the enzymes responsible have not been characterized in plants. Particular stages of processing may occur outside the nucleolus but are probably completed before the rRNA enters the cytoplasm.

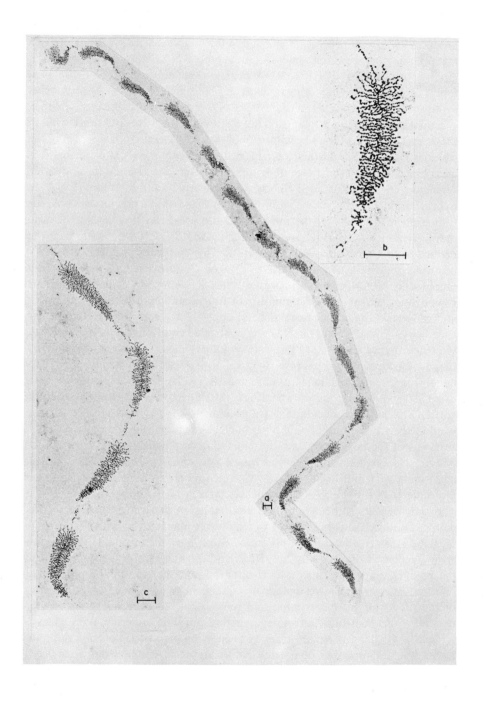

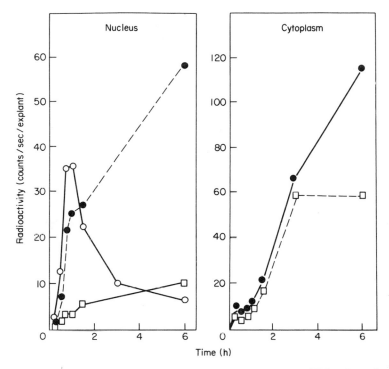

FIG. 2.19. Uptake of radioactivity into RNA fraction from artichoke (*Helianthus tuberosus*) tuber tissue. Tuber explants were incubated for 15 min in ^{32}P-orthophosphate, and then for a further 1–6 h in non-radioactive medium. Samples were taken at intervals. Nucleic acids were extracted and fractionated by gel electrophoresis and the radioactivity in each fraction was determined. Open circles, rRNA precursor; closed circles, 25*s* rRNA; squares: 18*s* rRNA (from ref. 391).

Results obtained with some plants cannot easily be accommodated with the above scheme. It is possible, therefore, that in some cases alternative processing pathways may operate. For example, in plants where two rRNA precursors are detected, such as carrot and sycamore, the larger precursor may be converted to

---

FIG. 2.18 Transcription of the genes coding for rRNA in *Acetabularia* as seen with the electron microscope. (a) Micrograph of spread nucleolar material, showing the regular pattern of "matrix" units and "spacer" segments. (b) Matrix unit at higher magnification, showing the dense packing of the lateral fibrils. The central axis is double-stranded DNA; the lateral fibrils are the growing rRNA precursor molecules which are attached to DNA by RNA polymerase enzymes. The rRNA precursor molecules increase in length as the enzymes move slowly along the gene, but the free ends of the RNA chains fold up, probably after association with proteins. Note the occasional groups of small fibrils associated with some of the spacer regions, suggesting that these regions may be transcribed to some extent (c) Scale lines indicate 0·5 μm (from ref. 481).

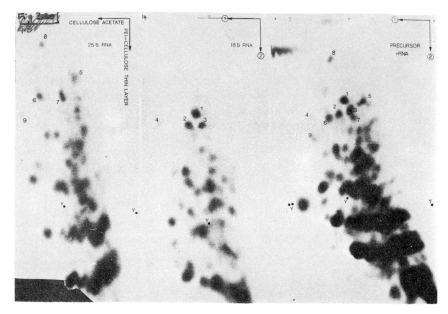

FIG. 2.20. "Fingerprint" patterns of rRNA precursor, 25s rRNA and 18s rRNA from rye (*Secale cereale*). The RNAs were subjected to partial hydrolysis with ribonuclease T1. The resultant oligonucleotides were fractionated by electrophoresis in cellulose acetate (direction 1) followed by chromatography on thin layers of poly(ethylene-imine) cellulose (direction 2). Some of the larger oligonucleotides are numbered to facilitate comparison of the three "fingerprint" patterns. The locations of the marker dye, Orange-G, in the second dimension are denoted by "Y" (from ref. 415).

the smaller one, or alternatively the two molecules may be synthesized and processed separately [90, 268]. There is no reason to suppose that particular processing events must necessarily occur in a given temporal sequence. For example, in some cases the 18s rRNA might be excised directly from the precursor molecule. This would mean that no 0.9×10^6 daltons RNA would be detected, but a "new" molecule with a molecular weight of about 1.7×10^6 would be produced and has indeed sometimes been observed (Fig. 2.22).

Chloroplasts and mitochondria also synthesize large RNA molecules which are probably rRNA precursors. A good demonstration of this comes from experiments with mitochondria isolated from the fungus *Neurospora crassa* where a precursor with a molecular weight of 2.4×10^6 daltons is processed to form the rRNAs of molecular weights 1.28 and 0.72×10^6 [146] (see also Chapter 4).

It is important to realise that at the same time as the post-transcriptional modifications of the rRNA precursor are taking place in the nucleolus, the rRNA portions of the molecule become associated with proteins to produce the

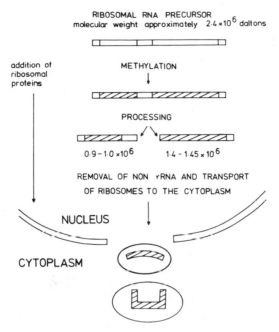

FIG. 2.21. Scheme for the post-transcriptional modification of the rRNA precursor. The molecular weights shown are approximate; small variations from these values occur frequently. The arrangement of the rRNA and excess RNA sequences within the precursor molecule is hypothetical, since this has not been determined in plants (from ref. 160).

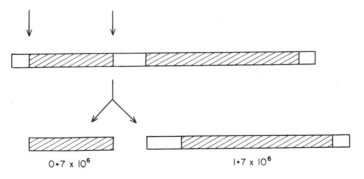

FIG. 2.22. Possible alternative processing scheme for the rRNA precursor.

ribosomal subunits. The non-rRNA portions of the precursor molecule have never been detected separately, suggesting that they are degraded during or immediately after cleavage. Their function is not clear. The excess RNA in different species is too variable in amount and sequence for it to act as an mRNA for the ribosomal proteins. It is possible that it may serve as some type of recognition sequence important in processing or transport of the RNA. More probably, the excess RNA performs a "folding function" by interacting with other parts of the molecule to produce double-stranded segments. Such interactions might be important in inducing the correct conformation of rRNA for the complex process of protein attachment during ribosome assembly.

The 5s RNA which forms part of the larger subunit of the ribosome appears to be transcribed separately because its synthesis can be observed in cells in which the production of 25s and 18s rRNA is artificially inhibited. As might be expected, there is a precursor to 5s RNA but the mechanism of its synthesis and processing in plants has not been extensively studied. Evidence from a number of experiments with plants and animals suggests that there are multiple genes coding for 5s RNA, but these do not appear to be located in the nucleolus [476].

C. The mechanism of rRNA synthesis in prokaryotes and eukaryotes

The synthesis of 25s and 18s rRNA as part of a polycistronic precursor molecule results in the production of both types of rRNA equimolecular proportions. However, co-ordinate synthesis is also observed in prokaryotes where no polycistronic precursor molecule is normally detected and one may, therefore, ask how the synthesis of both types of rRNA is regulated in such cases. A detailed analysis of the structure and organization of the rRNA genes in prokaryotes shows that, although they are not duplicated as many times as in eukaryotes, they are clustered on the chromosome and the 23s and 16s rRNA cistrons alternate. Detailed studies have shown that after initiating transcription at a 16s rRNA cistron, an RNA polymerase enzyme moves along the DNA and continues to transcribe the neighbouring 23s rRNA cistron [312]. This results in the synthesis of equal numbers of 23s and 16s rRNA molecules. A polycistronic precursor molecule is not normally formed because the two precursor RNA sequences are immediately separated by a processing enzyme. An exception to this occurs in a mutant of *E. coli* deficient in ribonuclease III. In these cells a rapidly-labelled RNA with a molecular weight of 1.8×10^6 daltons can be detected and competitive hybridization experiments have shown that it contains the sequences for both the larger and smaller rRNAs. Furthermore, it is processed *in vitro* by purified ribonuclease III to produce molecules which are very similar in size to bacterial rRNA [110]. This shows that the mechanism of synthesis of rRNA in prokaryotes is similar in principle to that which operates in eukaryotes. The main difference appears to be that in eukaryotes the polycistronic transcript exists for some time before it is processed. The significance of this is not clear; it may simply be related to the growth rate of the cells or it may have some more subtle implication for the mechanism of ribosome assembly. It is interesting to note that in chloroplasts and mitochondria, which are often compared with prokaryotes, polycistronic precursors to rRNA are detected and exist for some minutes before they are processed.

VIII. POLYDISPERSE RNA

A. The problem of studying mRNA

The study of mRNA metabolism in higher organisms is of particular interest in relation to the regulation of gene expression since much of our knowledge of the mechanism of protein synthesis comes from *in vitro* experiments using artificial mRNAs or those obtained from viruses. In order to understand how gene expression is controlled it is necessary to study the metabolism of endogenous mRNA. In ideal circumstances this requires that a particular mRNA sequence can be unequivocally identified so that the details of its synthesis and metabolism can be worked out. This is difficult to do for technical reasons. The major problem is in purifying a single species of mRNA from cells which normally contain a large number of different mRNA molecules, each present in comparatively small amounts. It is also necessary to show, normally by *in vitro* protein synthesis followed by chemical studies on the product formed, that the mRNA fraction directs the synthesis of a single, identifiable protein. It is therefore not surprising that this has only been achieved in a small number of instances. In most experiments it is only possible to establish rather general, and therefore, imprecise, criteria for the characterization of mRNA. Early workers in the field assumed that mRNA is rapidly synthesized (and therefore rapidly labelled), has a DNA-like base composition, becomes associated with polyribosomes shortly after synthesis, and is metabolically unstable. More recently it has become clear that other types of RNA also conform to one or more of these criteria while certain classes of mRNA do not. As with rRNA it is probable that at least some mRNAs are first transcribed as part of larger precursor molecules and that post-transcriptional modification of the RNA takes place before it becomes actively associated with polyribosomes. In the following section an account is given of the synthesis, properties and metabolism of polydisperse RNA, and its relationship to functional mRNA is discussed.

B. D-RNA

The use of polyacrylamide gels for the fractionation of high molecular weight RNA did not become possible until 1967 and for this reason early experiments on plant RNA synthesis involved the use of sucrose gradients and MAK columns. Experiments carried out by Ingle, Key and Holm with soya bean (*Glycine max*) hypocotyls led to the detection of a rapidly labelled RNA fraction very different from soluble and ribosomal RNA [208]. This fraction is called D-RNA because its base composition in some respects resembles that of DNA and is quite different from the majority of cellular RNA (see Table 2.IV). In sucrose gradients D-RNA sediments as a broad peak suggesting that it is composed of a population of different sized molecules (Fig. 2.23). Chromatog-

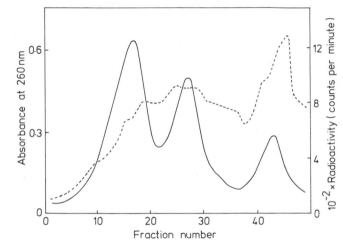

FIG. 2.23. Sedimentation of newly synthesized RNA from soya bean (*Glycine max*). The solid line shows the distribution of total RNA, measured by absorbance of u.v. light at 260 nm. The dotted line shows the distribution of radioactive RNA after incubation for 30 min in ^{32}P-orthophosphate. Part of this radioactive RNA is D-RNA (from ref. 208).

raphy on MAK columns on the other hand leads to the recovery of much of this RNA fraction as a single peak, eluting after the heavy 25*s* rRNA component (Fig. 2.24). This behaviour is probably explained by the high AMP content of the D-RNA which causes it to be quite firmly bound to the column. Considerable importance was attached to D-RNA because of its possible relationship to mRNA and this stimulated a great deal of work in a number of laboratories which showed that it occurs in a wide range of plant tissues. It is sometimes more readily detected in plant segments, probably because excision often leads to a reduced rate of rRNA synthesis, which makes it easier to detect the D-RNA. Ingle and Key showed that continued D-RNA synthesis was necessary for growth to be maintained although rRNA synthesis could apparently be inhibited without any effect on growth [240]. There is only a limited amount of evidence available concerning the subcellular distribution of D-RNA. The fact that it is necessary to employ detergents to extract DNA and D-RNA from cell

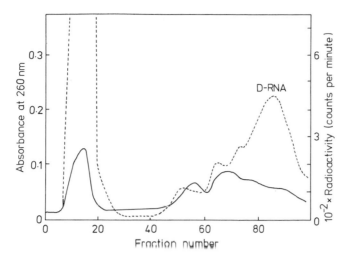

FIG. 2.24. Fractionation of D-RNA by chromatography on a column of methylated albumin-kieselguhr (MAK) (from ref. 208).

homogenates whereas milder procedures are sufficient to extract rRNA provides some evidence that the D-RNA is present in the nucleus. By following the disappearance of radioactivity from the D-RNA fraction during a pulse-chase experiment the half-life was estimated to be about 2 h. Subsequently, it was shown in Key's laboratory [280] that about 20% of the D-RNA becomes associated with polyribosomes active in protein synthesis. On the basis of this evidence it seems reasonable to conclude that at least part of the D-RNA fraction represents mRNA.

Although there is no doubt that plant tissues synthesize D-RNA it is now known that in some of the early experiments, bacteria growing on the surface of the plant material sometimes contributed to the labelling pattern of the nucleic acid samples [436]. In addition, the methods most commonly used for MAK fractionation often cause aggregation of rRNA and D-RNA and this alters the chromatographic behaviour of the various fractions. A further difficulty is that when the rRNA precursor is present in RNA samples it may chromatograph with the D-RNA. For these reasons any identification based solely on elution characteristics is unsound.

More recently Jackson and Ingle have extracted rapidly labelled RNA from the roots of aseptically cultured peas and fractionated it on MAK columns by an improved technique. These experiments compared the behaviour of various RNA fractions, including polydisperse RNA and the rRNA precursor, during MAK column chromatography and gel electrophoresis [213]. Under the conditions employed to run the MAK columns (lower salt concentrations and

D

higher pH values) aggregation of rRNA was reduced and the polydisperse
RNA separated into two fractions, one eluting with the rRNA and the other
remaining tightly bound to the column and requiring sodium dodecylsulphate
for elution. Both types of polydisperse RNA have a similar size distribution
judged by polyacrylamide gel electrophoresis (see Fig. 2.25) but the second
fraction bound to the column has a much higher AMP content and is assumed
to contain the D-RNA. The subcellular distribution and function of the
polydisperse RNA is not known. It seems likely that part consists of nuclear
RNA and part is present in the polyribosomes.

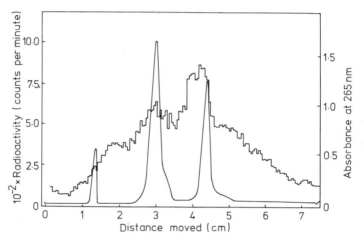

Fig. 2.25. Fractionation of "tightly bound" RNA from an MAK column by gel electrophoresis.
Pea (*Pisum sativum*) roots were incubated for 1 h in ^{32}P-orthophosphate. Nucleic acids were
extracted and fractionated by chromatography on a column of methylated albumin-kieselguhr.
The RNA which was not eluted with KCl (i.e. the "tightly-bound" RNA) was washed from the
column with detergent, mixed with non-radioactive "carrier" RNA, and fractionated by gel
electrophoresis (from ref. 213).

C. RNA containing poly-adenylic acid

More recent work has shown that at least some mRNAs contain poly-(adenylic
acid) sequences at the 3′ end of the molecule. Although the function of the poly-
(A) regions is not known, their presence can be exploited in order to purify any
RNA species containing poly-(A) and this has greatly stimulated research on
mRNA. The poly-(A) sequences can be detected in a variety of ways. They
were originally identified in plants by digesting RNA with pancreatic ribonuclease A, which specifically attacks phosphodiester bonds next to pyrimidine
residues, together with ribonuclease T1 which is specific for guanosine.
Analysis of the oligonucleotide fragments that remained showed that they were

composed largely of adenylic acid residues. The length of these fragments has been estimated by gel electrophoresis to be within the range 70–250 nucleotides by comparison with soluble RNA [398]. This method only provides an approximate estimate of the sequence length because of the large differences in properties, unrelated to length, between poly-(A) and the markers used. Hybridization of ^{3}H-poly-(uridylic acid) with RNA provides a method for detecting and quantitatively measuring the amount of poly-(A) present in RNA. The poly-(A) forms a hybrid molecule with ^{3}H-poly-(U) which is rendered resistant to low concentrations of ribonuclease because it is present in a double stranded structure. The T_m of the product suggests that it represents a true poly-(A)–poly-(U) hybrid and the amount of ribonuclease-resistant radioactivity is proportional to the quantity of poly-(A) present in the sample. Using this procedure it has been shown that between 0·01% and 0·1% of plant RNA is actually poly-(A); the quantity varies considerably with the age of the plant material and the stage of growth [89, 160]. The formation of hybrids between poly-(A) and sequences of either uridylic or deoxythymidylic acid is commonly exploited for the purification of poly-(A)-containing RNA. To achieve this, sequences complementary to poly-(A) are covalently attached to an insoluble support suitable for column chromatography and RNA solutions passed through the column. Poly-(U)-sepharose and oligo-(dT)-cellulose are commonly used materials and at moderate salt concentrations the poly-(A) forms a stable hybrid with the nucleotide sequence attached to the column. In contrast, the rRNA precursors, rRNA and soluble RNA have no affinity for the column and can be washed away (Fig. 2.26). The purified poly-(A)-containing RNA can be recovered by lowering the salt concentration, and perhaps raising the temperature, sufficiently to break the hydrogen bonds responsible for its attachment to the column. Only part of each molecule is composed of poly-(A) sequences as can be seen from the nucleotide composition of the RNA purified in this way (Table 2.IV). It is remarkably similar to and probably related to the polydisperse RNA from MAK columns. The fractionation by gel electrophoresis of poly-(A)-containing RNA from leaves of *Phaseolus aureus* pulse-labelled with ^{32}P-orthophosphate is shown in Fig. 2.26b. This sample contains RNA from both nuclei and polyribosomes. There are some problems associated with the estimation of molecular weights of poly-(A) containing RNA on gels because it may tend to aggregate and show other anomalous behaviour. Nevertheless, it is truly polydisperse in size and the estimated molecular weight ranges from several million down to a few hundred thousand daltons. There is no evidence for the "giant RNA" often said to be present in animal nuclei.

Part of the polydisperse RNA fraction associated with polyribosomes also contains poly-(A)-RNA. These molecules are on average smaller than those extracted from nuclei and undoubtedly represent functional mRNA molecules.

D. Grierson

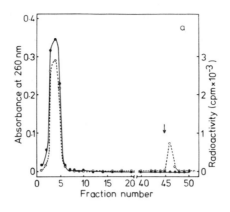

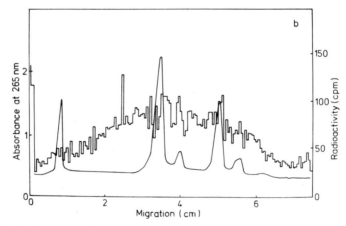

FIG. 2.26. The isolation and fractionation of poly(A)-containing RNA. (a) Pulse-labelled RNA from leaves of *Phaseolus aureus* (mung bean) was applied to an oligo(dT)-cellulose column. The column was washed with 0·3 M KCl in a buffer solution. The KCl concentration was then lowered (arrow) causing elution of the poly(A)-containing RNA. (b) The poly(A)-containing RNA from the column was mixed with non radioactive nucleic acids and fractionated by gel electrophoresis (from ref. 160).

The proof of this comes from experiments carried out on the synthesis of the plant protein leghaemoglobin. This protein is produced in substantial amounts in the nitrogen-fixing root nodules of legumes and for this reason the mRNA fraction from these cells would be expected to be enriched with mRNA for leghaemoglobin. Verma and colleagues have purified the poly-(A)-containing RNA from polyribosomes isolated from root nodules [498]. Analysis of this RNA on sucrose gradients shows that it contains polydisperse RNA but two discrete peaks sedimenting at 9s and 12s are also observed (Fig. 2.27). When these RNA fractions are supplied to an *in vitro* protein-synthesizing system

TABLE 2.IV. Base composition of various polydisperse RNA fractions

Type of RNA	Mol%					Ref.
	C	A	G	U	G+C	
Total RNA (soya bean)	23·9	23·2	31·9	21·1	55·8	208
D-RNA (soya bean)	20·2	28·5	25·8	25·7	45·9	208
Bound RNA from MAK column (pea)	18·9	38·0	24·6	19·0	43·6	213
Poly(A)-containing RNA from polysomes (French bean)	19·0	38·4	21·5	21·2	40·5	161

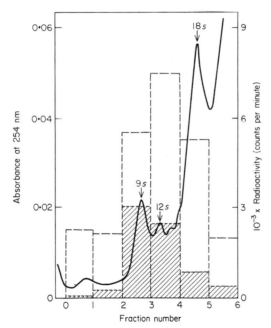

FIG. 2.27. Sedimentation characteristics and coding properties of poly(A)-containing RNA from polysomes of soya bean (*Glycine max*) root nodules. Purified poly(A)-containing RNA was mixed with non-radioactive total RNA and fractionated in a sucrose gradient (sedimentation left to right). Fractions were collected from the gradient (smooth curve) and assayed, in an *in vitro* protein-synthesizing system, for their ability to stimulate amino acid incorporation into total protein (histogram) and into protein precipitable by antibodies to leghaemoglobin (shaded histogram) (from ref. 498).

they stimulate the incorporation of amino acids into protein. Furthermore, part of the protein product formed is precipitated by an antibody specific for leghaemoglobin and analysis of this protein fraction by gel electrophoresis shows that it has the same mobility as leghaemoglobin.

The presence of discrete peaks representing single species of mRNA coding for individual proteins is to be expected in cells synthesizing large amounts of one protein. In most cells this does not occur and the mRNA fraction codes for large numbers of different proteins and is polydisperse in size [193]. This is shown for the mRNA from polyribosomes of leaves of *Phaseolus aureus* (Fig. 2.28). Calculations based on the degree of resistance to digestion with ribonuclease T 1 and pancreatic ribonuclease A suggest that only a small percentage of the RNA is poly-(A). This is consistent with mRNA molecules many hundreds of nucleotides in length, each containing one poly-(A) sequence. An RNA molecule of 1500 nucleotides, sufficient to code for 450 amino acids, has a molecular weight of 0.45×10^6 which is in reasonable agreement with the average value calculated from the electrophoretic mobility of the RNA shown in Fig. 2.28.

Poly-(A)-containing RNA can also be extracted from nuclei. These mole-

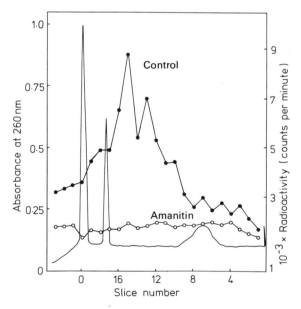

FIG. 2.28. Gel electrophoresis of poly(A)-containing RNA from polysomes. Poly(A)-containing RNA was purified from polysomes isolated from leaves of mung bean (*Phaseolus aureus*) pulse-labelled in the presence or absence of α-amanitin. It is clear that α-amanitin inhibits the synthesis of poly(A)-containing RNA (from ref. 193).

cules are generally larger in size than mRNA from polyribosomes and it is probable that at least some of them represent macromolecular precursors of mRNA. For plants very little evidence is available on this point but in animal systems it has been shown that the mRNAs for a number of proteins are transcribed as part of larger molecules which are subsequently processed and modified before they enter the cytoplasm [51, 300]. One interesting observation is that the poly-(A) sequences appear to be added after transcription. The evidence for this is that during pulse-labelling experiments the poly-(A) tract becomes radioactive more rapidly than the remaining portion of the mRNA [300]. In addition, there do not seem to be sequences present in DNA sufficiently long to code for the transcription of poly-(A) of the appropriate length [161, 420] although shorter stretches of oligo-(A) may be transcribed [215]. It must be emphasized that although the poly-(A) sequences make it easier to study mRNA metabolism, not all mRNAs contain such regions. Poly-(A) is absent from a number of animal mRNAs [309] and from the mRNA fraction from yeast mitochondria [165] and spinach chloroplasts [514a]. Estimates suggest that as much as 60% of mRNA from cultured plant cells may lack poly-(A) [161]. Its function remains obscure. It does not appear to be essential for the transport of mRNA to the cytoplasm or for the effective translation of the mRNA [17] and there is no evidence that it is ever translated. One possibility is that it may be involved in extending the lifetime of mRNAs in the cytoplasm, by inhibiting the action of ribonuclease [277].

There are a number of important questions that remain unanswered. Very little is known about the function of the repeating sequences that occur in nuclear RNA or the mechanism of association of RNA with proteins and the transport of mRNA, probably as messenger-ribonucleoprotein particles, to the cytoplasm. The most important tasks for future research will probably be to determine what proportion of the genome is transcribed into RNA and to elucidate the metabolic relationship between nuclear polydisperse RNA and mRNA.

IX. VIRAL RNA

The economic importance of some plant viruses is sufficient justification for them to be mentioned here. In addition, they provide a potentially useful source of mRNA for studies on protein synthesis *in vitro*. They are also of considerable biological interest.

Most plant viruses consist of single-stranded RNA molecules surrounded by a coating of protein. Infection of susceptible plants normally leads to the subversion of host metabolism and the production of new virus particles. In most cases the molecular weight of the RNA is within the range $1-4 \times 10^6$

daltons [303], sufficient to code for several proteins including the coat protein and a replicase enzyme for synthesizing new viral RNA.

Tobacco mosaic virus (TMV), which featured prominently in some of the classical experiments of molecular biology, is a rod-shaped particle about 300 nm by 17 nm. It is composed of more than 2000 protein subunits surrounding a single-stranded RNA which, in the particle, is arranged in a spiral and has a molecular weight of 2×10^6 daltons [303]. While protected by the coat protein the viral RNA is insensitive to ribonuclease. Infection of susceptible plants probably occurs by entry or the virus particles via a wound. After entering the cells viral RNA is liberated from the protein coat and although some of it may be degraded by cellular ribonucleases the remainder is conserved [379] and is responsible for directing the synthesis of the protein and RNA components of new virus. This is commonly observed in stems and leaves where the infection gives rise to pale green or yellow patches (the mosaic). The growth of the leaves is inhibited and the accelerated breakdown of chloroplast and cytoplasmic rRNA may be observed although the severity of the symptoms depends upon the strain of TMV [142]. Very high concentrations of virus may be produced and the synthesis and accumulation of TMV RNA in the cells can be detected by gel electrophoresis (Fig. 2.29). In some cases the TMV RNA accounts for as much as 75% of the total leaf RNA [143].

Evidence that the RNA of the infecting virus particles functions as a mRNA for the synthesis of viral proteins is as follows. When purified RNA is added to an *in vitro* protein-synthesizing system it stimulates polyribosome formation and the incorporation of amino acids into a large protein. No native coat protein is produced, probably because this normally requires a specific proteolytic cleavage of the larger protein. Nevertheless, Roberts and his colleagues have shown that among the protein products of *in vitro* synthesis is a polypeptide positively identified as being part of the coat protein [388]. The RNA is also responsible for directing the synthesis of new viral RNA. This probably involves the formation of a replicative intermediate and evidence for this has been obtained from detailed studies on the infection cycle in tobacco (*Nicotiana tabacum*) protoplasts. Takebe has shown that in addition to the infecting RNA strand, two other species of RNA can be detected [465]. The appearance of these molecules occurs a short time after infection and before the accumulation of virus particles takes place. One of these molecules is totally double stranded judging from its physical properties. It has a molecular weight of 4×10^6 daltons and can be converted to molecules with a similar electrophoretic mobility to the single-stranded RNA by denaturation. Furthermore, the nucleotide composition is typical of a double-stranded RNA [465]. It probably represents a hybrid structure composed of a molecule of RNA identical to the infecting strand complexed with a complementary strand synthesized after infection. The other type of molecule observed appears to be a single strand of

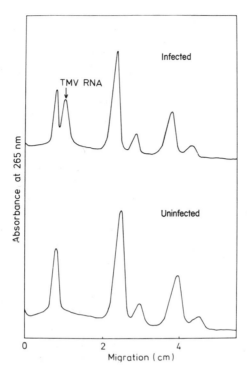

FIG. 2.29. Gel electrophoresis of nucleic acids from uninfected tobacco (*Nicotiana tabacum*) leaves and from leaves infected with tobacco mosaic virus (from ref. 161).

RNA in the process of being copied at several points along its length (compare with the transcription of the rRNA genes, section VIIC). From this evidence it appears that the first stage in the production of viral RNA is the synthesis by an RNA replicase enzyme of complementary strands to the infecting RNA. These would then be the templates for the synthesis of new infecting strands which subsequently become encapsulated to form virus particles.

Tobacco mosaic virus exists as only one type of virus particle. Other viruses, however, may have their genomes divided into several fragments, each present in a separate particle, and at infection the presence of each type of particle may be necessary for multiplication of virus. One example is brome mosaic virus (BMV) and studies on the capacity of the RNAs from this virus to direct the synthesis of proteins *in vitro* have provided evidence for a mechanism regulating viral protein synthesis during infection. BMV has four components or virions each containing a single-stranded RNA molecule. The smallest, RNA_4, is not needed for infection but is always produced during the infection cycle. Part of its sequence is identical to RNA_3, from which it is probably formed by cleavage. RNA_3 is the second smallest and it and RNA_4 are known both to contain the coat protein cistron. RNA_4 directs the synthesis of the coat protein *in vitro* but RNA_3 is translated into a protein larger than the coat protein. The other two types of RNA also stimulate the synthesis of other, unknown, proteins *in vitro*. An intriguing observation is that when RNA_4 is mixed with the other RNAs only the coat protein is made and the synthesis of all other proteins is inhibited. Extrapolating from these results to the *in vivo* situation, it seems that early in the infection cycle $RNAs_{1, 2 \& 3}$ may be translated, presumably to produce proteins required for viral replication. At a later stage in the infection cycle it is proposed that RNA_4 is formed from molecules of RNA_3 by nucleolytic cleavage. This would provide coat protein synthesis and at the same time inhibit further synthesis of the proteins coded for by the other RNAs [421].

Another extremely interesting class of pathogenic RNA molecules includes the potato spindle tuber agent, which causes epinasty and stunting of potato (*Solanum tuberosum*) and tomato (*Lycopersicum esculentum*) plants, and the chrysanthemum stunt agent. Much of the work on these viroids has been performed by Diener [105]. All the available evidence suggests that the causative agents are very small naked RNA molecules without protein coats. The molecular weight of the RNA responsible for the potato spindle tuber disease is estimated to be 80 000 daltons. It appears to be single stranded, with hairpin loops, and has a theoretical coding capacity for a peptide of about 80 amino acids in length. However, it does not appear to direct the synthesis of proteins *in vitro* and may not be translated. The mechanism of action of this highly potent RNA is unknown.

X. RNA STRUCTURE

A. Double-stranded regions in RNA

Different types of RNA exhibit varying degrees of hyperchromicity. Totally double-stranded molecules such as those associated with the "killer" character in *Saccharomyces cerevisiae* have a very sharp thermal transition with a high T_m (Fig. 2.30). In contrast, single-stranded molecules such as rRNA have a broader melting profile and the T_m is much lower (Fig. 2.30). The nucleotide composition of double-stranded molecules shows an equivalence of AMP with UMP and GMP with CMP. Conventional base pairing leads to the formation of a double helix but the bases are tilted relative to the long axis of the molecule. The result of this is that eleven base-pairs are accommodated in one complete turn of the helix whereas in the B form of DNA there are ten (see Chapter 1). In 0·15 M sodium chloride, double-stranded RNA is resistant to digestion with ribonuclease A, although it is readily degraded at lower salt concentrations.

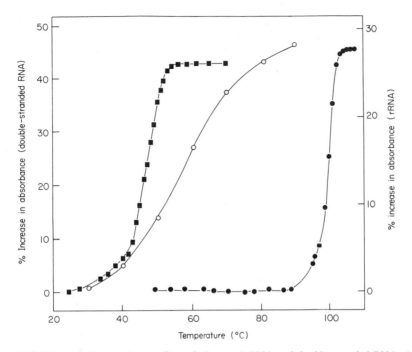

FIG. 2.30. Thermal denaturation profiles of ribosomal RNA and double-stranded RNA. 18s rRNA from *Spirodela oligorhiza* (a species of duckweed) was thermally denatured *in situ* in a polyacrylamide gel (open circles) (Redrawn from ref. 392). Double-stranded "killer" RNA from yeast was thermally denatured in the presence of 0·017M Na⁺ plus 67% formamide (squares) or in the presence of 0·017M Na⁺ with no formamide (closed circles) (Redrawn from ref. 32).

The secondary structure of single-stranded RNA results partly from the interaction between adjacent bases in the polynucleotide chain (base stacking) and also partly from the presence of short double-stranded regions of variable length consisting of "hairpin loops". From the degree of resistance of purified rRNA to digestion by ribonuclease it is calculated that as much as 60% of the molecule may be involved in the formation of such double-stranded segments. Treatment of ribosomes with ribonuclease *in vitro* shows that the rRNA is afforded additional protection presumably because much of it is buried within the ribosomal particle and is protected by the ribosomal proteins. Certain regions, however, must be exposed near the surface because such treatment results in the cleavage of the rRNA at specific sites in the molecule [353].

B. The fragmentation of rRNA

Fragmentation of rRNA also occurs *in vivo* and this has in the past led to some misunderstandings about the nature of rRNA. A particular fragment of wide-spread occurrence is the 5·8s RNA. It appears to be synthesized as part of the rRNA precursor molecule and it is later found in the larger subunit of the ribosome hydrogen bonded to the 25s rRNA to which it remains firmly attached. It can be released by breaking the hydrogen bonds by heat treatment or with formamide or urea, and is then detected separately by gel electrophoresis as it migrates more slowly than 5s RNA, and has a molecular weight of about 50 000 daltons (Fig. 2.31). Bacteria, blue-green algae and chloroplasts

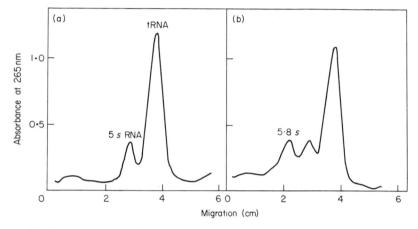

FIG. 2.31. The detection of 5·8s RNA. (a) Low molecular weight RNA fractionated in a 7·5% polyacrylamide gel. (b) Low molecular RNA, after dissociation of the 5·8s RNA from the 25s rRNA by heating (from ref. 158).

all lack the 5·8s RNA and it appears to be confined to eukaryotes [352]. The extraction and fractionation of chloroplast rRNA under normal conditions gives rise to other very characteristic fragments, derived by breakdown of the 23s rRNA and first described by Ingle (ref. 210 and Chapter 4.) Whether or not the fragments separate during RNA extraction depends upon the T_m of the double-stranded regions. This can be quite low for relatively short helical regions which explains why the chloroplast 23s rRNA only remains intact at temperatures below 5°C [267]. It is known that magnesium is particularly effective in stabilizing helices, thus increasing the T_m, and this is undoubtedly the reason for the apparent integrity of the 23s RNA in the presence of magnesium ions. A similar situation seems to occur with the 25s cytoplasmic rRNA from *Euglena* [374], and with mitochondrial rRNA [266]. The phenomenon is probably widespread but in most cases the "nicks" are masked because the double-stranded regions are longer than in chloroplast rRNA and therefore stable at normal temperatures. If the RNA is heated or denatured with urea or formamide then the fragments of rRNA dissociate and are easily detected separately [158]. The functional significance of the nicks is not known.

C. The properties of rRNA

Loening showed that the molecular weight of RNA is inversely proportional to its mobility in polyacrylamide gels [284]. This provides a very convenient method for determining molecular weights of unknown RNAs by co-electrophoresis with markers of known size. The presence of double-stranded regions can influence the mobility as can the base composition of the RNA [283], but in general these factors do not have a large effect. Occasional anomalies do occur and the exact molecular weight calculated may depend upon the buffer used, the concentration of the gel or the temperature [156, 381]. It is probably important to use markers with similar overall properties to the RNA being examined but this cannot always be done. The most satisfactory method is to compare RNAs under completely denaturing conditions so that length is the only property that exerts a significant effect on electrophoretic mobility.

Three types of ribosomes with different s-values are found in higher plants, located in the cytoplasm ($80s$), the chloroplasts ($70s$) and the mitochondria ($78s$). Each type of ribosome generally contains three classes of rRNA as structural components. The molecular weights of these RNAs are shown in Table 2.I (see also Chapter 4). There is a striking similarity between the size of the ribosomes and rRNAs of chloroplasts, bacteria and blue-green algae. This, together with the fact that chloroplast ribosomes are susceptible to the same range of antibiotics as the prokaryotic ribosomes, has been interpreted as evidence in favour of the endosymbiont hypothesis for the origin of chloroplasts. Mitochondria contain a third class of ribosomes with properties intermediate between those of the cytoplasm and the chloroplast (Table 2.I). This makes speculation about their origin more difficult. Furthermore, there are major differences in the properties of mitochondrial ribosomes and rRNAs from animals, fungi and higher plants [266].

Among eukaryotes the molecular weight of the $18s$ rRNA is, with very few exceptions, remarkably constant, whereas the largest rRNA varies in size, from $1 \cdot 3 \times 10^6$ daltons in plants to $1 \cdot 75 \times 10^6$ daltons in humans [283]. There is speculation that this increase in size is related to the more complex control mechanisms thought to govern protein synthesis in higher eukaryotes [283]. In addition, there is a general trend for the size of the rRNA precursor to be related to the degree of taxonomic "advancement" [163, 285, 358] but the wide variation in molecular weights of these RNAs in different plant species suggests that other factors may also govern the size of these transcription units.

Despite the similarity in size between rRNA molecules from different plants there are quite large differences in nucleotide composition (Table 2.V). This is consistent with the view that there are certain overall requirements for the correct functioning of rRNA which are related to size and secondary structure

TABLE 2.V. Base composition of ribosomal RNA from different organisms

Organism	rRNA	mol%					Ref.
		C	A	G	U	G+C	
Yeast (*Saccharomyces*	25s	19·2	26·4	28·4	26·0	47·6	135
cerevisiae)	25s	19·1	26·6	26·1	28·1	45·2	
Mushroom (*Psalliota*	25s	20·1	27·7	27·8	24·3	47·9	363
campestris)	18s	18·0	24·9	30·1	27·1	48·1	
Artichoke (*Helianthus*	25s	21·6	28·1	31·5	18·9	53·1	391
tuberosus)	18s	20·5	27·6	28·1	23·9	48·6	
Pea (*Pisum sativum*)	25s	21·8	24·9	32·2	21·1	54·0	391
	18s	21·1	25·9	28·5	24·4	49·6	
Mung bean (*Phaseolus*	25s	23·0	24·5	32·4	20·1	55·4	162
aureus)	18s	20·2	25·4	30·2	25·2	50·4	
Maize (*Zea mays*)	25 s	28·0	21·3	33·9	16·9	61·9	363
	18 s	25·07	22·0	33·1	20·2	58·17	

but that considerable variation may be accommodated. There does seem to have been a tendency for the G + C content of rRNA to increase during evolution [12]. A higher percentage of G–C pairs might be expected to confer greater stability on the rRNA but little is known about the significance of this trend. Specific regions of rRNA may be less open to change. The complete nucleotide sequence of plant rRNA is not yet available but a comparison of the sequences of the 5.8s RNA from a number of species suggests that the sequence of this particular rRNA molecule has been highly conserved during the evolution of the flowering plants [521]. The sequence of the 5s rRNA seems to have been even more highly conserved [350]. Readers interested in an account of sequencing methods should see reference 344.

Ribosomal RNA appears to function as a framework for the attachment of the many ribosomal proteins. Nomura has shown that it is possible to separate and purify all the RNA and protein components of *E. coli* ribosomes and to reassemble them *in vitro* to produce functionally active particles [331]. These studies have shown that there are specific attachment sites in the rRNA for individual ribosomal proteins. It is possible that rRNA molecules may also participate in other functions such as the association of the ribosomal subunits or in the binding of mRNA or tRNA to ribosomes. It remains to be established whether or not this is the case.

D. The structure of tRNA

The nucleotide sequences of a large number of different tRNAs from a variety of organisms are known. They contain an unusually high proportion of modified bases. These modifications are made by special enzymes after the tRNA precursors are transcribed and during the maturation process. Preparations of tRNA show considerable hyperchromicity, indicating a high degree of secondary structure, and they also form mixed crystals which suggest that the main structural features must be common to all of the tRNAs. If the nucleotide sequences are arranged so as to permit the formation of the maximum number of hydrogen bonds it is found that all the molecules can adopt a common configuration—the characteristic "clover-leaf" pattern. This consists of a number of base-paired regions, including the stem, which contains the 3' and 5' ends of the molecule, and several arms which have loops at the ends (Fig. 2.32). The "acceptor stem" terminates in a purine followed by the sequence CCA—OH which is at the 3' terminus. During charging of the tRNAs by the synthetase enzymes, amino acids are added to the 3' A residue (see Chapter 3). In pulse-labelling experiments the CCA sequence becomes radioactive more rapidly than the remaining portion of the tRNA because it is continually removed and resynthesized in the cytoplasm. Other common features include the "dihydro-U loop", which in a few cases may not actually contain any dihydro-U, and the "T-pseudo-U-C loop" which universally contains this sequence. There is, in addition, a "variable arm" which, as its name suggest, is the most variable part of the general structure.

The anti-codon is exposed at the end of a loop and is always flanked by a purine or a modified purine on the 3' side. The anti-codon itself frequently contains a rare or modified base. This is thought to allow more flexible base-pairing with the mRNA codons so that where more than one codon exists for an amino acid these may be recognized by a single tRNA. This is the basis of Crick's "Wobble Hypothesis" [93]. For example, the codons for phenylalanine

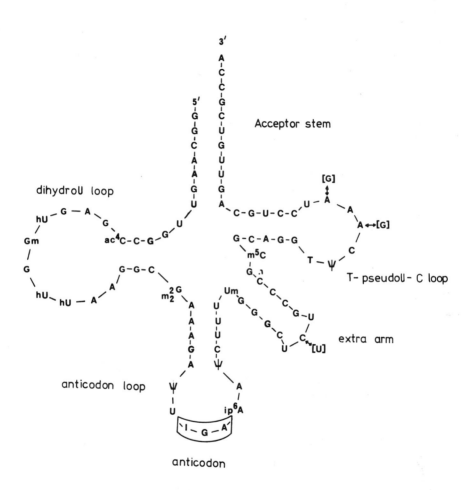

dihydroU loop

Acceptor stem

T- pseudoU- C loop

extra arm

anticodon loop

anticodon

a.

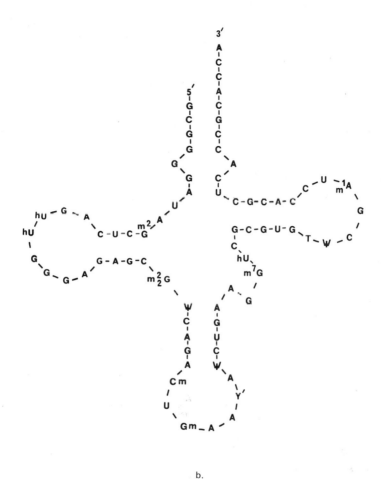

b.

FIG. 2.32. Nucleotide sequences of rRNAs. The sequences are arranged in the clover leaf pattern. Parallel sequences represent regions of base pairing. (a) Serine tRNA, from yeast. Serine tRNA$_2$ differs in only three positions (indicated in brackets) (from ref. 538). (b) Phenylalanine tRNA from wheat germ (from ref. 108).

Abbreviations: **A** = adenine, **C** = cytidine, **G** = guanosine, **I** = Inosine, **T** = thymidine, **U** = uridine. Prefixes indicate the presence and position of base modifications: **M** = methyl, **ac** = acetyl, **ip** = isopentyl. **Gm** and **Cm** refer to nucleosides with 2′ methyl ribose. ψ = pseudouridine, **hU** = dihydrouridine and **Y′** is an unidentified base.

are UUU and UUC and the anti-codon in wheat germ tRNA^{phe} is GmAA (Fig. 2.32). Since mRNA interacts with the anti-codons in an anti-parallel manner this means that the AA of the anti-codon pairs with the UU of the codons. The base G can form a base pair with either of the bases U or C (the wobble) and thus tRNA^{phe} can respond to both phenylalanine codons. The modified purine next to the anti-codon is of particular interest because in many cases it has biological activity as a cytokinin. Ribosylzeatin, isopentyladenine and their 2-methylthio derivatives have been identified in plant tRNA preparations and similar molecules appear to be present in tRNAs from most organisms [433]. They commonly occur in tRNAs that respond to codons containing U as the first letter but are not present in tRNAs recognizing other initial letters. Conversely many of the tRNAs that recognize A as the initial letter contain another modified purine, N-(purin-6-ylcarbomoyl) threonine, which has no cytokinin activity [433].

Gefter and Russell demonstrated that the presence of a modified purine next to the anti-codon is of great functional significance [149]. They studied the properties of the tRNA^{tyr} produced by $E.\ coli$ infected with defective trans-ducing phase. Under these conditions three different forms of tRNA^{tyr} are produced in large amounts. They are all capable of being charged with tyrosine and differ only in whether or not the A residue next to the anti-codon is modified. Those tRNA molecules in which the A residue is modified bind to ribosomes and participate effectively in protein synthesis whereas rRNAs without the modified A bind to ribosomes much less efficiently. These findings have led to speculation about the possible role of cytokinins in regulating protein synthesis. Although these modified purines are of great importance to the functioning of tRNA molecules, a direct physiological relationship with freely-occuring cytokinins in fact seems doubtful. Firstly, most evidence suggests that the modifications to the appropriate A residue in tRNA occurs by the addition of a side chain only [433]. This is incompatible with a mode of action involving the incorporation of the intact cytokinin molecule into tRNA. Secondly, the hydroxyl group of ribosylzeatin in plant tRNA appears to be in the cis position whereas the hydroxyl of the freely occurring base is in the $trans$ configuration [433].

The cloverleaf representation of tRNA is incomplete in the sense that it takes no account of tertiary structure. Attempts to modify chemically residues not involved in base-pairing in the cloverleaf model show that certain bases are unexpectedly resistant. This suggests that they are not readily accessible and is interpreted to mean that they are masked in some way by the tertiary structure of the molecule. There are a number of possible three-dimensional arrange-ments involving folding of the various arms but there is no decisive evidence in favour of one structure and some experimental evidence suggests that there may be two alternative conformations [344].

XI. THE REGULATION OF RNA METABOLISM

There seem to be four points at which the regulation of RNA metabolism may occur. One can envisage control mechanisms governing transcription, the post-transcriptional modification and transport of RNA, the regulation of mRNA translation in the cytoplasm and the degradation of RNA. In eukaryotes very little is known about the relative importance of each of these processes or the molecular mechanisms involved.

A. Control of transcription

Clearly the factors that determine which genes are transcribed are of consider able interest. A mechanism involving template availability is one obvious possibility. In its most familiar form this involves repressor proteins which reversibly bind to particular DNA sequences thereby regulating their transcription. Such molecules are known to operate in bacteria but as yet no specific repressor protein has been identified in higher plants. Although it is clear that histones exert some influence on the availability of DNA for transcription it is quite clear that they lack the appropriate specificity for individual gene sequences normally associated with bacterial repressor proteins (see Chapter 1). Further research may reveal more specific repressors among the acidic proteins of the nucleus, although to date the evidence suggests that the acidic proteins are de-repressors (see Chapter 1). However, it should be remembered that although the λ and *lac* repressor proteins in bacteria are specific for unique DNA sequences, other transcriptional control elements, such as the cyclic AMP binding protein, regulate a large number of genes [348]. This precedent and the fact that each gene or operon regulated by a repressor protein would require a second gene to code for the repressor itself suggests that specific repressor proteins may not be very common in eukaryotes.

Positive control is an alternative mechanism. This requires a certain specificity for RNA synthesis to be imparted to RNA polymerase enzymes. For example in *E. coli* the RNA polymerase consists of a number of subunits including a "sigma factor". Evidence from *in vitro* studies has shown that the basic enzyme can function in the absence of the sigma factor which reversibly associates with the "core" enzyme and induces it to initiate transcription at specific loci in the DNA template [66]. Eukaryotes contain multiple RNA polymerases with different properties and subcellular locations. In higher plants, the mitochondria and chloroplasts have distinct enzymes for copying organelle DNA to RNA and the nucleus has at least two separate RNA polymerases. These can be distinguished by their differential sensitivity to certain antibiotics and metal ions and by their chromatographic behaviour. One of these enzymes (RNA polymerase I) is localized in the nucleolus and is presumably responsible for transcribing the rRNA genes [50]. There is at least one other RNA polymerase enzyme, RNA polymerase II (and perhaps more than one), present in the remaining part of the nucleus. RNA polymerase II is inhibited by the antibiotic α-amanitin and is assumed to be responsible for the synthesis of nuclear polydisperse RNA and mRNA. The characterization of these enzymes and their template specificities is still at an early stage but there is evidence for a polypeptide which stimulates the initiation of transcription of RNA polymerase [319] and in addition growth substance–receptor proteins appear capable of modifying RNA synthesis [497].

B. Post-transcriptional control

One important feature of RNA metabolism that was first detected in eukaryotes and only later in bacteria is the phenomenon of post-transcriptional modification of larger precursor molecules. This is potentially of great significance because RNA sequences may be conserved and selected for transport to the cytoplasm or they may be degraded without leaving the nucleus. As a regulatory step it has one important argument in its favour: it has been shown to occur for certain in the case of the metabolism of rRNA, 5s RNA and tRNA. Although not all the details are clear, the production of mRNA probably also involves post-transcriptional modifications (see section VIII). Following evidence obtained from his studies on RNA synthesis in mammals, Scherrer suggested that after synthesis of mRNA precursor molecules and during the stages of processing, certain RNA sequences are selected for transport to the cytoplasm; this is the hypothesis of "cascade control" [402]. Cascade control would provide a second level of regulation of mRNA production, with the specificity for template selection residing in the processing enzymes and transport proteins which presumably respond to specific recognition sequences in the RNA molecules. It therefore follows that a variation in the activity of different processing enzymes or transport proteins could alter the pattern of mRNA molecules directed to the cytoplasm.

C. Control in the cytoplasm

Once in the cytoplasm three possible fates await a messenger RNA molecule. Firstly it may become associated with ribosomes and direct the synthesis of a protein, secondly it may be stored in an inactive form until some suitable signal triggers its translation or thirdly it may be degraded by ribonuclease. If the first mechanism operates, i.e. immediate translation, this would suggest that the control of gene expression must occur before the protein synthesis step. In fact this is not quite true because factors regulating the rate of translation of individual mRNAs, and therefore the amount of protein produced, could have profound effects on development. One extreme type of translational control involves the storage of mRNA in an inactive form and evidence suggesting that such a mechanism has been obtained from experiments with a variety of organisms. For example, the unicellular green alga *Acetabularia* will produce a characteristic reproductive "cap" under suitable conditions. If the cap is surgically removed then the cell is capable of regenerating a new one, even if the nucleus is removed from the cell (ref. 174 and Chapter 6). This implies that the enucleated cell still contains genetic information for the production of a cap. Regeneration of a new cap by enucleated cells takes place in the presence of actinomycin D, which inhibits DNA-dependent RNA synthesis, and this observation appears to rule out the possibility that the DNA present in the chloroplasts or mitochondria is responsible for determining cap formation. The evidence from these experiments is consistent with the view that stable mRNAs coding for proteins essential for cap formation persist in the cytoplasm and are translated when the old cap is removed.

Similar conclusions can be drawn from experiments with slime-moulds. Under constant conditions a characteristic sequence of morphological and biochemical events takes place during the development of the fruiting bodies in the slime-mould *Dictyostelium* and the activities of particular enzymes are observed to increase at different times in a sequential manner. When actinomycin D is applied immediately before the expected increase in activity of an enzyme it has very little effect, suggesting that RNA synthesis is not essential for the change in enzyme activity. However, if actinomycin D is applied for a short time some hours before the enzyme change normally takes place and the antibiotic is then removed to allow RNA synthesis to continue, this often results in the failure of the enzyme to appear [458]. If it is assumed that mRNAs for particular enzymes are synthesized some time before they are translated this would explain why actinomycin D has an inhibitory effect if applied several hours in advance but is ineffective immediately prior to the enzyme change.

Evidence in favour of the "masked messenger" concept in higher plants comes from studies on germination. During cotton (*Gossypium*) seed germination, as with other seeds, there are characteristic changes in the activity of enzymes in the cotyledons. These enzymes such as isocitrate lyase and carboxypeptidase, are responsible for breaking down and mobilizing stored reserves for use by the growing embryo axis. The increase in activity of these enzymes occurs even if RNA synthesis is inhibited by actinomycin D [207]. In barley (*Hordeum vulgare*) the technique of density labelling (see Chapters 6 and 7) has been used to demonstrate that many enzyme changes are probably caused by *de novo* synthesis rather than activation of inactive enzyme [537]. Other experiments suggest that the increase in the number of polyribosomes that takes place during the early stages of germination does not depend on new mRNA synthesis. The most satisfactory interpretation of these results is that during seed maturation certain mRNA molecules are synthesized and stored in readiness to be translated during germination. Although the precise mechanisms involved are not understood it is clear that plant growth substances in some way influence these processes (see Chapters 6 and 7).

It should be pointed out that the evidence for stored or long-lived mRNA depends heavily on the assumption that actinomycin D prevents mRNA synthesis. Recently it has been found that although RNA synthesis is greatly reduced, actinomycin D does not completely inhibit mRNA synthesis in the slime-mould *Dictyostelium* [216]. This raises some doubts about the evidence for long-lived mRNA, and it is clear that more detailed evaluation of the effects of actinomycin D on plant cells is necessary.

The number of protein molecules that are synthesized from a particular mRNA depends on the frequency with which ribosomes initiate and complete translation of the template. The length of time that the messenger exists in the cytoplasm before it is degraded can therefore have a profound effect on the amount of enzyme produced. In bacteria, for example, the processes of translation and degradation of mRNA are often tightly linked. Inhibition of mRNA synthesis therefore results in rapid cessation of enzyme synthesis. Very little is known about the mechanism of RNA turnover in plants but much of the cellular RNA appears to be comparatively stable. The half-life of rRNA in *Lemna* has been estimated to be 4 days for the cytoplasmic rRNA and 15 days for chloroplast rRNA [484] and it is probable that masked mRNA persists in some cells for a very long time. Nevertheless, there are a number of reasons for believing that functionally active mRNA is often quite rapidly degraded although accurate measurements in higher organisms are rare, due mainly to the difficulty of mRNA identification (see Chapter 6).

SUGGESTIONS FOR FURTHER READING

DAVIDSON, J. N. (1972) "The Biochemistry of the Nucleic Acids" (7th edition). Chapman and Hall, London.

LEWIN, B. (1974). "Gene Expression", Volume 2: "Eukaryotic Chromosomes". John Wiley, London and New York.

PARISH, J. H. (1972). "Principles and Practice of Experiments with Nucleic Acids". Longmans, London.

STEWART, P. R. and LETHAM, S. S., eds (1973). "The Ribonucleic Acids". Springer Verlag, Berlin.

WATSON, J. D. (1975). "The Molecular Biology of the Gene" (3rd edition). W. A. Benjamin, New York.

3. Protein Synthesis

C. M. BRAY

I. INTRODUCTION

In the cell, gene expression is ultimately manifested by specification of the amino acid sequence and hence structure of all cellular proteins. A multitude of cellular proteins, having many diverse biological functions, are all synthesized by a common mechanism involving the participation of over 100 different macromolecules. The overall process requires RNA synthesis (transcription) followed by polymerization of a combination of the 20 naturally occurring amino acids into an ordered sequence determined by the sequence of bases in the messenger RNA template. This latter process is known as translation. Translation occurs both in the cell cytoplasm and in intracellular organelles such as mitochondria and chloroplasts. The discussion in this chapter is limited to cytoplasmic protein synthesis with particular emphasis on plant systems, although reference is made to other eukaryotic systems and to the well studied prokaryotic system wherever necessary (see also refs 43, 182, 539).

The protein synthetic process can be conveniently divided into several distinct steps: amino acid activation and aminoacyl-tRNA synthesis, peptide chain initiation, peptide chain elongation, chain termination and peptide release. Simultaneously with or following peptide release the ribosome dissociates into its subunits thus allowing the cyclic process of protein synthesis to begin another round. Much of our knowledge concerning these processes has been derived from studies of cell-free protein-synthesizing systems isolated from prokaryotes, and in particular studies on the system isolated from the micro-organism *Escherichia coli*. It has been demonstrated that despite the drastically different life-styles of bacteria, plants and animals the genetic code, although degenerate in the sense that several codons may code for one amino acid, is apparently universal in that the same codons code for the same amino acid in different organisms. The incidence of degeneracy can be predicted by Crick's "Wobble hypothesis" [93].

In the following sections the aim has been to try to give an overall picture of the present state of knowledge of the mechanism and possible sites of regulation of the protein synthetic process in plant systems and to compare these processes to the corresponding stages of protein synthesis in other eukaryotic systems which in most cases have been studied in more detail than plant systems.

II. AMINO ACID ACTIVATION: AMINOACYL-tRNA SYNTHESIS

The first step in protein biosynthesis involves the activation of amino acids in the presence of ATP, Mg^{2+} ions and aminoacyl-tRNA synthetases (amino acid: tRNA ligases) to produce an enzyme-bound aminoacyl-adenylate or activated amino acid. This same synthetase enzyme then catalyses the transfer of the amino acid from the aminoacyl-adenylate to a specific tRNA molecule. The overall reaction can be represented as a two-step mechanism:

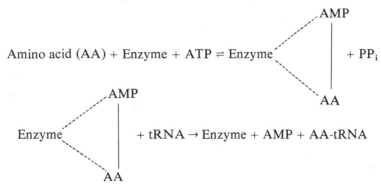

$$\text{Amino acid (AA)} + \text{Enzyme} + \text{ATP} \rightleftharpoons \text{Enzyme} \Big\langle \begin{array}{c} \text{AMP} \\ | \\ \text{AA} \end{array} + \text{PP}_i$$

$$\text{Enzyme} \Big\langle \begin{array}{c} \text{AMP} \\ | \\ \text{AA} \end{array} + \text{tRNA} \rightarrow \text{Enzyme} + \text{AMP} + \text{AA-tRNA}$$

In the aminoacyl-tRNA molecule, the amino acid is esterified to the tRNA via the 2′- or 3′-hydroxyl of the terminal (3′-) adenosine residue.

However, Loftfield [287] has cast doubt upon whether this is the physiological mechanism of amino acid activation. He suggests that in the cell the aminoacyl-tRNA synthetase is almost totally associated with tRNA and he proposes a concerted reaction for aminoacyl-tRNA synthesis in which tRNA, amino acid, synthetase and ATP react to form aminoacyl-tRNA, AMP, PP_i and free enzyme in the absence of any discrete intermediates. In this case the production of the enzyme-bound aminoacyl adenylate intermediate would occur only in the non-physiological absence of tRNA.

Whichever mechanism is the correct physiological mechanism for aminoacyl-tRNA synthesis, it must possess the property of specificity. A characteristic of protein synthesis is the absolute specificity by which the correct amino acids are incorporated into each position in the polypeptide chain. Hence the accuracy of the aminoacyl-tRNA synthetases in discriminating amongst both potential amino acid substrates and the many tRNA species in the cell must be at least as great as that of the overall protein synthetic process. A considerable amount of work has been done on the specificity of binding of tRNA by the synthetase but no conclusive evidence regarding the nature and position of the binding sites has been obtained. Present evidence indicates that all tRNA molecules are of approximately the same size (75–80 nucleotides) and appear to have similar three-dimensional configurations. The "clover leaf" model of tRNA is shown in Fig. 2.32 (Chapter 2). The distance from the anticodon to the 3′ (–C–C–A) terminus of the tRNA molecule appears virtually the same for all tRNA species.

In order to achieve the high degree of specificity in the tRNA-synthetase interaction, the synthetase must recognize much more essential parts of the tRNA than the –C–C–A (3′) terminus (which is common to all tRNA molecules). However, apart from the anticodon and terminal (3′-) adenosine, the vast majority of bases, including the "minor" bases found in tRNA, appear to be withdrawn into the interior of the tRNA molecule and thus play no specific role in synthetase-tRNA recognition [287]. This would suggest a structure for tRNA consisting of a molecule whose outer surface is a regular repetition of ribose and phosphate with most of the bases turned inwards, possibly in many cases forming Watson-Crick type base pairs. If this is the case then a particular tRNA molecule would best be distinguished from its companions by difference in shape. Hence the shape of the tRNA molecule may be the major factor responsible for the unique attraction between tRNA and its cognate synthetase, there being few or no unique bases in exposed positions.

There are no reports of naturally co-existing amino acids being attached to the wrong tRNA by aminoacyl-tRNA synthetases under physiological conditions and in general these enzymes exhibit a much reduced specificity against amino acids not normally present in the cell. However, there are certain plants which contain naturally occurring amino acid analogues and in these plants it is found that the aminoacyl-tRNA synthetase discriminates against these amino acids in the amino acid activation reaction, whereas the corresponding synthetase from plants in which the amino acid analogue is absent exhibits no such discrimination, activates the analogue and in some cases transfers the activated analogue to tRNA. Azetidine carboxylic acid, a proline analogue (Fig. 3.1) is activated and transferred to tRNA^{Pro} by the proline enzyme from *Phaseolus aureus* (mung bean), in which the analogue is naturally absent, but not by the corresponding enzyme from *Polygonatum multiflorum* (Solomon's seal) which contains high levels of azetidine carboxylic acid. Similarly the arginine-tRNA synthetase of *Canavalia ensiformis* (jack bean) discriminates against the activation of canavanine, an arginine analogue (Fig. 3.1) occurring naturally in this plant. The phenylalanine-tRNA synthetases of *Mimosa* and *Leucaena* discriminate against the activation of mimosine, a naturally occurring phenylalanine analogue (Fig. 3.1), found in these plants but highly toxic to all other species.

The phenomenon of species specificity between aminoacyl-tRNA synthetases and tRNA preparations from different organisms has been investigated and in some instances it is found that an enzyme from a heterologous source will charge (i.e. will effect the combination of an amino acid with its tRNA) plant tRNA either partially or fully whilst in other instances no charging of tRNA is observed. Allende [4], using wheat embryo synthetases and tRNA, demonstrated that yeast seryl-tRNA synthetase was completely effective in charging the appropriate wheat embryo tRNA whilst the corresponding synthetase from wheat embryo was only partially effective in charging *E. coli* tRNA. The *E. coli* synthetase, however, was totally inactive with wheat tRNA^{Ser}. In the

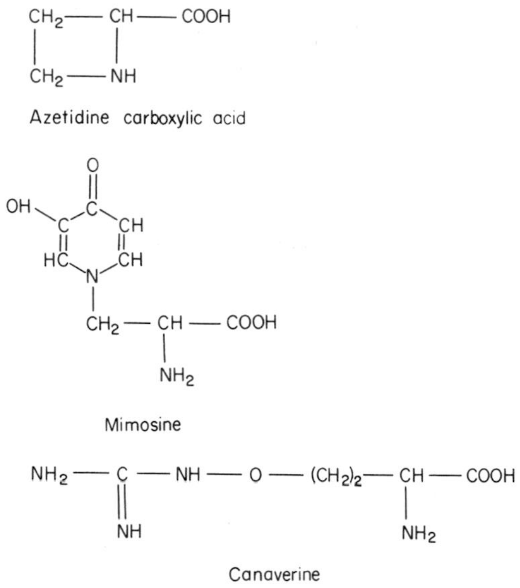

$$CH_2 \!-\! CH \!-\! COOH$$
$$CH_2 \!-\! NH$$

Azetidine carboxylic acid

Mimosine

$$NH_2 \!-\! C \!-\! NH \!-\! O \!-\! (CH_2)_2 \!-\! CH \!-\! COOH$$
$$\| \qquad\qquad\qquad\qquad |$$
$$NH \qquad\qquad\qquad\qquad NH_2$$

Canaverine

FIG. 3.1. Examples of amino acid analogues which occur in plants.

case of methionyl-tRNA synthetase, the enzyme from *E. coli* would charge only one of the two species of $tRNA^{Met}$ present in wheat embryo.

There is little information available concerning the number and specificity of aminoacyl-tRNA synthetases present in an organism which has a number of iso-accepting tRNA molecules (i.e. different species of tRNA which will accept the same amino acid). Anderson and Cherry [8] demonstrated that soya bean cotyledons contained six $tRNA^{Leu}$ species which could be charged by the cotyledon synthetase preparation but the corresponding soya bean hypocotyl system, utilizing hypocotyl tRNA and synthetase preparations, appeared to possess only four major leucyl-tRNA species although traces of the other two (leucyl tRNA$_5$ and $_6$) were found. Kanabus and Cherry [231] provided an explanation for the differences when they fractionated soya bean cotyledon leucyl-tRNA synthetase into three fractions and demonstrated that soya bean hypocotyls were deficient in one of these synthetase fractions (the fraction responsible for specific aminoacylation of $tRNA_5^{Leu}$ and $_6$). Further discussion of this topic can be found in section VII.

III. INITIATION OF PROTEIN SYNTHESIS

A. The prokaryotic system

Studies on protein-synthesizing systems isolated from *E. coli* demonstrated that the amino acid most commonly occurring at the N-terminus of $E.$ $coli$ proteins was methionine, suggesting that methionine is the first amino acid incorporated into nascent protein. Subsequently it was found that bacterial cells initiate protein synthesis by incorporating N-formyl-methionine using an initiating transfer RNA, $tRNA_F^{Met}$. Methionine is first bound to this transfer RNA and subsequently formylated enzymically before incorporation at the N-terminus of the protein while internally situated methionine is introduced into the growing polypeptide chain by a different transfer RNA, $tRNA_M^{Met}$, methionyl $tRNA_M$ being incapable of accepting a formyl group.

The process of initiation of protein synthesis in prokaryotes has been studied extensively and is thought to require at least three protein initiation factors (IF–1, IF–2 and IF–3), $30s$ and $50s$ ribosomal subunits, messenger RNA, GTP and N-formyl-methionyl tRNA. The first step in the process involves the formation of a $30s$ ribosomal subunit–mRNA–f-met-$tRNA_F$ complex which is then converted into a $70s$ initiation complex by the addition of a $50s$ ribosomal subunit, this complex now being able to participate in protein synthesis. The exact role of the three initiation factors is not clear but it is known that IF–3 is required for the binding of the mRNA to the $30s$ ribosomal subunit and also had a role in ribosome dissociation. IF–2 is involved in the binding and transfer of the initiator tRNA to the complex involving IF–3, mRNA and the $30s$ ribosomal subunit, while IF–1 promotes the catalytic activity of IF–2.

B. The eukaryotic system

1. Intracellular Organelles

Studies on the initiation of protein synthesis in mitochondria and chloroplasts of eukaryotes demonstrated that f-met-$tRNA_F$ was also the initiator tRNA in these intracellular organelles. The protein-synthesizing apparatus of these organelles also utilizes ribosomes of the $70s$ type and it has been proposed that f-met-$tRNA_F$ is the universal initiator of protein synthesis on $70s$ ribosomes (see Chapter 4).

2. The Cytoplasmic System

Elucidation of the mechanism of peptide chain initiation in the cytoplasm of eukaryotes has lagged behind that of prokaryotes, but recently, data regarding this process have become available. Present evidence indicates that f-met-$tRNA_F$ is absent from the cytoplasm of eukaryotes but the cytoplasmic $tRNA^{Met}$ from mammals, yeast, wheat germ and $Vicia faba$ does consist of two major methionine-accepting tRNAs. The cytoplasm of mammals and yeast does not contain an active transformylase and so neither of the two $tRNA^{Met}$ species is formylated under normal circumstances. However, the transformylase from $E.$ $coli$ will formylate one of these species, designated $tRNA_{F*}^{Met}$, which despite primary structural differences will substitute for the *E. coli* initiator in the *E. coli* cell-free protein synthesizing system indicating that the $tRNA_{F*}^{Met}$ of mammals and yeast has retained the structural features of f-met-$tRNA_F$ necessary for this initiator action. The other methionyl-tRNA (met-$tRNA_M$) which cannot be formylated, donates methionine only into internal positions of the growing peptide chain. Therefore, initiation of protein synthesis on 80*s* ribosomes in yeast and mammalian systems appears to involve a non-formylated met-tRNA which can be formylated *in vitro* by a transformylase from *E. coli*.

Peptide chain initiation has been studied in some detail in two 80*s* plant systems, that from wheat germ [296] and that from *Vicia faba* (broad bean) [533]. The cytoplasm of these plants was found to contain two major and one minor $tRNA^{Met}$ species of which the minor and one of the major species could function in initiating protein synthesis. In contrast to the mammalian and yeast systems, both plant systems contain an active transformylase but this may originate from the cell organelles, as may the minor $tRNA^{Met}$ species which comprises only 4% of the total $tRNA^{Met}$ of wheat embryo. Wheat and bean transformylases will formylate the minor, presumably plastid, $tRNA^{Met}$ but neither of the two major $tRNA^{Met}$ species is formylated by homologous transformylase or by *E. coli* transformylase, in direct contrast to the mammalian and yeast systems. Hence initiation in plant cytoplasmic systems appears to involve an initiating $tRNA^{Met}$ which is not formylatable. The other major $tRNA^{Met}$ species of bean and wheat embryo cytoplasm introduces methionine into internal positions in the growing polypeptide and is not involved in the initiation process.

(*a*) *Formation of the initiation complex.* The translation of natural messenger RNAs or viral RNAs possessing the initiator codon AUG or GUG at or near their 5′-end requires the participation of the initiating tRNA, GTP, ribosomes, ATP (in the case of wheat embryo) and several protein initiation factors. The sequence of addition of these components to form an "initiation complex" has been studied in both prokaryotic and eukaryotic systems. The eukaryotic systems studied in most detail are the reticulocyte system [271, 407] and the wheat embryo system [506]. The available evidence suggests that each system exhibits a degree of individuality with regard to the sequence of addition of components involved in the formation of the initiation complex.

Formation of the initiation complex in prokaryotic systems has been described previously. Chain initiation in the reticulocyte system involves binding of the initiator met-$tRNA_{F*}$ to the 40*s* ribosomal sub-unit in the presence of initiation factors IF–E_2 and IF–E_3 independently of the messenger RNA to be translated (Fig. 3.2a). This process appears to require GTP which may be bound to the complex [271]. This first intermediate then interacts with messenger RNA to produce a 40*s* subunit–mRNA–met-$tRNA_{F*}$ intermediate

C. M. Bray

(a) In mammals (rabbit reticulocytes):

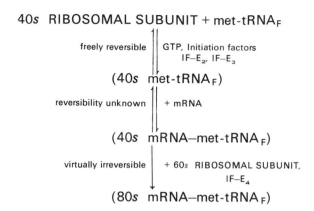

$40s$ RIBOSOMAL SUBUNIT + met-tRNA$_F$

freely reversible ‖ GTP, Initiation factors IF–E$_2$, IF–E$_3$

($40s$ met-tRNA$_F$)

reversibility unknown ‖ + mRNA

($40s$ mRNA–met-tRNA$_F$)

virtually irreversible | + $60s$ RIBOSOMAL SUBUNIT, IF–E$_4$

($80s$ mRNA–met-tRNA$_F$)

(b) In plants (wheat embryo):

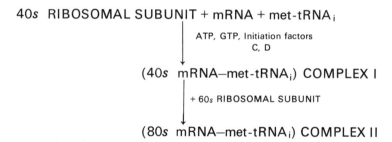

$40s$ RIBOSOMAL SUBUNIT + mRNA + met-tRNA$_i$

| ATP, GTP, Initiation factors C, D

($40s$ mRNA–met-tRNA$_i$) COMPLEX I

| + $60s$ RIBOSOMAL SUBUNIT

($80s$ mRNA–met-tRNA$_i$) COMPLEX II

Fig. 3.2. Sequence of reactions involved in the initiation of protein synthesis in eukaryotes.

which, in the presence of initiation factor IF–E$_4$, combines with a $60s$ ribosomal sub-unit to produce the $80s$ ribosome–mRNA–met-tRNA$_F$* complex active in protein synthesis. This model of eukaryotic chain initiation suggests that the initiator tRNA is bound first and as a result assists in the binding and correct phasing of the messenger RNA rather than vice versa.

The only plant system in which "initiation complex" formation has been studied in detail is the wheat embryo system in which tobacco mosaic virus (TMV) RNA or satellite tobacco necrosis virus (STNV) RNA acts as messenger [505]. These studies indicated that chain initiation in the wheat embryo system differs from both the prokaryotic and reticulocyte systems (Fig. 3.2b). Chain initiation proceeds via formation of the $40s$ subunit–mRNA–met-tRNA$_i$ complex (Complex I) which requires the presence of GTP, ATP and two initiation factors. This complex subsequently combines with the $60s$ ribosomal subunit to form an $80s$ monoribosome (Complex II) functional in protein synthesis. One major difference between initiation complex formation in the wheat embryo protein-synthesizing system and that in the prokaryotic and other eukaryotic systems studied is the requirement for the presence of ATP in the wheat system which is highly specific and absolute. ATP cannot be replaced by GTP or any other nucleoside triphosphate tested. This is consistent with the suggestion of Obendorf and Marcus that the level of ATP in wheat embryos during early germination may be a significant factor in the regulation of the germination process [338]. However, this difference between the plant system and other eukaryotic systems may be more apparent than real, since recent work by Staehelin has indicated that ATP (along with six other initiation factors) may also be specifically involved in the initiation of protein synthesis in mammalian cells [445].

The above comparison serves to illustrate the danger of assuming the universality of all aspects of the protein synthetic mechanism not only between prokaryotes and eukaryotes but also between different eukaryotic systems.

(b) *Ribosomes and messenger RNA*. The presence of an initiating codon at the 5′ end of a messenger RNA molecule is an essential prerequisite for *in vitro* mRNA translation under physiological conditions. The initiating triplet need not be the terminal 5′ triplet of the messenger RNA; indeed in polycistronic mRNA molecules one or more of the initiating triplets must occur intramolecularly. Studies on bacteriophage RNA have demonstrated long sequences of untranslated nucleotides (between 90 and 130 nucleotides) at the 5′ terminus of these molecules. The presence of untranslated sequences of nucleotides in eukaryotic messenger RNA has been demonstrated in the case of haemoglobin mRNA which has approximately 200 more nucleotides than are required for coding of the globin chains. Weeks and Marcus [505] have isolated a subcellular fraction from dry wheat embryos (messenger or MF fraction) which has the characteristics of messenger RNA in an *in vitro* amino acid incorporating system. This MF fraction contains a component, other than RNA, which is necessary for messenger-like activity and Weeks and Marcus suggest that the functional form of the messenger fraction may be a ribonucleoprotein rather than RNA alone. Concerning the problem of whether ribosomes can attach to any initiation site in a polycistronic message or whether translation begins only from the 5′ terminus of the mRNA molecule, evidence gained from prokaryotic systems is inconclusive, suggesting that either process may occur.

There is further uncertainty concerning the site or position on the ribosome to which the initiating tRNA binds primarily. It is suggested that the ribosome contains two sites, the A site involved in the binding of aminoacyl-tRNA and the P site involved in peptidyl-tRNA binding (i.e. binding of the growing polypeptide chain). These sites are not thought to be localized on any one ribosomal subunit [52], but have elements on both 30s and 50s subunits. The initiating tRNA (met-tRNA$_i$) may bind to the ribosome at the A site and subsequently be transferred to the P site in a process which could involve GTP hydrolysis or, unlike other aminoacyl-tRNAs, bind directly to the P site. Whichever of these alternatives represents the true physiological situation, it would appear that the initiating tRNA must be bound at the P site before a second aminoacyl-tRNA can bind to the ribosome at the A site. When this situation exists then chain initiation is complete and peptide chain elongation may begin.

IV. ELONGATION OF THE PEPTIDE CHAIN

Most of the detailed knowledge concerning the elongation process has been gained from studies on prokaryotic systems. Wheat embryo is the only higher plant system in which this aspect of protein synthesis has been studied in any detail and current evidence suggests that the mechanism of polypeptide chain elongation is basically similar in both prokaryotic and eukaryotic systems.

A. Requirements for elongation

In vitro studies have shown that in a system in which initiation of protein synthesis has already occurred, potassium or ammonium ions, magnesium ions, GTP, aminoacyl-tRNA and several protein factors isolated from the particle-free supernatant fraction are required for the elongation process to occur. These cytoplasmic protein factors are collectively known as transfer factors and can be further resolved into binding factors, which are responsible for the binding of aminoacyl-tRNA to the acceptor site on the ribosome, and translocase factors which are responsible for transferring peptidyl-tRNA from the acceptor to the donor site on the ribosome after peptide bond formation has occurred. The actual process of peptide bond formation is a property of the enzyme peptidyl transferase which is a component of the large ribosomal subunit in both prokaryotic and eukaryotic systems.

B. The elongation process in prokaryotes

The peptide chain is synthesized from its N-terminal end by the sequential addition of amino acids. This process of peptide chain elongation can be conveniently divided into four phases (Fig. 3.3):

(i) The starting phase where peptidyl-tRNA occupies the donor (P) site on the ribosome and the acceptor (A) site is unoccupied.

(ii) The aminoacyl-tRNA whose anticodon matches the mRNA triplet

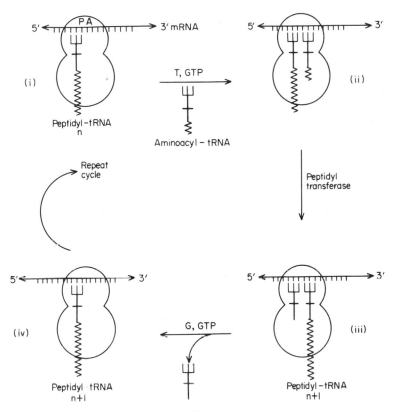

FIG. 3.3. Elongation of the polypeptide chain.

codon adjacent to the peptidyl-tRNA now binds to the A site in a reaction requiring binding factor T and GTP.

(iii) Peptidyl transferase now catalyses peptide bond formation resulting in an elongated peptidyl-tRNA bound at the A site and an uncharged tRNA, which had initially bound the peptidyl moiety in stage i, at the P site.

(iv) The final phase involves removal of the uncharged tRNA from the P site and translocation of the elongated peptidyl-tRNA from the A site to the now vacant P site. Translocation requires the presence of factor G and GTP is hydrolysed in the process.

This situation (stage iv) is now analogous to that appertaining in stage i except the mRNA molecule has moved along the ribosome by the length of one codon thus exposing a new codon under the acceptor site which can now bind a new aminoacyl-tRNA. The elongation steps are repeated many times during protein synthesis. During translation the ribosome moves along the mRNA molecule from the 5′ end to the 3′ end and in so doing once again exposes the initiation site on the message, allowing another ribosomal subunit to form a new initiation complex with the mRNA. Repetition of this process means that each mRNA molecule may be translated simultaneously by several ribosomes each of which will carry a growing polypeptide chain at different stages of completion. This complex of mRNA and several ribosomes is known as a polysome or polyribosome.

C. Properties of transfer factors

1. Prokaryotic Transfer Factors

Binding factor T isolated from bacterial sources can be further sub-divided into two components T_s and T_u which remain associated in the cytoplasm. T_u isolated from *E. coli* has a molecular weight of 40 000 whilst the molecular weight of T_s is 19 000. T-factor is implicated in aminoacyl-tRNA binding to the ribosome; it is found that it is specifically the T_u component that possesses this property. The (T_u-T_s) complex is thought to react with GTP and a charged tRNA molecule to produce a ternary complex, aminoacyl-tRNA-GTP-T_u; which then binds to the ribosome. After binding has occurred GTP is hydrolysed and GDP leaves the ribosome as a GDP–T_u complex. T_s functions in the regeneration of T factor (i.e. the T_u-T_s complex) thus allowing further reaction of T with GTP and aminoacyl-tRNA. The overall process is represented in Fig. 3.4a. Aminoacylated tRNA must be used to form a complex with T factor since uncharged tRNA, *N*-acetylated tRNA and f-Met-tRNA are all inactive in this process.

After aminoacyl-tRNA has bound to the ribosome the α-amino group on the amino acid undergoes a rapid condensation reaction with the terminal carboxyl residue of the peptidyl-tRNA on the adjacent P site resulting in the formation of a peptide bond. Peptidyl transferase, a constituent protein of the 50s ribosomal subunit, catalyses this peptide bond formation in a reaction requiring neither supernatant factors nor GTP. After formation of the peptide bond the newly extended peptidyl-tRNA is attached to the acceptor (A) site on the ribosome and must now be transferred back to the P site before the A site can bind another aminoacyl-tRNA and permit elongation to proceed. This process of translocation is catalysed by translocase factor G in conjunction with GTP which is hydrolysed to GDP and P_i. It is possible that G factor is involved in the removal from the P site of the uncharged tRNA remaining on the ribosome after donation of its peptidyl moiety to the adjacent aminoacyl-tRNA on the ribosomal A site during peptidyl transfer.

2. Transfer Factors in Plant Systems

In the absence of purified plant messenger RNAs the elongation process in plant systems has been studied using a synthetic messenger, poly-uridylic acid,

(a) in bacteria

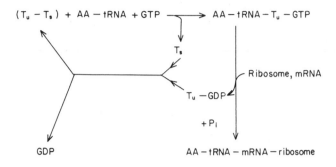

(b) in plants (wheat embryo)

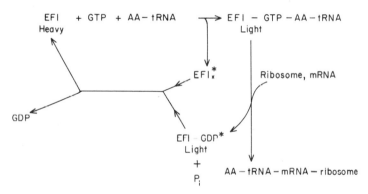

Fig. 3.4. Reactions involved in the elongation of the polypeptide chain in bacteria and wheat embryo. An asterisk represents postulated intermediates in the elongation process in wheat embryo analogous to those found in the prokaryotic elongation process.

which codes for the synthesis of poly-phenylalanine in *in vitro* systems. Poly-phenylalanine synthesis in an initiated wheat embryo system requires the presence of GTP and two soluble aminoacyl transfer factors, a binding factor (EF1) and a translocase factor (EF2) [270]. Resolution of these transfer factors followed by their subsequent purification has allowed some of their properties to be studied in detail. Similarly Yarwood *et al.* have isolated two thermolabile elongation factors from developing bean cotyledons, one factor having the properties of a binding factor similar to EF1 while the other factor is involved in polymerization [534]. Yeast has also been shown to possess two soluble transfer factors with characteristics similar to those of EF1 and EF2 of wheat embryo.

(a) *Wheat Embryo Binding Factor EF1.* Golinska and Legocki have purified wheat germ EF1 70-fold [153]. The purified preparation contained three active forms of EF1 with molecular weights of 250 000, 180 000 (the most stable form), and 60 000. It was suggested that the species having a molecular weight of 250 000 was a tetramer with $4 \times 60\,000$ subunits. The active forms of EF1 could interact directly with GTP and aminoacyl-tRNA and subsequently bind the aminoacyl-tRNA to the ribosome. Lanzani and associates, studying the same EF1 from wheat embryos, found that several forms of the factor having different molecular sizes could be isolated but only one "heavy" form with a molecular weight in the range 240 000–500 000 could join directly in a binary complex with GTP [261]. Furthermore, this heavy form had to be converted into a lighter species of molecular weight about 67 000 before the ternary complex aminoacyl-tRNA-GTP-EF1 could be formed. The light form itself could form a binary complex with GTP but was unable to form a ternary complex, suggesting that transformation of the heavy form into the light form was necessary to obtain a stable ternary complex. By analogy with the bacterial elongation process involving T_u and T_s, and in the absence of any data on GDP binding by EF1 protein species, the action of EF1 in the wheat embryo system can be represented by the sequence of interactions shown in Fig. 3.4b where hypothetical intermediates analogous to those found in bacterial systems are designated by an asterisk.

(b) *Wheat Embryo Translocase Factor EF2.* The wheat embryo translocase factor has not been studied as extensively as the binding factor. Twardowski and Legocki [487] found that EF2 was required for the translocation of newly formed peptidyl-tRNA from the acceptor site on the ribosome to the donor site during the elongation process. EF2 was purified 350-fold and found to be a single protein with a molecular weight of 70 000, which was similar to that of the bacterial and mammalian translocase factors. Sulphydryl groups were implicated in translocase activity and it was estimated that one EF2 molecule was bound per ribosome during each elongation cycle. In addition to EF2, magnesium ions, potassium or ammonium ions and GTP were required for the polymerization of phenylalanine in the wheat embryo *in vitro* system.

V. TERMINATION AND RELEASE OF POLYPEPTIDE CHAINS

The termination process has been studied in most detail in bacterial systems whilst comparatively few studies have been performed on eukaryotes. Of the work done on eukaryotic systems it is the mammalian system that has received most attention while this final step in protein synthesis has not been studied at all in higher plants. However, as the other steps of protein synthesis in plants appear similar to those in prokaryotes and other eukaryotes, and bearing in mind the apparent universality of the genetic code, then it might be expected that the termination process in plant systems will be similar to that found in other eukaryotic systems.

A. The termination process in prokaryotes

Termination occurs on the ribosome and the completed protein is released as a free polypeptide rather than as peptidyl-tRNA. There are three triplets which, when present in a messenger RNA molecule and read in phase, will code for chain termination. These triplets, UAG, UAA and UGA, must be recognized in the termination process since the presence of an unreadable triplet in the messenger RNA molecule prevents chain termination. This recognition does not involve a specific tRNA or tRNA-like molecule but requires codon-specific protein factors to hydrolyse the completed polypeptide chain from the tRNA to which it is esterified and thus free the polypeptide from the ribosomal complex.

In *Escherichia coli* three release factors, R_1, R_2 and S, have been identified. R_1 and R_2 are acidic proteins with molecular weights of approximately 44 000 and 47 000 respectively and differing in the terminator codons which they recognize. R_1 is necessary for recognition of the codons UAA and UAG while R_2 recognizes UAA and UGA but neither the activity of R_1 nor of R_2 is stimulated by the presence of the other R factor. Factor S has no release activity itself but stimulates the rate of release dependent on a particular R factor and appropriate terminator codon. The presence of another factor (factor Z), which appears to be essential for the completion of mRNA translation during or subsequent to recognition of the signal for polypeptide chain termination, has been demonstrated in *E. coli*. Work on *E. coli* mRNA indicates that terminator codons often occur in pairs, hence this tandem arrangement of "nonsense" codons may represent the termination signal. Just as the initiator codon need not be the first triplet codon at the 5'-end of the mRNA molecule, the terminator codon(s) although lying toward the 3'-end of the message need not form the 3'-terminus. Substantiating evidence for this is found in sequence studies on the terminal region of bacteriophage −R17−RNA which have shown that a substantial portion of the terminal region remains untranslated.

There is a lack of detailed knowledge regarding the actual mechanism of polypeptide chain termination and release, but *in vitro* studies support the idea of a two-step mechanism for the termination process involving the formation of an R factor-terminator codon−70s ribosome intermediate. This two-step reaction mechanism proposes that R-factor binds to a specific terminator codon in the first step, followed by a peptide release step involving a hydrolytic reaction in which R-factor converts the peptidyl transferase of the large ribosomal subunit into a hydrolase, resulting in the transfer of the peptidyl moiety of peptidyl-tRNA to water rather than to aminoacyl-tRNA, as would happen in the elongation phase of polypeptide synthesis. It should be noted here that the evidence for the role of peptidyl transferase in this hydrolase reaction is inconclusive. In the absence of translocation, removal of the uncharged tRNA which remains bound to the ribosome after release of the polypeptide chain is thought to be achieved by a factor TR, this being the first step in ribosome dissociation which follows termination (any tRNA remaining bound to the ribosome would tend to keep the subunits together). This dissociation step is discussed in more detail in section VC.

B. The termination process in eukaryotes

Studies on systems isolated from yeast and mammalian sources have indicated that the same three terminator codons UAA, UAG and UGA which act in prokaryotes also act in eukaryotes. A release factor (R factor) purified from rabbit reticulocytes has a molecular weight of about 255 000, and, in contrast to bacterial release factor, cannot be separated into components. The purified reticulocyte release factor has an associated ribosome-dependent GTPase activity and, again in contrast to the bacterial system, GTP hydrolysis appears to be a prerequisite for the chain termination process in mammals. However, GTP hydrolysis can occur in the absence of peptidyl-tRNA hydrolysis; thus the requirement for GTP hydrolysis appears to be essential to but not directly coupled to the hydrolysis of peptidyl-tRNA. No S-like factor has been found in eukaryotes and the role of peptidyl transferase in eukaryotic chain termination, just as in prokaryotes, remains controversial. Involvement of GTP in the initiation, elongation and termination steps in eukaryotes may indicate its involvement in a single intermediate event which is common to all three steps in the synthesis of the polypeptide chain. In conclusion, in the absence of any evidence from plant systems, and although evidence concerning polypeptide chain termination in other eukaryotic systems is incomplete, the eukaryotic chain termination process appears to be basically similar to that of bacteria apart from the discrepancy regarding the involvement of GTP.

C. The ribosome cycle

In order that the ribosome having just completed a translation cycle may now form a new initiation complex and begin the translation process again, it must first be dissociated into subunits in both prokaryotes and eukaryotes. The small ribosomal subunits so produced are now free to form new initiation complexes in the manner previously described. This subunit–polysome–ribosome cycle, commonly known as the ribosome cycle, appears to have the same general features in eukaryotes and prokaryotes and is represented diagrammatically in Fig. 3.5.

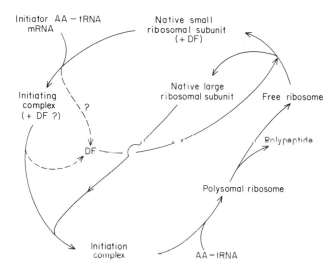

FIG. 3.5. The prokaryotic ribosome cycle. The scheme involves a proposed role for dissociation factor DF (IF–3) in the ribosome cycle. The other two prokaryotic initiation factors IF–1 and IF–2 also become part of the native 30s subunit and are released in the course of its incorporation into a polysomal ribosome.

In bacteria the dissociation of the ribosome into subunits requires the stoichiometric complexing of the ribosome with a ribosome dissociation factor (DF) present in the cell in a limiting supply thus regulating the level of the subunit pool in the cell, even though the level of the free ribosome pool may vary according to the rate of protein synthesis in the cell at any particular time. The discovery in bacteria that dissociation factor DF was in fact initiation factor IF–3 provided a link between chain termination and the initiation of protein synthesis in prokaryotes. Initiation factor IF–3 was demonstrated to have a dual function in the ribosome cycle, not only possessing the ability to dissociate ribosomes into subunits by decreasing the affinity of the small subunit for the large subunit thus ensuring a supply of 30s subunits at or after chain termination (IF–1 and IF–2, the other bacterial initiation factors, were inactive in this process) but also increasing the affinity of the small subunit for the mRNA molecule. Whether the ribosome is released as a free ribosome at chain termination and subsequently dissociated or is released directly as subunits remains an area of controversy.

Jacobs-Lorena and Baglioni have found that in a reticulocyte cell-free protein-synthesizing system, the ribosomal subunits continually produced by dissociation of the ribosomes actively involved in translation were preferentially re-utilized in the initiation of a new round of protein synthesis [215]. They suggested that this may occur for topographical reasons, i.e. the ribosomes completing translation may dissociate rapidly and the ribosomal subunits so formed will be located closer to an initiation site on the messenger RNA than other ribosomal subunits. This represents a significant difference between eukaryotic and prokaryotic ribosome cycles in that the eukaryotic ribosome pool does not equilibrate rapidly with added subunits or freshly run-off ribosomes whereas in bacteria it appears that free ribosomes are in rapid equilibrium with any subunits which may be present. In addition to this work, subunit exchange has also been demonstrated in yeast and, furthermore, a dissociation factor equivalent to DF (IF−3) has also been isolated from yeast.

Hence there is good evidence that similar ribosome cycles operate in bacteria and eukaryotic cytoplasm despite qualitative differences in the components of the protein-synthesizing systems. However, certain questions still remain to be answered, including the role of eukaryotic initiation factors in the ribosome cycle and whether any modification of dissociated ribosomal subunits is necessary after termination in order that they may be re-utilized in the initiation step.

Using the evidence gained from *in vitro* studies on plant systems, in particular from the protein-synthesizing system isolated from wheat embryo, and by analogy to other eukaryotic systems when evidence for plant systems is lacking, e.g. in the termination step, it is possible to construct a tentative model for the mechanism of protein synthesis in plant cells. This model is shown in Fig. 3.6. The model lacks much of the detail of the models for bacterial and mammalian protein synthesis with which it has many basic similarities, but this may be remedied in the near future as interest in this aspect of plant biochemistry increases. Very recently the wheat embryo *in vitro* protein-synthesizing system has been demonstrated to carry out efficient translation of a mammalian mRNA—rabbit globin 9*s* RNA [388]—and studies on mammalian mRNA translation combined with studies on natural plant messenger RNAs should assist in the elucidation of the mechanism of protein synthesis in plants.

VI. ASPECTS OF CELLULAR PROTEIN SYNTHESIS

A. The Ribosome

The ribosome is the "workbench" upon which the process of translation is carried out. Cells of higher organisms contain cytoplasmic ribosomes that have sedimentation coefficients of about $80s$ and molecular weights of $4-5 \times 10^6$ while their intracellular organelles such as mitochondria and chloroplasts possess ribosomes resembling the bacterial type, having sedimentation coefficients of about $70s$ and molecular weight of approximately 3×10^6 (see Chapter 4). The $80s$ eukaryotic cytoplasmic ribosome consists of two unequal ribonucleoprotein particles and can be dissociated into these sub-particles by lowering the magnesium ion concentration of the medium whereupon $40s$ and $60s$ ribosomal sub-particles appear. Plant ribosomes consist of approximately 40–50% RNA and 50–60% protein and their shape approximates to that of a prolate ellipsoid with a size in the dry state of $17 \times 17 \times 20$ nm.

Two high-molecular-weight RNA components are found in $80s$ ribosomes and are characterized by sedimentation coefficients of approximately $25-28s$ and $18s$, corresponding to molecular weights of $1\cdot3 \times 10^6$ and $0\cdot7 \times 10^6$ respectively. In addition to these two high-molecular-weight components the $80s$ ribosome has two low-molecular-weight RNA components, usually termed $5s$ and $5\cdot8s$ RNA. Ribosomal RNA contains the four main nucleotides of RNA plus small amounts of minor nucleotides which include methylated bases, pseudouracil and nucleotides with methylated ribose moieties making the ribosomal RNA similar in this respect to tRNA rather than to mRNA, although the proportion of minor nucleotides in ribosomal RNA is much less than in tRNA. The reader is referred to Chapter 2 for a more detailed discussion of the RNA components of ribosomes.

The ribosomal protein components of $80s$ ribosomes have been shown to consist mainly of basic proteins with few cysteine residues and apparently no disulphide bridges. Ultimately it is the interaction between this ribosomal protein and ribosomal RNA which is responsible for the final compact shape of the ribosome. The ribosomal proteins isolated from *Pisum sativum* (pea) ribosomes have been fractionated into at least 52 components [474], 24 of which can be found in the small ribosomal subunit and 28 in the large ribosomal subunit. The molecular weights of these proteins are in the range $10-35 \times 10^3$. When the arrangement of RNA and protein within the ribosome is considered it appears that the high molecular weight ribosomal RNA forms numerous short helical regions successively linked by single-stranded regions forming a flexible "rod" [443]. The protein molecules interact with the RNA mainly along the non-helical regions and the resultant ribonucleoprotein strand is folded into the final compact ribosomal sub-particle. The most characteristic and universal feature of the ribosome, whose shape varies little from species to species, is the presence of a groove or cleft which divides the two ribosomal sub-particles perpendicular to the long axis of the complete ribosome.

In addition to containing RNA and protein, the ribosome also contains some "low-molecular-weight" components which are permanently bound to the

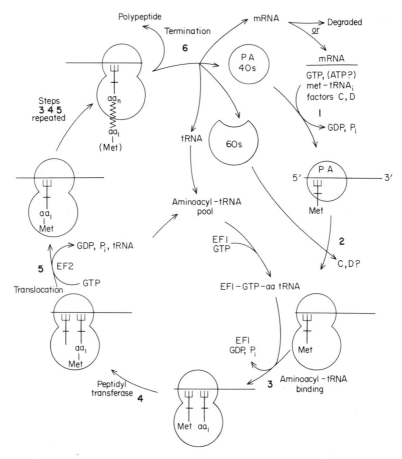

FIG. 3.6. Possible mechanism of protein synthesis on 80s plant ribosomes.

INITIATION STEPS

Step 1. The messenger RNA binds to the 40s ribosomal subunit at or near the 5′ end of
the molecule in a reaction requiring GTP, initiation factors C and D, and, in the
case of the wheat embryo system, ATP. Methionyl-tRNA$_i$, the initiating
tRNA, also binds to the 40s subunit either directly at the P site or initially at the
A site followed by translocation to the P site. The order in which the initiation
factors, met-tRNA$_i$ and mRNA bind to the plant 40s ribosomal subunit during
the formation of the initiation complex is unknown.

Step 2. The 60s ribosomal subunit binds to the initiation complex formed in Step 1,
probably with the subsequent release of initiation factors C and D, to produce a
functional ribosome.

ELONGATION STEPS

Step 3. Chain elongation begins as the EF 1-aa-tRNA complex donates the aatRNA to
the A site of the functional ribosome. GTP is hydrolysed in this process.

ribosome. Of these components, magnesium ions, calcium ions and di- and polyamines such as putrescine, cadaverine and spermidine appear to be the most essential. There is an absolute requirement for the presence of divalent metal ions, magnesium in particular, in order that the structural integrity of the ribosome is maintained. Magnesium ions are thought to stabilize the ribosomal structure by suppressing intraribosomal electrostatic repulsion due to the charged groups present and possibly by forming a system of intraribosomal cross-links (magnesium bridges within the ribosome).

B. The role of cell structure in protein synthesis

It has been estimated that the efficiency of an *in vitro* protein-synthesizing system is only 1% of that of the corresponding *in vivo* system. This is an indication that the disintegration of cell structure has resulted in the loss of some vital factor(s) necessary for efficient protein synthesis. As the importance of "cell architecture" is now being realized it seems appropriate at this point to discuss the related topic of cytoplasmic (free) and membrane-bound (mb) ribosomes and their role in protein synthesis *in vivo*.

FIG. 3.6. *contd*

Step 4. Peptide bond formation, catalysed by peptidyl transferase, occurs, resulting in the nascent protein chain subsequently produced being located at the A site on the ribosome.

Step 5. Translocation of the peptidyl-tRNA from the A site to the P site with the concomitant displacement of the discharged tRNA from the P site requires the presence of EF2 and the hydrolysis of GTP.
A repetition of steps 3, 4 and 5 is required for subsequent chain elongation.

TERMINATION STEPS

Step 6. The termination process has not been studied in plant systems but by analogy with other eukaryotic and bacterial systems the probable requirements are the presence of specific release factors and a terminator codon in the mRNA molecule. The completed polypeptide is released from the ribosome together with mRNA and tRNA while ribosome dissociation, a prerequisite for the re-initiation of protein synthesis, may also occur.

(i) met-tRNA$_i$ is shown entering at the P site initially (step 1) but it may first enter at the A site and then be translocated to the P site.
(ii) the polypeptide may be modified during synthesis so that the N-terminal amino acid may not be methionine.
(iii) the termination process (step 6), if analogous to that of mammalian systems, may involve the hydrolysis of GTP and the active participation of release factors.
(iv) chain initiation in the wheat embryo system requires the presence of ATP. Similarly it has been demonstrated recently that ATP is also a requirement for chain initiation in the mammalian system.

Early studies were performed on animal cells in particular using tissues which secreted much of the protein that they synthesized. For example, liver cells secrete serum proteins which are synthesized on membrane-bound polysomes while non-serum proteins, which are not secreted, are synthesized on free polysomes. This observation led to the suggestion that protein synthesized in a cell but utilized elsewhere, i.e. protein for "export", was synthesized on polysomes bound to membranes. In contradiction to this generality it was found that cells which did not secrete protein, e.g. rat cerebral cortex, skeletal muscle, HeLa cells, still had a proportion of their polysomes bound to the membrane and the products of protein synthesis on these membrane-bound polysomes was utilized intracellularly, not secreted into the endoplasmic reticulum as was the case with secretory tissues. Thus present evidence suggests but does not conclusively prove that in both secretory and non-secretory cells different classes of protein are synthesized on polysomes which are topographically separated.

An analogous situation exists in plant cells where polysomes may be free or membrane-bound, but the only detailed investigation into the types of protein synthesized by these different ribosome populations involves legume storage protein synthesis. Electron microscopic studies have demonstrated a correlation between the development of endoplasmic reticulum and the transfer of storage protein from ribosomes to the protein bodies during seed development. This correlation has been substantiated by the studies of Boulter and co-workers on developing seeds of *Vicia faba* where globulin synthesis has been demonstrated to occur on rough endoplasmic reticulum and the protein transferred to the developing protein bodies via the endoplasmic reticulum, not across the cytoplasm [16]. Hence, in plants it appears probable that free and membrane-bound polysomes specifically synthesize different types of protein.

This conclusion raises the question of whether two distinct classes of ribosome exist within the cell and, if there are two classes, whether they differ solely with regard to whether or not they are attached to membranes or if they possess specific distinguishing features which confer membrane-attachment or non-attachment. Conclusions drawn from studies of developing *Vicia faba* seeds at the stage at which storage protein is being synthesized suggest that while membrane-bound ribosomes can give rise to free ribosomes the converse is not true and that any increase in the proportion of membrane-bound ribosomes arises through *de novo* ribosome synthesis. However, these results were obtained in a system undergoing a change in levels and type of proteins synthesized and this may be atypical; thus the possibility of ready interchange of free and membrane-bound ribosomes in other systems cannot be excluded.

Some support for the idea of a lack of exchange of membrane-bound and free ribosomes is found from studies on animal cells by Baglioni *et al.* [15]. Their observations suggested that 60s ribosomal subunits bind directly to the membrane, probably at a specific binding site, while in the cytoplasm initiation complex formation occurs between the 40s ribosomal subunit, mRNA and initiating tRNA. If membrane-bound and free polysomes do synthesize different classes of protein then the different initiation complexes produced in the cytoplasm must distinguish in some unknown way between 60s subunits in the cytoplasm and those which are membrane bound.

C. Protein turnover

In the cell, when the synthesis of a protein has been completed, that protein is in most cases available to participate in the metabolic processes of the cell. However, the lifetime of cellular proteins is finite and both enzymes and structural proteins undergo continual degradation and resynthesis. This process is known as protein turnover. Discussion of this topic is largely outside the scope of this chapter but two pertinent points will be mentioned here (see also Chapters 6 and 7). Firstly, in plants, just as in other living systems, cellular proteins participate in the dynamic processes of the cell, the exception being storage proteins which are synthesized by the plant at certain developmental stages and removed from active participation in cellular metabolism, remaining undegraded until required for such processes as seed germination. Secondly, protein turnover is a characteristic of all actively metabolizing cells and occurs even when cellular protein content is decreasing as, for example, in the cells of pea cotyledons during seed germination, and in leaves during senescence.

VII. REGULATION OF POLYPEPTIDE SYNTHESIS AT THE TRANSLATIONAL LEVEL

It seems probable that much of the regulation of protein synthesis occurs at the transcriptional level but this does not exclude the possibility that the translational process is controlled or modulated in some manner. The previous sections have demonstrated that translation may be viewed as occurring in several distinct steps, each step being a possible site for regulation. Amino acid activation and aminoacyl-tRNA synthesis may be controlled via the availability of particular transfer RNAs, amino acids and the control of aminoacyl-tRNA synthetase activity. The rate of initiation of peptide synthesis may be controlled by the availability of both quantity and types of mRNA present at any one time and also by the availability of other components required in initiation complex formation e.g. ribosomal subunits, specific initiation factors, etc. The rate of movement of the ribosome along the message during the elongation process will control the rate of protein synthesis, while the termination of polypeptide synthesis and subsequent release of completed polypeptide chains from the ribosome offers another possible regulatory site. Certain types of post-transcriptional control such as those operative in unfertilized eggs or quiescent seeds appear quite general because they inhibit virtually all cellular protein synthesis.

The best evidence for translational control in prokaryotic systems is to be found in viral systems and the evidence for this has been reviewed elsewhere [182]. Solid evidence for translational control of protein synthesis in eukaryotes is lacking, one prime difficulty being the shortage of purified eukaryotic mRNA species which are required in order to investigate aspects of translational control in *in vitro* systems. Four eukaryotic *in vitro* protein-synthesizing systems have been studied in detail, namely those isolated from reticulocytes, various mammalian livers, Krebs II ascites cells and wheat embryo, while the messenger RNAs whose translation has been studied in such systems are those coding for haemoglobin, lens α-crystallin, histones, immunoglobulin, myosin and viral proteins. These systems have been used in the search for tissue-specific and species-specific factors that might affect translation rates. As in most aspects of eukaryotic protein synthesis most of the studies on translational control have been performed on animal cells, and plant systems have been almost entirely neglected except in studies of the processes of germination and senescence.

Although fairly extensive studies have been performed on mammalian systems, in particular with regard to the postulated existence of tissue-specific translation factors, the presence of these factors has not been conclusively demonstrated. Conclusive proof of the existence of such factors can only be obtained by fractionation and isolation of the individual factors followed by suitable assays with messenger RNAs under optimal conditions. The observation that mRNA can be translated in a heterologous system has been generally interpreted as evidence for a lack of specificity in recognition of the mRNA by the protein-synthetic machinery. However, evidence that mammalian initiation factors can distinguish between different messages has been provided in the translation of EMC viral RNA, globin mRNA and myosin mRNA by heterologous systems. Furthermore, it was suggested that if optimal conditions were not used in the assay then the translation process became independent of the specific factors, hence the absence of any evidence for specific factors in mRNA translation by heterologous systems may be due to the sub-optimal conditions of the assay [393]. It seems likely that the number and relative amounts of each of these factors vary in different cells and tissues and that this variation is responsible for the observed cell and tissue specificity in the translation of messenger RNA.

In addition to the control of the types of mRNA translated preferentially by a cell, both the rate of initiation and elongation of peptide chains may be subject to regulation. Investigations into the synthesis of α- and β-chains of haemoglobin in rabbit and human reticulocytes demonstrated that the polysomes on which the β-chains were synthesized were 30–40% larger than the polysomes in which α-chains were synthesized [281]. In order that equal numbers of α- and β-globin chains may be synthesized, the reticulocytes must contain either 30–40% more α-chain mRNA than β-chain mRNA or, if equal quantities of α- and β-chain mRNA are present, the α-chain mRNA must be translated at a rate 30–40% faster than β-chain mRNA. Evidence suggests that reticulocytes contain more α-chain mRNA than β-chain mRNA and that although there is no difference in the rate of elongation or termination of protein synthesis on either mRNA species there could be a difference in the relative rates of chain initiation.

Several studies have been performed on the possible involvement of phosphorylation of ribosomal proteins in the control of protein synthesis but even though the phosphorylation of these proteins has been demonstrated to occur both *in vivo* and *in vitro* there is no evidence that this process is involved in the regulation of ribosome activity.

In plants the process of seed germination has been studied in relation to the control of protein synthesis and although, in contrast to mammalian systems, little is known of the finer control mechanisms of the translational process, nevertheless some interesting observations have been made. During seed germination, water imbibition is followed by biochemical changes resulting in a transformation from a dormant or quiescent state, where very little protein synthesis occurs, to a state of rapid biosynthetic activity. In wheat embryo at least, the lack of protein-synthetic activity in the quiescent state is not due to any deficiency in the translation machinery and DNA transcription appears not to be involved in the "trigger" mechanism of germination; rather, the interaction of a pre-existing ("masked") mRNA in the dry seed with ribosomes appears to constitute this "trigger", at least in part. Masked mRNA is discussed in more detail later.

A further postulate regarding the regulation of protein synthesis is an extension of the idea put forward originally by Ames and Hartman [5]. The point of control in this theory is the rate of translation of the mRNA molecule which could be regulated if cells contain different amounts of the various tRNA species. In this case ribosomes would wait for a longer time period at the codons corresponding to the charged tRNAs in short supply than at codons for which a plentiful supply is available. Not only the concentration of specific tRNA molecules but also the activity of the corresponding aminoacyl-tRNA synthetases may play a critical role in limiting translational capacities. It has been proposed that the initiating events in the process of cell senescence consist of the loss of certain translational capacities in the cell. The limiting quantities of specific tRNA species and/or synthetases may restrict a cell's capacity to read certain genetic code words and the loss in specific code-reading abilities concerned with the synthesis of essential cellular components will eventually lead to deleterious effects on long-term cell function. Studies on soya bean cotyledon senescence, however, have not demonstrated any loss of specific synthetase activity in older cotyledons and the observed deficiency in charging of tRNA^{Leu} species in older cotyledons is attributed to the accumulation in the cells of a repressor substance that complexes with the aminoacyl-tRNA synthetase complex and prevents synthetase molecules from entering into further rounds of charging [33].

Studies on the levels of total and individual aminoacyl-tRNA synthetase activities in various tissues of wheat (*Triticum*) during seed maturation and germination demonstrated, with a few exceptions, very little change in the relative levels of the individual synthetases in the various tissues [332]. Similar studies on the aminoacyl-tRNA synthetases of *Phaseolus vulgaris* during germination demonstrated that these enzymes are present in the dry seed and that their synthesis is not a limiting factor in seed germination [6]. Furthermore, measurement of synthetase activity in the cotyledons during germination indicated that the synthetase enzymes did not control the rate of protein synthesis in this senescing tissue. Synthetase activity in the plumule and radicle, which are actively growing tissues, exhibited a spectacular increase during the first 2–3 days of germination with the unexplained exception of histidine and asparagine-tRNA synthetase activities in the plumule. The variation in synthetase activity during growth bore a striking resemblance to changes in dry weight.

Control of transcriptional and translational processes during the germination of wheat embryo has been studied in detail by many workers, in particular by Marcus and his colleagues. The initiation of protein synthesis in the early phases of seed germination is correlated with a rapid conversion of the monoribosome population into polyribosomes. In the germination of wheat embryo, this conversion takes place in the apparent absence of mRNA transcription during the first 12 h of germination [79], although ribosomal RNA synthesis does occur during this period. These observations suggest that the conversion of monoribosomes to polyribosomes in germinating wheat embryo involves the activation of a "preformed" messenger RNA which was synthesized during some stage of wheat embryo development. Weeks and Marcus obtained a subcellular fraction (messenger fraction or MF) from dry wheat embryos which stimulated *in vitro* amino acid incorporation in a messenger-free wheat embryo system [505]. Inhibitor studies demonstrated that there was an obligatory requirement for MF to attach to ribosomes and produce polysomes for this stimulation of amino acid incorporating activity to be observed in the cell-free system. RNA purified from MF would stimulate amino acid incorporation to only a limited extent and MF appears to function only when in the form of a ribonucleoprotein particle. MF was found in the dry embryo in substantial amounts but disappeared rapidly as germination proceeded and there was a striking correlation between MF disappearance and polyribosome formation. Unfortunately the nature of the MF protein product of translation is unknown, but the isolation of MF represents the best evidence available at present concerning the existence of an "endogenous messenger" which has been postulated to exist in dry seeds and unfertilized sea urchin eggs; this postulate is an attempt to explain the occurrence of translation in the apparent absence of transcription during the onset of germination or zygote development respectively. One unexplained major problem concerns the way in which the temporal separation of MF transcription from MF translation is controlled. Although studies on this aspect are lacking in the case of MF from wheat embryo, control of translation in germinating cotton seed has shed some light on possible control mechanisms.

Ihle and Dure, working on cotton seed embryogenesis and germination, have postulated that the synthesis of a group of enzymes required for germination is under the same co-ordinated regulation involving activation of preformed mRNA (i.e. mRNA transcribed but not translated during embryogenesis) in cotton seed cotyledons during the early stages of germination [207]. The development of two particular enzyme activities, a carboxypeptidase (i.e. protease) and isocitrate lyase, were studied during cotton seed germination, these enzymes being synthesized *de novo* during germination even when RNA synthesis was inhibited. Making use of the fact that young immature cotton embryos could be excised from the mother plant and induced to germinate (precocious germination), Ihle and Dure demonstrated that the mRNA species responsible for carboxypeptidase and isocitrate lyase production was synthesized at a stage in embryogenesis when the embryos had reached 65% of their final size, equivalent to 20 days before reaching maturity. Translation of the mRNA during the maturation period was prevented by abscisic acid, a plant growth substance synthesized by the ovule tissue surrounding the developing embryo. This inhibition of translation by abscisic acid also required the simultaneous synthesis of non-ribosomal RNA. On maturation of the seed and during the subsequent desiccation process the ovule tissue sclerifies and dies, thus removing the source of the mRNA translation inhibitor, vivipary being prevented at this stage by rapid desiccation of the embryo which is no longer protected by the living ovule tissue. When subsequent hydration and germination occurs the synthesis of isocitrate lyase and carboxypeptidase takes place. The mechanism of this hormone-mediated control of mRNA translation is not understood but possibly involves the synthesis of a regulatory protein (see Chapter 7 for further discussion). This separation in time of transcription from translation appears to be a widespread phenomenon in developmental systems. However, in the cotton embryo, in contrast to the wheat embryo, the nature of the "stored" mRNA is unknown.

VIII. POST-TRANSLATIONAL CONTROL OF ENZYME ACTIVITY

Regulation of enzyme activity at the post-translational level is one which changes readily and quickly in response to physiological changes and involves a variety of reactions and cell locales. All enzymes are proteins which differ in properties such as molecular weight, conformation, substrate specificity, etc. Some enzymes perform their catalytic role as simple proteins while others contain non-protein components as integral parts of the catalytic apparatus, these enzymes being termed "conjugated" enzymes. The active conjugate enzyme, the holoenzyme, consists of a protein component, the apoenzyme, and a coenzyme or prosthetic group. Several coenzymes are obligatory reactants in a large number of widely different reactions and any increase in the activity of one reaction could, therefore, deplete the coenzyme pool of the cell, resulting in an effect on the rate of other cellular enzymic reactions requiring this coenzyme.

In bacterial cells which are rapidly dividing, the concentration of enzyme in the cell is controlled mainly by a control of the rate of enzyme synthesis. However, under conditions such as spore development when the bacterium undergoes elaborate intracellular changes without extensive cell division, control of enzyme activity via a control of enzyme degradation appears to play a major role. In higher organisms the cell population often exists for relatively long periods between cell divisions and during the cell cycle may have to undergo many physiological changes in response to a changing environment (see Chapter 5). Consequently, if cells are unable to destroy enzymes specifically, then their ability to respond to environmental changes would consist solely of sequentially adding enzyme activities, which would severely limit the capabilities of a cell compared to the virtually unlimited range possible if enzymes can be selectively destroyed. However, very little is known about the factors controlling selective enzyme degradation *in vivo* but it is known that in certain cases the presence of substrate stabilizes enzyme activity, e.g. glucose stabilizes yeast hexokinase against trypsin attack, possibly by altering the conformation of the enzyme, hence making the enzyme–substrate complex unavailable for proteolysis. This could provide a basis for the selective degradation of enzymes by proteases, depending upon the availability of substrate for the enzyme at any particular time (see also Chapters 6 and 7).

Post-translational changes in conformation may also affect the stability and activity of some enzymes. Most of the hydrogen bonds in a protein are formed during chain assembly on the ribosome; after assembly, amino acid residues can be covalently substituted with groups which affect tertiary structure and function, although in some cases these substitutions may have already begun during assembly. Indeed many proteins are strictly dependent on this kind of substitution, i.e. sulphation, glucosylation, acetylation or lipid addition for functional activity. In addition to these substitutions, the protein conformation may be altered by the formation of disulphide bridges between cysteine residues in the protein molecule, the resulting covalent bonds having strong stabilizing effects on structure.

The quaternary structure of a protein is based on the association of subunits through hydrogen and covalent bonds. Failure of the subunits to associate as well as the unavailability of essential prosthetic groups can lead to a feedback inhibition on translation. Some proteins undergo physiological proteolytic cleavage resulting in the conversion of an inactive form of the enzyme produced during translation into an active form. Good evidence exists that this is a widespread phenomenon in many plant systems, particularly at a time during which dramatic physiological changes are taking place, e.g. during seed development and germination. The amylopectin-1, 6-glucosidase (debranching enzyme) of germinating pea (*Pisum sativum*) cotyledons is found in the dry seed in an inactive form, present in zymogen-like granules; activation of enzyme activity occurs by limited proteolysis of the precursor form during the early stages of germination [417]. This activation process may involve a change in conformation of the enzyme. The β-amylase of cereal grains occurs bound to reserve protein via disulphide bonds which are cleaved on germination of the seed to release active enzyme. Several seeds have been shown to develop trypsin-like proteolytic activity during germination, not as a result of *de novo* enzyme synthesis but through removal of an endogenous inhibitor or limited proteolysis to activate an inactive precursor form, or by a combination of both mechanisms.

Spatial separation of enzyme from available substrate is another means by which a cell can regulate enzyme activity at the post-translational level. Evidence for subcellular compartmentalization of enzyme activities in plant cells has been demonstrated by the discovery of such intracellular particles as glyoxysomes which, in the castor bean (*Ricinus*) have been shown to possess all the enzymes of the glyoxylate cycle and β-oxidation pathways plus a very active catalase. Similarly the aleurone grains in cotton seed which contain reserve stores of protein and phytate also contain acid phosphatase and acid proteinase activities and hence the ability to degrade themselves [535]. These bodies can be considered as autolytic repository organelles that have a role in intracellular digestion, and in this respect are similar to the lysosomes of animal cells.

Thus the flow, storage and concentration of protein through the cytoplasm and intracellular structures in a cell can have feedback effects on translation. Overall, at the post-translational level the specific control of enzyme degradation may be a significant process in the control of cellular metabolism in higher organisms second only in importance to control of enzyme synthesis (see also Chapters 6 and 7).

IX. CONCLUDING REMARKS

It is apparent from the preceding sections that many questions concerning the mechanism and control of protein synthesis in eukaryotes, and in particular in plant systems, remain unanswered. Much more interest is now being shown in the mechanism of the cytoplasmic protein-synthetic process in plants, particularly as the requirements of the corresponding mammalian system become more fully understood. Thus we might anticipate that over the next few years the state of knowledge regarding protein synthesis in plants will approach that of other eukaryotic systems.

SUGGESTIONS FOR FURTHER READING

ALLENDE, J. E. (1970). Protein biosynthesis in plant systems. *In* "Techniques in Protein Biosynthesis" (P. N. Campbell and J. R. Sergent, eds), Vol. II Academic Press, London and New York.

BOULTER, D., ELLIS, R. J. and YARWOOD, A. (1972). Biochemistry of protein synthesis in plants. *Biol. Rev.* **47**, 113–175.

GOODWIN, T. W. and SMELLIE, R. M. S., eds (1973). Nitrogen metabolism in plants. *Biochem. Soc. Symp.* **38**.

HASELKORN, R. and ROTHMAN-DENES, L. B. (1973). Protein synthesis. *A. Rev. Biochem.* **42**, 397–438.

IHLE, J. N. and DURE, L. S. (1972). The developmental biochemistry of cotton seed embryogenesis and germination. III. Regulation of the biosynthesis of enzymes utilised in germination. *J. biol. Chem.* **247**, 5048–5055.

MATHEWS, M. B. (1973). Mammalian messenger RNA. *In* "Essays in Biochemistry" (P. N. Campbell and F. Dickens, eds), pp. 59–102. Academic Press, London and New York.

ZALIK, S. and JONES, B. L. (1973). Protein biosynthesis. *A. Rev. Pl. Physiol.* **24**, 47–68.

4. Nucleic Acids and Protein Synthesis in Chloroplasts and Mitochondria

J. A. BRYANT

I. INTRODUCTION

The suggestion that plastids and mitochondria exhibit hereditary continuity was first made in the late nineteenth century. Hereditary continuity demands that new plastids and mitochondria arise from pre-existing plastids and mitochondria, and the most straightforward mechanism that satisfies this demand is a simple binary fission of the organelles. Division of plastids has been widely observed in green algae [403], and has been occasionally observed in higher plants. These observations of chloroplast division have recently been supported by the claim that higher plant chloroplasts can undergo division *in vitro* [385]. Mitochondrial division is harder to observe because of the pleiomorphic nature of mitochondria. Nevertheless, it is well established that mitochondria, like plastids, do undergo division [403]. A note of caution has been sounded by Bell, who has suggested that in ferns, plastids and mitochondria may arise from evaginations of the nuclear membrane [24, 25]. However, the biochemistry of the organelles, and in particular the nature of their protein-synthesizing system, make such an origin extremely unlikely. Accordingly, Bell's suggestion has received little support.

The hereditary continuity of plastids and mitochondria raises the possibility that they possess their own genetic systems, and the phenomenon of cytoplasmic inheritance supports this suggestion. Cytoplasmic inheritance was first detected by Correns and Baur in the early years of the twentieth century [23, 88]. Correns obtained a mutant of *Mirabilis* which was deficient in a factor affecting chloroplast development. Plants carrying the mutant allele possessed very pale plastids. Crosses between mutant and wild-type plants revealed strict maternal inheritance; that is, all the progeny resembled the female parent, whether the female parent was mutant or wild-type (Fig. 4.1). Baur worked with *Pelargonium*; he also found a mutant lacking a factor affecting chloroplast

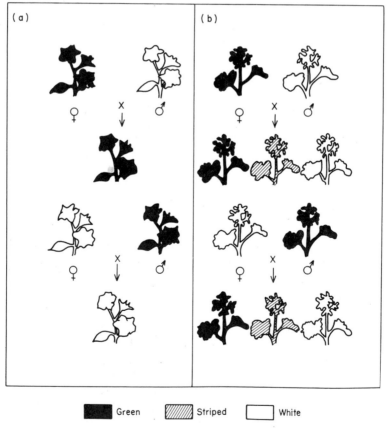

FIG. 4.1. Cytoplasmic inheritance in higher plants. (a) Maternal inheritance in *Mirabilis*. (b) Biparental non-Mendelian inheritance in *Pelargonium* (from ref. 397).

development. In crosses between the wild-type and the mutant forms, inheritance was biparental. However, the ratios of different types of progeny were non-Mendelian and very variable. Further, in progeny which were heterozygous for the factor, somatic segregation occurred, so that the plants were striped or sectored (Fig. 4.1). Since 1909, a large number of instances of cytoplasmic inheritance have been documented. These have involved either maternal inheritance or biparental inheritance with variable ratios and somatic segregation. Strict maternal inheritance may be explained by supposing that the male gamete contributes no cytoplasmic genes to the progeny. In instances of biparental inheritance, it is likely that male and female parents contribute variable numbers of cytoplasmic genes to the progeny.

In green plants, many of the features transmitted by cytoplasmic genes are associated with chloroplast development and function. This led to the idea that the genes concerned may be wholly or partly located in the chloroplast [307, 382, 406]. Similarly, it was proposed that cytoplasmic genes affecting mitochondrial development in fungi were actually located in the mitochondria [122].

Thus, the phenomenon of hereditary continuity of plastids and mitochondria, taken with the phenomenon of cytoplasmic inheritance, has led to the supposition that these organelles contain their own genes. This supposition has been confirmed by the discovery and characterization of DNA and RNA in plastids and mitochondria, and by the demonstration that both organelles are capable of protein and nucleic acid synthesis. The first report of organelle DNA was published by Ris and Plaut in 1962 [387]. Since then progress in this area has been rapid, and in some respects more is known about the molecular biology of the organelles, particularly chloroplasts, than is known about the nucleo-cytoplasmic system. This chapter is devoted to a consideration of the structure and function of the organelle genomes and protein-synthesizing systems in green plants.

II. CYTOPLASMIC INHERITANCE

Two of the features which facilitate detection of cytoplasmic genes, namely maternal inheritance or biparental inheritance with variable non-Mendelian ratios, make classical genetic studies of cytoplasmic inheritance very difficult. In particular, it is generally not possible to study linkage and recombination of cytoplasmic genes. However, Sager has established an elegant experimental system in which it has proved possible to establish linkage data for cytoplasmic genes in the unicellular green alga *Chlamydomonas reinhardi* [397].

Chlamydomonas is a haploid organism with a single large chloroplast. It normally reproduces vegetatively, but it also has a sexual reproductive cycle in

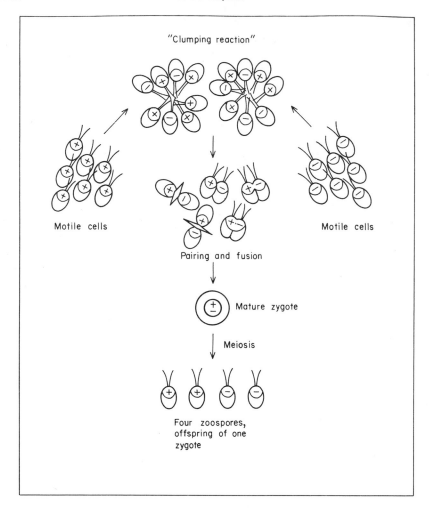

FIG. 4.2 The sexual cycle of *Chlamydomonas* (from ref. 397).

which two individuals, one of each mating type (mt^+ and mt^-) fuse to form a diploid zygote (Fig. 4.2). The mating is isogamous, with each individual contributing equal cell contents to the zygote. The mature zygote then undergoes meiosis to produce four haploid zoospores. Nuclear genes, including those controlling mating type (mt), segregate in these zoospores in a 2:2 ratio. Extensive analysis has been carried out with the nuclear genes, and the presence of 16 nuclear linkage groups has been established [276].

   In addition to the genes which segregate in the normal 2:2 ratio, Sager and her colleagues have detected a number of genes which segregate in a 4:0 ratio in the progeny, with a strict preference for the allele carried by the mt^+ mating type [397]. This is therefore equivalent to the strict maternal inheritance first reported by Correns in *Mirabilis,* and is evidence for the presence of cytoplasmic genes in *Chlamydomonas.* The genes which show a 4:0 segregation pattern include genes affecting photosynthetic capacity, and genes conferring resistance to antibiotics. Initially, the existence of strictly uniparental inheritance in *Chlamydomonas* was surprising. Maternal inheritance is widely thought to be caused by the female gamete contributing a much larger amount of cytoplasm to the zygote than the male gamete. However, some other mechanism must cause uniparental inheritance in *Chlamydomonas,* since mating is isogamous in this organism. This is discussed more fully later in the chapter.

The 4:0 uniparental segregation pattern exhibited by cytoplasmic genes in *Chlamydomonas* can be converted to a 2:2 biparental segregation pattern by irradiating the mt^+ parent with ultraviolet light just before mating [397]. Further, when cytoplasmic genes are inherited biparentally, they undergo recombination during the mitosis following zoospore formation. This feature has facilitated the study of linkage of cytoplasmic genes, and the presence of a single circular linkage group has been established. Vegetative cells which are heterozygous for cytoplasmic genes (cytohets) occur in the progeny following u.v.-induced biparental inheritance. Very occasionally, cytohets arise spontaneously in populations of normal cells. The genetic behaviour of cytohets is entirely consistent with the cytoplasmic linkage group being diploid [397].

Sager suggests that the most likely basis for the circular cytoplasmic linkage group is chloroplast DNA [397]. Support for this view comes from a study of the behaviour of chloroplast DNA during zygote maturation. By growing colonies of a single mating type in the presence of ^{15}N, Sager was able to density-label the DNA of either the mt^+ or the mt^- cells. The behaviour of the chloroplast DNA from each mating type was then followed by density gradient centrifugation of the DNA at different stages of zygote maturation, following mating between ^{15}N-labelled and unlabelled cells. These experiments have shown that chloroplast DNA from mt^- cells is broken down during zygote maturation [397]. This loss of mt^- chloroplast DNA seems the most likely explanation of uniparental inheritance of cytoplasmic genes in an isogamous organism.

Sager's work has thus provided strong evidence for chloroplast genes, located in chloroplast DNA in *Chlamydomonas.* As yet, no equivalent genetic evidence exists for the mitochondria. However, linkage data have been obtained for mitochondrial genes in yeast [86] and changes in mitochondrial DNA have been identified as the basis of certain mutations affecting mitochondrial development [31]. It seems probable that similar work will soon be carried out with unicellular algae. No linkage data exist for the cytoplasmic genomes of higher plants, although Rhoades has obtained genetic evidence for the existence of two separate cytoplasmic linkage groups in *Zea mays* (maize) [383]. In view of the wealth of information concerning organelle nucleic acids in higher plants (sections III and IV), it is reasonable to suggest that the two cytoplasmic linkage groups in *Zea* are located in chloroplast and mitochondrial DNA respectively.

III. CHLOROPLAST NUCLEIC ACIDS

A. DNA

The first definitive report of chloroplast DNA was made in 1962 by Ris and Plaut [387]. With the electron microscope they detected two areas of low electron density, containing fibrils 2·5–3·0 nm diameter in the chloroplasts of *Chlamydomonas*. The fibrils were very similar in appearance to the DNA observed in bacterial cells. The fibrils were not visible in the micrographs if the sections were treated with deoxyribonuclease. Similar observations were made on the chloroplasts of higher plants. Following this discovery, a large number of investigators attempted to extract DNA from isolated chloroplasts, and to characterize the extracted DNA. Sager and Ishida were among the earliest investigators to achieve success with this technique, again working with *Chlamydomonas* [396]. When centrifuged in gradients of caesium chloride, the DNA of *Chlamydomonas* separates into two bands, the major band (nuclear DNA) having a greater buoyant density than the minor band (tentatively identified as chloroplast DNA) (Fig. 4.3). If the DNA extracted from a crude chloroplast preparation is centrifuged in caesium chloride, the microdensitometer trace shows a reduction in the amount of the heavier species and a great increase in the amount of the lighter species. Thus, the DNA tentatively identified as chloroplast DNA is enriched in a crude chloroplast preparation. By using marker DNA species of known density in the caesium chloride gradients, it was shown that nuclear DNA from *Chlamydomonas* has a buoyant density of $1·724$ g cm^{-3}, whilst the chloroplast DNA has a buoyant density of $1·695$ g cm^{-3}. Following this work, there have been many further definitive demonstrations and partial characterizations of chloroplast DNA in unicellular green algae, in particular in *Euglena gracilis* and in *Chlorella* spp. As in *Chlamydomonas*, the chloroplast DNA has a much lower buoyant density than nuclear DNA (Table 4.I); the two DNA species may therefore be readily separated by centrifugation in caesium chloride.

The investigators who carried out the early work on chloroplast DNA in higher plants were not as successful as their colleagues who worked with algae.

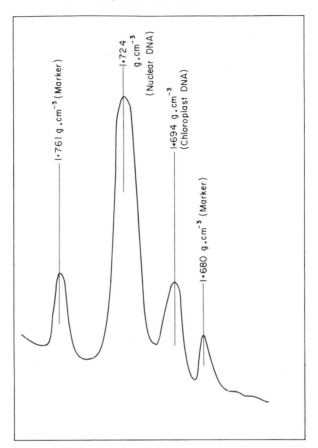

FIG. 4.3. The DNAs of *Chlamydomonas*. Microdensitometer trace from u.v. photograph of ultracentrifuge cell following equilibrium density gradient centrifugation of *Chlamydomonas* and marker DNAs in caesium chloride (from ref. 397).

Firstly, the chloroplast DNA of most higher plant cells makes up a smaller proportion of the total DNA than in green algae. Secondly, it has proved much more difficult to determine the true buoyant density of chloroplast DNA in higher plants. The first attempt to characterize the DNA of higher plant chloroplasts by use of caesium chloride gradients was reported in 1963 [83]. The chloroplast DNA prepared from *Spinacea oleracea* (spinach) or from *Beta vulgaris* (beet) separated into three bands. The major band had a buoyant density very similar to that of nuclear DNA ($1 \cdot 694$–$1 \cdot 695$ g cm^{-3}). One of the minor bands had a density of $1 \cdot 705$ g cm^{-3} while the other had a density of $1 \cdot 719$ g cm^{-3}. The major band was attributed to contamination of the chloroplasts by nuclear fragments, while the dense minor bands were identified as

J. A. Bryant

TABLE 4.I. Buoyant densities of chloroplast DNA and nuclear DNA from green algae and higher plants .[a]

Species	Density in caesium chloride (g cm^{-3})	
	Chloroplast DNA	Nuclear DNA
Chlorella ellipsoidea	1·692	1·716
Euglena gracilis	1·685	1·707
Chlamydomonas reinhardi	1·695	1·723/4
Vicia faba	1·697	1·695
Spinacea oleracea	1·697	1·694
Nicotiana sp.	1·697	1·695
Allium cepa	1·696	1·691
Triticum sp.	1·697	1·702

[a]Taken from the compilation by Ellis and Hartley [120].

chloroplast DNA. The identification of the dense DNA species as chloroplast DNA was widely accepted, and indeed was supported by results obtained with a wide variety of higher plants, although the number of DNA species attributed to the chloroplast varied from one to two in different reports.

The view that chloroplast DNA from higher plants has a very much higher buoyant density than nuclear DNA was challenged by Kirk [246]. His challenge was based on two findings. Firstly, he found by chemical analysis that the chloroplast DNA of *Vicia faba* (broad bean) has a G–C content of 39·4%, as compared with 37·4% for nuclear DNA. These base compositions are equivalent to buoyant densities of 1·6965 g cm^{-3} and 1·6955 g cm^{-3} respectively (allowing for the presence of 5-methyl-cytosine in nuclear DNA). Two DNA species of such similar buoyant densities would be extremely difficult to separate by centrifugation in gradients of caesium chloride. Secondly, chloroplast isolation techniques have improved markedly since 1963 and the dense DNA species (mentioned above) are not detectable in highly purified chloroplast preparations. It is probable that their appearance in earlier chloroplast DNA preparations was caused by the presence of contaminating bacteria and/or mitochondria. Chloroplast DNA from a wide variety of higher plants is now known to have a buoyant density of 1·696–1·697 g cm^{-3}, while the nuclear DNA of higher plants varies in density between 1·691 and 1·702 g cm^{-3} (Table 4.I). Thus in many higher plant species, the chloroplast DNA and the nuclear DNA have very similar buoyant densities. This confirms the data derived by chemical analysis of chloroplast and nuclear DNA from *Vicia faba* and gives an explanation for the failure of earlier investigators to identify correctly the chloroplast DNA in higher plants.

The large difference in the buoyant densities of chloroplast and nuclear DNA in unicellular algae indicates that the base compositions of the two DNA species are very different. A buoyant density of $1·724$ g cm^{-3} (*Chlamydomonas* nuclear DNA) is equivalent to a G–C content of 67%. A density of $1·695$ g cm^{-3} (*Chlamydomonas* chloroplast DNA) represents a G–C content of 39%. This difference in overall base composition is reflected in significant differences in the actual base sequences of the DNA molecules as indicated by nearest neighbour frequency analysis. Figure 4.4 shows that the nearest-neighbour

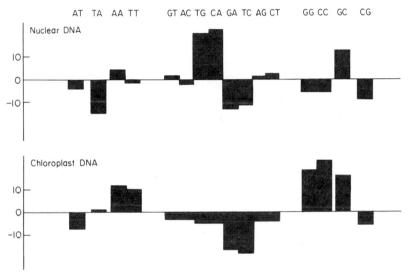

FIG. 4.4. Nearest-neighbour frequency analysis of *Chlamydomonas* chloroplast and nuclear DNA. The increase or decrease in frequency of each doublet compared with random frequencies (adjusted for the overall base composition of each DNA) is indicated by the bars above and below the base line (from ref. 397).

frequencies of *Chlamydomonas* chloroplast and nuclear DNA differ markedly from each other. Nearest-neighbour frequency analysis has not been carried out with chloroplast DNA from higher plants. Such a study could well prove very interesting, since it has been suggested that the constant buoyant density observed for higher plant chloroplast DNAs is indicative of a highly conserved base sequence [512].

A further indication of the difference between chloroplast DNA and nuclear DNA is the absence of 5-methyl-cytosine in chloroplast DNA [473]. In nuclear DNA, up to 25% of the cytosine residues are methylated (Chapter 1). Some authors have suggested that this is one of the most reliable criteria for distinguishing chloroplast DNA from nuclear DNA [120].

Nuclear DNA exists *in vivo* as a nucleic acid–protein complex (chromatin).

The bulk of the protein in chromatin is made up of histones, which confer a good deal of configurational stability on the chromatin (Chapter 1). By contrast, chloroplast DNA is not complexed with histones, and in this respect, chloroplast DNA is similar to bacterial DNA [473].

Another feature which is useful in establishing the identity of chloroplast DNA is the ease with which the two strands of the molecule reanneal after denaturation (Fig. 4.5). This contrasts markedly with the renaturation behaviour of nuclear DNA. The fact that chloroplast DNA reanneals completely

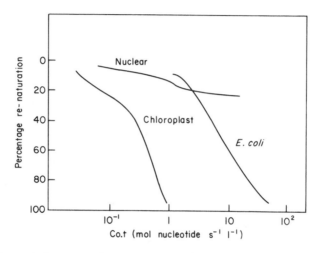

FIG. 4.5. Reassociation curves of nuclear and chloroplast DNA from *Nicotiana* and of *Escherichia coli* DNA (from ref. 473).

within a few hours means that the rate of renaturation may be used to compute the size of the chloroplast genome. At a given concentration of DNA, the second order reaction rate (Fig. 4·6) is proportional to the number of copies of the genome present (Chapter 1). The genomic sizes of several different chloroplast DNAs have been measured by this method, using bacteriophage T_4 DNA or *Escherichia coli* DNA (unique sequences of known genomic size) as standards. Some experiments of this type have provided evidence for two types of chloroplast DNA, one of which reannealed very rapidly, and the other of which reannealed more slowly [511, 513]. The rapidly reannealing DNA was ascribed to reiterated sequences in the chloroplast genome, while the more slowly reannealing fraction was assumed to represent the unique sequences. Other authors, however, failed to detect the rapidly reannealing fraction [253, 453, 473]. Tewari and Wildman have suggested that the very rapid phase of reannealing may be an artefact, caused by Type 1 renaturation or "snap-back" of strands incompletely separated during denaturation [473]. Although there is no direct evidence for this view, it is widely accepted. According to this view, chloroplast DNA contains no reiterated sequences. Table 4.II gives values for genomic size (kinetic complexity) of a variety of chloroplast DNAs. The most marked feature of these data is that the kinetic complexities of chloroplast DNAs from plants as different as *Chlamydomonas* and *Nicotiana* all fall in the

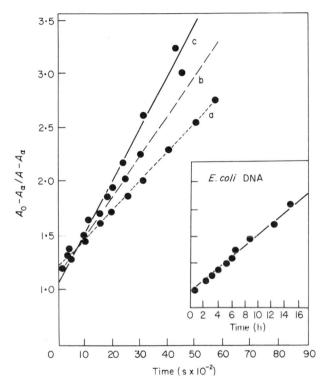

FIG. 4.6. Second-order rate plots of the reassociation of *Nicotiana* chloroplast DNA, and of *E. coli* DNA. Three different concentrations of chloroplast DNA were used: (a) 17 μg ml⁻¹; (b) 21 μg ml⁻¹; (c) 23 μg ml⁻¹ (from ref. 473).

range $0 \cdot 9 - 1 \cdot 0 \times 10^8$ daltons. The significance of this is not clear, but it has been suggested that the information content represented by a genomic size of $0 \cdot 9 \times 10^8$ daltons may be the minimum possible consistent with normal chloroplast function [120].

Osmotic lysis of isolated chloroplasts releases the DNA in a form suitable for examination with the electron microscope. This enables estimates to be made of the length and hence the molecular weight of the chloroplast DNA.

TABLE 4.II. Kinetic and analytical complexities of chloroplast DNA.[a]

Species	Kinetic complexity (daltons × 10⁸)	Analytical complexity (daltons per chloroplast × 10⁹)
Euglena gracilis	0·9	6
Chlamydomonas reinhardi	0·99	4–5
Nicotiana sp.	0·93	3
Lactuca sativa	0·98	2
Pisum sativum	0·95	?

[a]Recalculated by Ellis and Hartley [120] from previously published data, using the most recent determination of the molecular weight of the bacteriophage T4 DNA as a standard.

Initially, the DNA released from chloroplasts by osmotic lysis appeared linear; the size of the molecules seemed very variable [473]. It is now known that these features can be explained in terms of shearing of the DNA molecule during the lysis procedure [120]. With the use of gentler methods of lysing chloroplasts, it has been shown that chloroplast DNA is in fact circular. The contour lengths of the chloroplast DNA molecules from a number of higher plants and from *Euglena gracilis* are all between 37 and 44 μm, corresponding to molecular weights of $0·85-1·1 \times 10^8$ daltons [253, 294, 295]. These estimates are in very good agreement with the molecular weights (kinetic complexities) calculated from renaturation rates.

Many of the features of chloroplast DNA are consistent with it being the site of the cytoplasmic genes detected and analysed by Sager (section II, above). The existence of chloroplast DNA as a circular molecule, corresponding to the circular linkage group, is particularly noteworthy. However, in one respect the physico-chemical data concerning chloroplast DNA are at variance with the genetic data. Sager's work showed that the chloroplast genome in *Chlamydomonas* behaves as if it is diploid. It is clear, however, that there are many more than two copies of the chloroplast DNA in each chloroplast. Estimates of the total DNA per chloroplast ("analytical complexity") vary from 2×10^9 to 6×10^9 daltons (Table 4.II). Since the molecular weight of chloroplast DNA is of the order of 1×10^8 daltons, there must be between 20 and 60 copies of the chloroplast DNA molecule in each chloroplast. Further, there is evidence that the amount of DNA per chloroplast in a given plant species can vary according to the size of the chloroplast [191]. Clearly, these features are difficult to reconcile with the view that the chloroplast genome is diploid. It is possible that the chloroplast contains two "master" copies of the genome which govern the behaviour of chloroplast DNA as detected by genetic analysis, and that the other copies are "slaves" whose sequence can be checked against the master copies.

The synthesis of chloroplast DNA is mediated by DNA polymerase, which has been detected in isolated chloroplasts of higher plants and green algae [413, 441, 442, 472]. The product of the polymerase reaction is hydrolysable by deoxyribonuclease, has the same buoyant density as chloroplast DNA, and reanneals rapidly after denaturation. The polymerase enzyme is tightly bound to the chloroplast membranes. To date, only one chloroplast DNA polymerase, that of *Euglena gracilis*, has been successfully solubilized and purified [236]. A soluble DNA polymerase has been detected in chloroplasts of spinach (*Spinacea oleracea*) [442]. It is not clear whether this is a different enzyme from the bound enzyme, or whether the bound and soluble polymerases represent two populations of the same enzyme.

The synthesis of chloroplast DNA is clearly under at least a partly separate control from the synthesis of nuclear DNA, since the two events may take place at different times in the cell cycle [81, 87, 293]. Further, chloroplast replication often occurs in cells which are not dividing, suggesting that chloroplast DNA synthesis can occur in the absence of nuclear DNA synthesis [104, 424]. By use of the density transfer technique (see Chapter 1). a number of investigators have shown that the replication of chloroplast DNA, like that of nuclear DNA, is semi-conservative [81, 293] (Fig. 4.7).

Manning and Richards [293] observed that in exponential cultures of *Euglena gracilis*, chloroplast DNA is replicated 1·5 times as fast as nuclear DNA. Despite this, the ratio of chloroplast DNA to nuclear DNA does not change. This suggests that turnover of the chloroplast DNA occurs. The occurrence of DNA turnover was confirmed by use of a double labelling technique. Cells were supplied with ^{3}H and ^{32}P. Tritium was then withdrawn from the medium, and the ratio of ^{3}H to ^{32}P in chloroplast and nuclear DNA was measured through several cell cycles. The results indicated that *Euglena* chloroplast DNA has a half-life of 1·6 cell generations. The significance of this turnover is not understood at present.

B. RNA

In 1962 Lyttleton [290] reported the isolation of ribosomes from chloroplasts of *Spinacea oleracea* (spinach). Very soon after this, Bogorad's group reported the detection in electron micrographs of ribonuclease-sensitive ribosome-like particles in the plastids of *Zea mays* [217]. Following these reports, it has been confirmed by a large number of investigators that chloroplasts contain ribosomes. Indeed, in leaves of higher plants, up to 50% of the ribosomes may be contained in the chloroplasts [38, 118]. When Lyttleton first reported the

F

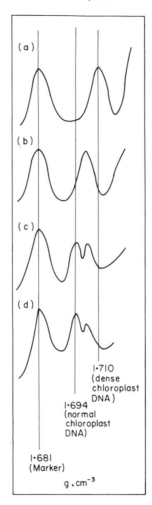

FIG. 4.7. Semi-conservative replication of chloroplast DNA in *Chlamydomonas*. Ultracentrifugation of chloroplast DNA. (a) After many generations of growth in ^{15}N: all the chloroplast DNA is dense. (b) After one round of DNA replication in the presence of ^{14}N: all the chloroplast DNA is of intermediate density. (c) After two rounds of DNA replication in the presence of ^{14}N: half the DNA is of normal density and half is of intermediate density. (d) After three rounds of DNA replication in the presence of ^{14}N (from ref. 81).

isolation of chloroplast ribosomes, he also showed that the chloroplast ribosomes have a lower sedimentation constant than cytoplasmic ribosomes. Again, this has been widely confirmed by subsequent work. Table 4.III shows the *s*-values for chloroplast ribosomes and ribosome subunits from a variety of plants. It is very clear that chloroplast ribosomes are all of the 70*s* type, with

TABLE 4.III. Sedimentation coefficients (s values) of chloroplast ribosomes

Species	s value of ribosome	s value of ribosome subunits	Reference
Euglena gracilis	70	50, 30	412
Spinacea oleracea	70	50, 30	439
Nicotiana tobaccum	70	50, 35	38
Pisum sativum	70	45, 32	460
Triticum vulgare	69	49, 31	226

subunits of 50s and 30s. This contrasts with the 80s cytoplasmic ribosomes with 60s and 40s subunits. Further, the ribosomes of chloroplasts are more readily dissociated into subunits by a lowering of the divalent cation concentration than are cytoplasmic ribosomes [38, 226]. Thus, chloroplast ribosomes resemble those of prokaryotic cells, rather than those of the cytoplasm of eukaryotic cells. Another vary marked similarity between chloroplast ribosomes and the ribosomes of prokaryotic cells is their response to antibiotics and inhibitors. Protein synthesis on chloroplast ribosomes and on bacterial ribosomes is inhibited by D-*threo*-chloramphenicol, spectinomycin, erythromycin and lincomycin, but not cycloheximide [118]. The converse is true for protein synthesis on 80s ribosomes.

The distribution of ribosomes within chloroplast to some extent parallels the distribution of ribosomes in the cytoplasm (Chapter 3). Some of the chloroplast ribosomes are "free" within the stroma, whereas others are membrane-bound [134]. The membrane-bound ribosomes may be released from the thylakoids by treatment with detergents. Both types of ribosome participate in protein synthesis, but whether they synthesize different types of protein is not known [80].

Early estimates of the size of the major RNA species in chloroplast ribosomes were made using sucrose density gradient techniques. Although the results obtained were not always unequivocal, they did suggest sedimentation coefficients of 22–23s for the RNA of the large subunit of the ribosome, and 16s for the RNA of the small subunit [198, 454]. These values are very similar to those accepted for the ribosomal RNAs of *Escherichia coli* (23s, 16s). With the introduction by Loening [282] of the use of polyacrylamide gel electrophoresis for the fractionation of RNA, a much more sensitive method became available for estimating the size of RNA molecules (Chapter 2). Use of this technique has shown that in a very wide variety of higher plants and algae, the 23s chloroplast ribosomal RNA has a molecular weight of $1 \cdot 1 \times 10^6$ daltons, whereas the 16s species has a molecular weight of $0 \cdot 56 \times 10^6$ daltons [120,

286]. These figures again emphasize the similarity between chloroplast ribosomes and bacterial ribosomes. In *Escherichia coli,* the molecular weights of the two RNA species are also $1 \cdot 1$ and $0 \cdot 56 \times 10^6$ daltons. The corresponding RNA species from cytoplasmic ribosomes have sedimentation coefficients of 25*s* and 18*s* and molecular weights of $1 \cdot 3$ and $0 \cdot 7 \times 10^6$ daltons (Chapter 2).

The base composition of ribosomal RNA from chloroplasts has been compared with that of cytoplasmic ribosomal RNA and with that of bacterial ribosomal RNA (Table 4.IV). Ellis and Hartley have pointed out a number of interesting features arising from such comparisons [120]. Firstly, the overall base composition of chloroplast ribosomal RNA differs from that of cytoplasmic ribosomal RNA. Secondly, the interspecific variation with respect to the base composition of the chloroplast ribosomal RNA is less than the variation with respect to the composition of cytoplasmic ribosomal RNA. Thirdly, in terms of base composition, chloroplast ribosomal RNA also differs from the ribosomal RNA of prokaryotic cells.

In the great majority of plant studied, the 23*s* chloroplast ribosomal RNA molecule shows some degree of instability [210, 267]. If chloroplast RNA is extracted and fractionated on polyacrylamide gels in the absence of magnesium ions, the optical density peak corresponding to a molecular weight of $1 \cdot 1 \times 10^6$ daltons is much smaller than expected. Further, there are "extra" optical density peaks corresponding to molecular weights of $0 \cdot 4 – 0 \cdot 5 \times 10^6$ daltons

TABLE 4.IV. Base compositions of ribosomal RNAs.[a]

Source of ribosomal RNA	Moles %			
	Adenine	Uracil	Guanine	Cytosine
Euglena gracilis				
Chloroplast	25·5	23·0	31·1	20·4
Cytoplasm	21·6	22·3	32·0	24·1
Hedera helix				
Chloroplast	23·4	21·1	34·1	21·6
Cytoplasm	22·2	22·0	31·7	24·1
Brassica rapa				
Chloroplast	24·4	19·8	32·1	22·7
Cytoplasm	21·5	22·7	32·2	23·6
Beta vulgaris				
Chloroplast	23·7	20·3	33·6	22·3
Cytoplasm	22·0	23·4	31·4	23·2
Escherichia coli				
23 *s* RNA	25·5	21·0	32·5	21·0
16 *s* RNA	24·2	21·3	32·1	22·3

[a]Taken from the compilation by Ellis and Hartley [120].

and 0.6–0.7×10^6 daltons (Fig. 4.8). This occurs even when ribonuclease is completely inhibited during extraction. If extraction and fractionation are carried out in the presence of magnesium ions, the optical density peak corresponding to 1.1×10^6 daltons is not diminished, and the "extra" peaks do not appear (Fig. 4.8). It has been suggested that *in vivo*, the 23s (1.1×10^6 daltons) RNA contains a "hidden break" [267]; the molecule is held together

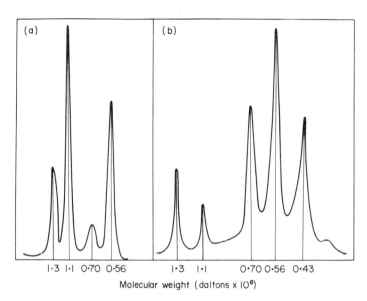

Molecular weight (daltons x 10^6)

FIG. 4.8. Instability of chloroplast ribosomal RNA in the absence of magnesium. Ribosomal RNA was extracted from a crude chloroplast preparation from *Raphanus* (radish). The ribosomal RNA was fractionated by polyacrylamide gel electrophoresis, (a) in the presence of magnesium, and (b) in the absence of magnesium (from ref. 210).

by virtue of its secondary structure and its integration into the ribosome. During extraction, the secondary structure of the RNA, and hence the overall size of the molecule, may be preserved by the presence of divalent cations such as magnesium. A further interesting feature is that newly synthesized 23s RNA, as detected by radioisotope labelling, does not show any "hidden breaks" [210]. Thus the introduction of "hidden breaks" occurs sometime during the working life of the RNA molecule. The significance of this instability *in vivo* of the 23s RNA is difficult to ascertain, particularly in view of the finding that the ribosomal RNA of chloroplasts turns over considerably more slowly than that of the cytoplasm [484].

In addition to the large ribosomal RNA molecules described above, the cytoplasmic 80s ribosomes of eukaryotic cells contain two further RNA molecules. These are 5s RNA, which is integrated into the large subunit of the ribosomes, and 5·8s RNA, which is hydrogen-bonded to the 25s RNA and which some investigators regard as being part of the 25s RNA molecule (Chapter 2). In common with all other ribosomes so far examined, chloroplast ribosomes contain 5s RNA [351]. In *Vicia faba*, the chloroplast 5s RNA has a slightly higher GNC content and a slightly greater molecular weight than the cytoplasmic 5s RNA [351]. However, in contrast to cytoplasmic ribosomes, chloroplast ribosomes do not contain 5·8s RNA, and in this respect, they again resemble the ribosomes of prokaryotic cells [352].

In 1969, Barnett and his colleagues reported that illumination of dark-grown *Euglena* cells caused the appearance of a new population of transfer RNA molecules [20]. Since this did not happen with bleached *Euglena* mutants which contained no plastids, it was suggested that the light-induced transfer RNAs were located in the chloroplasts. Similar observations have since been made with higher plants, and it is now well established that chloroplasts contain transfer RNAs which are distinct from those of the cytoplasm [539]. Transfer RNA molecules corresponding to all the usual protein amino acids have been detected in chloroplasts, and there is evidence for the presence of iso-accepting transfer RNA species for isoleucine, leucine and valine [67, 248]. The presence of at least two methionine-accepting transfer RNAs in chloroplasts has been conclusively shown [168, 272, 408]. The chloroplastic transfer RNA molecules bear much more resemblance to prokaryotic transfer RNAs than they do to the cytoplasmic transfer RNAs of eukaryotic cells, particularly with respect to overall base composition [133].

In addition to the chloroplast-specific transfer RNAs, there are also chloroplast-specific amino-acyl-transfer RNA synthetases. These are present at very low levels in etioplasts or proplastids, but increase dramatically in activity following illumination of dark-grown cells [346, 539]. The organelle enzymes show some differences from the cytoplasmic enzymes, particularly with regard to magnesium optimum and the action of inhibitors [67]. There is a reciprocal specificity between chloroplast transfer RNAs and amino-acyl-transfer RNA synthetases. With some chloroplast transfer RNAs, the specificity is absolute so that the chloroplast transfer RNA can only be charged by the chloroplast synthetase and the chloroplast synthetase can only charge the chloroplast transfer RNA [249]. With other transfer RNAs, the specificity is less marked, so that some charging takes place in heterologous systems, where one component is chloroplastic and the other is cytoplasmic [249]. In cases where the specificity is absolute, amino-acyl-transfer RNA synthetases from *Escherichia coli* can substitute for those from the chloroplast [539].

The existence in chloroplasts of iso-accepting transfer RNAs for methionine has already been mentioned. One of the methionine-accepting transfer RNAs, amounting to up to 30% of the total chloroplast methionine-accepting transfer

RNA, is capable of formylation after amino-acylation, to make N-formyl-methionyl-transfer RNA_F [272, 408]. Formylation is mediated by a transformylase enzyme, which has been detected in isolated chloroplasts [273]. The initiation of protein synthesis on chloroplast ribosomes by N-formyl-methionyl-transfer RNA_F has been demonstrated in both green algae and higher plants [273, 408]. This represents yet one more similarity between chloroplasts and prokaryotic cells. Other than the involvement of formyl-methionyl-transfer RNA_F in initiation, very little is known about the mechanism of chloroplast protein synthesis. However, it is likely to be similar to bacterial protein synthesis, since protein synthesis on chloroplast ribosomes is supported by bacterial supernatant factors [539].

The presence of RNA polymerase in chloroplasts was first reported in 1964 [244]. Since then, the presence of chloroplast RNA polymerase has been demonstrated in a number of plants. As with many other chloroplast enzymes, the activity is low in dark-grown cells, and increases markedly during light-induced greening [39, 119]. The enzyme is tightly bound to the chloroplast membrane system and is therefore difficult to purify. However, solubilization, purification and partial characterization of the chloroplast RNA polymerase of *Zea mays,* maize, have been achieved by Bogorad and Bottomley and their colleagues [39]. The chloroplast enzyme differs from the nuclear enzymes in the size of the major subunits and in cofactor and template requirements. It is particularly interesting that the enzyme markedly prefers denatured chloroplast DNA as its template, since this raises the possibility that chloroplast DNA contains sequences that are specifically recognized by chloroplast RNA polymerase.

Drugs of the rifamycin group are inhibitors of RNA synthesis in prokaryotic cells; the drugs act by blocking the initiation of transcription. These drugs also inhibit the *in vivo* synthesis of chloroplast RNA in higher plants and in green algae, while not affecting the synthesis of cytoplasmic RNA [39, 275, 457]. Experiments with rifamycin *in vitro,* using isolated chloroplasts or (partially) purified chloroplast RNA polymerase have given conflicting results [39, 522]. The situation has been partially resolved by Bogorad who suggests that *in vitro,* the inhibitory action of the rifamycins is dependent on the presence of ammonium or potassium ions [39]. It may be concluded that chloroplast RNA polymerase, like that from *Escherichia coli,* is sensitive to the rifamycins, but that observation of the inhibition *in vitro* requires the maintenance of particular conditions.

The other widely used inhibitor of RNA polymerase is α-amanitin. Bacterial polymerases are insensitive to α-amanitin, as is the nucleolar RNA polymerase of eukaryotic cells. Nucleoplasmic RNA polymerases are strongly inhibited by α-amanitin, and this has provided one means by which the nucleolar and nucleoplasmic enzymes may be distinguished [39]. In this respect, the chloroplast enzyme resembles the nucleolar and the bacterial polymerases, since it is insensitive to α-amanitin.

In addition to RNA polymerase, the presence in wheat chloroplasts of poly-(adenylic acid)polymerase and poly-(guanylic acid)polymerase has been established [68]. Using pre-existing RNA molecules as primers, these enzymes add chains of AMP and GMP molecules respectively. The poly-(A) polymerase may be involved in the processing or maturation of chloroplast messenger RNA, although the presence of poly-(A) in chloroplast messenger RNA has not been demonstrated. The function of the poly-(G) polymerase is not known.

C. Functions of the chloroplast genome and protein-synthesizing system

It is clear that chloroplasts contain all the necessary equipment to code for and to synthesize RNA and proteins. A number of experimental techniques have been used in attempts to ascertain the function of the genes contained in the chloroplast DNA molecule. These techniques are: (1) Studies of the inheritance patterns of chloroplast components; (2) DNA–RNA hybridization; (3) investigation of the effects of mutations causing the loss of chloroplast ribosomes; (4) use of specific inhibitors of chloroplast RNA and protein synthesis; (5) identification of the RNAs and proteins synthesized *in vitro* by isolated chloroplasts.

1. *The inheritance of chloroplast components*

The study of inheritance patterns of chloroplast components requires four conditions to be fulfilled. Firstly, organisms must be used in which cytoplasmic inheritance is readily distinguished from Mendelian inheritance. Secondly, there is a need for the existence of different strains, varieties or species between which there are reliably identifiable differences in chloroplast structure or function. Thirdly, it is necessary that each difference is caused by a change in a singe gene. Fourthly, hybrids between the different forms must be viable, in order that progeny may be scored for the character under consideration. This requirement is easily met when the two forms are simply different strains or varieties of the same species. It is also met by inter-specific hybrids in genera such as *Nicotiana,* where the different species are very closely related.

Mutations involving the functioning of the photosynthetic electron transport chain are initially detected by the partial or complete non-functioning of the chain [274]. The component involved may then be identified by the "cross-over point" technique, which reveals the non-functional component as the point at which the redox state of the electron transport chain changes abruptly. It is then necessary to distinguish the non-functional component from the normal functional component by differences in features such as redox behaviour, absorption spectrum or electron spin resonance. Finally, in some instances, identification has been confirmed by the restoration of electron transport activity when the normal wild-type component is added to the nonfunctional chloroplast preparation. Using these techniques, Levine and his colleagues have identified mutations in *Chlamydomonas* which affect cytochromes b_{559} and b_{553}, plastocyanin and P700 (photosystem I). In each case, the mutation is inherited in a Mendelian fashion. The genes coding for these components of the electron transport chain may therefore be assigned to the nucleus.

inhibitors of 70s ribosome protein synthesis, such as streptomycin and erythromycin [39, 397]. In bacteria, such resistance is conferred by changes in the structure of one or more ribosomal proteins [509]. In *Chlamydomonas,* one of the mutations which confers resistance to erythromycin has been shown to affect the structure of at least one chloroplast ribosomal protein [39]. It therefore seems likely that in chloroplasts, as in bacteria, changes in ribosomal proteins are responsible for resistance to antibiotics. Study of the inheritance and linkage of such mutations in *Chlamydomonas* has enabled Levine [276] and Sager [397] to assign one gene involved in streptomycin resistance to the nucleus and three to the chloroplast. Using the same organism, Bogorad had identified two nuclear genes and one chloroplast gene as being implicated in erythromycin resistance [39]. It is therefore probable that both the nuclear and chloroplast genomes are involved in coding for chloroplast ribosomal proteins.

Extensive use has been made of interspecific hybrids in the genus *Nicotiana* for studies of the inheritance of the primary structure of chloroplast proteins. For example, the primary structure of the protein in the photosystem II chlorophyll–protein complex of *N. tabacum* differs from that of *N. glauca*. The differences may be revealed by two-dimensional electrophoresis of the peptides released by the tryptic digestion of the purified complex. In hybrids between the two species the primary structure of the protein is inherited in a Mendelian manner, showing that the genetic code for this protein resides in nuclear DNA [258]. Similar results have been obtained for the small subunit of fraction I protein [233]. The primary structure of the large subunit (ribulose diphosphate carboxylase), however, is inherited maternally, suggesting that this polypeptide is coded for by chloroplast DNA [75]. This suggestion is supported by the finding that the structure of the substrate binding site of ribulose diphosphate carboxylase (as detected by measurements of K_m) is also inherited maternally [429]. In addition to the use of tryptic digests, the mobilities of proteins on polyacrylamide gels have been used to detect differences between species or varieties. This has been the basis of the findings that the chloroplast aldolase is coded for in the nucleus [7], as are at least two of the proteins in the 50s subunit of the chloroplast ribosome [44].

2. DNA–RNA hybridization

The technique of DNA–RNA hybridization facilitates establishment of the location of DNA sequences which are complementary to given RNA molecules (Chapter 2). The technique has been widely used to establish whether chloroplast RNAs are coded for in the nucleus or in the chloroplast. The most thorough investigation of this type has been that by Ingle and his colleagues [210]. They fractionated both chloroplast and nuclear DNA of Swiss chard (*Beta vulgaris*) in gradients of caesium chloride; fractions from the gradients were then incubated with chloroplast ribosomal RNA or with cytoplasmic ribosomal RNA. The results (Fig. 4.9) clearly show that cytoplasmic ribosomal RNA hybridizes only with nuclear DNA, whereas chloroplast ribosomal

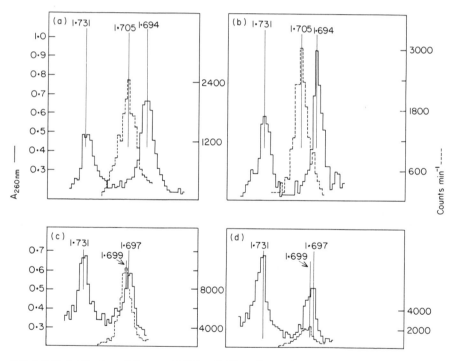

FIG. 4.9. Hybridization of ribosomal RNA to DNA in *Beta vulgaris* (Swiss chard). Preparations of chloroplast and nuclear DNA were centrifuged in gradients of CsCl in a preparative ultracentrifuge. DNA from *Micrococcus lysodeikticus* (density 1.731 g cm^{-3}) was used as a marker. The gradients were fractionated and each fraction was hybridized with radioactive ribosomal RNA. Following ribonuclease treatment (to remove unbound RNA) the amount of RNA hybridizing to each fraction was determined by estimation of radioactivity. (a) Nuclear DNA and cytoplasmic ribosomal RNA. (b) Nuclear DNA and chloroplast ribosomal RNA. (c) Chloroplast DNA and chloroplast RNA. (d) Chloroplast DNA and cytoplasmic RNA (from ref. 210).

RNA hybridizes with both chloroplast and nuclear DNA. The most obvious conclusion from these results is that sequences coding for chloroplast ribosomal RNA are located in both chloroplast and nuclear DNA [120, 473]. However, Ingle maintains that this is unlikely [210]. Experiments in which nuclear DNA was incubated with both chloroplast ribosomal RNA and cytoplasmic ribosomal RNA showed that the effect of the two RNAs is not additive. The two types of RNA therefore hybridize with the same sequences in the nuclear DNA. It is known that in several species of algae and higher plants, the two types of ribosomal RNA (chloroplast and cytoplasmic) differ in overall base composition [120]. It is therefore impossible that they should be coded for by the same DNA sequences. Ingle has suggested that the hybridization between nuclear DNA and chloroplast ribosomal RNA is non-specific, and perhaps relies on only a partial matching of sequences [210]. This suggestion is supported by results obtained by Scott [410] who, working with *Euglena*, showed that hybrids between chloroplast ribosomal RNA and nuclear DNA are denatured at a lower temperature than homologous hybrids, and further, do not show a sharp melting profile (Fig. 4.10). An additional possible explanation

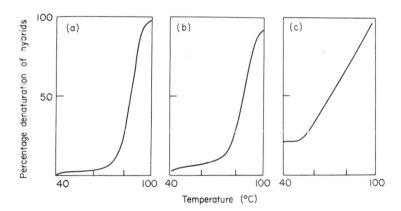

FIG. 4.10. Thermal denaturation of hybrids between ribosomal RNA and DNA from *Euglena gracilis*. (a) Nuclear DNA and cytoplasmic ribosomal RNA. (b) Chloroplast DNA and chloroplast ribosomal RNA. (c) Nuclear DNA and chloroplast RNA. Thermal denaturation was carried out in the presence of 0.17 M Na^+ (from ref. 410).

for the hybridization of chloroplast ribosomal RNA with nuclear DNA may be the contamination of preparations of chloroplast ribosomal RNA by small amounts of partially degraded cytoplasmic ribosomal RNA. It may therefore be concluded that the sequences coding for chloroplast ribosomal RNA are in chloroplast DNA.

Hybridization has also been used to give estimates of the number of cistrons coding for chloroplast ribosomal RNA. There is general agreement that each copy of the chloroplast genome contains between one and three copies of the ribosomal RNA cistrons [210, 410, 455, 473]. There is, therefore, no large-scale reiteration of the chloroplast ribosomal RNA cistrons, contrasting with the high degree of reiteration or amplification shown by the nuclear cistrons coding for cytoplasmic ribosomal RNA. However, because each chloroplast contains many copies of the chloroplast genome, there is in effect a "functionaal reiteration" of the chloroplast ribosomal RNA cistrons. This feature is of course true for all the genes contained in the chloroplast genome.

Similar hybridization experiments have been carried out with transfer RNAs from chloroplasts and cytoplasm [67, 473]. In order to avoid artefacts associated with contamination of transfer RNA preparations by fragments of ribosomal RNA, only RNAs which could be charged by amino-acyl-tRNA synthetases were used in these experiments. The results show that cytoplasmic transfer RNA specifically hybridizes with nuclear DNA and chloroplast transfer RNA specifically hybridizes with chloroplast DNA. Very little hybridization takes place in heterologous systems. Further, the sequences in chloroplast DNA which are complementary to chloroplast transfer RNA are different from those which are complementary to chloroplast ribosomal RNA, since the two RNAs do not compete with each other in hybridization [473]. The chloroplast DNA therefore contains specific sequences which code for chloroplast transfer RNA. The saturation level of hybridization represents $0.4–0.7\%$ of the chloroplast genome, which is enough to account for one cistron for each transfer RNA species per copy of the genome [473].

3. *Mutations causing the loss of chloroplast ribosomes*

In *Chlamydomonas,* mutant strain *ac-20* has a very much reduced chloroplast ribosome content [457]. The primary cause of the loss of chloroplast ribosomes is not known, but it is known that the *ac-20* gene is in the nucleus. Despite the lack of ribosomes, the plastids show some degree of development and possess many of the normal chloroplast components. Analysis of the mutant chloroplasts has been carried out in order to distingush between proteins synthesized on chloroplast ribosomes (which should be absent in strain *ac-20*) and those synthesized on cytoplasmic ribosomes (which should be present in strain *ac-20*). Using this approach, the synthesis of ribulose diphosphate carboxylase and at least one cytochrome (b_{559}) has been assigned to chloroplast ribosomes [457]. It is also likely that at least one membrane protein (in addition to the cytochrome) is synthesized on chloroplast ribosomes, since the chloroplasts in strain *ac-20* have a very much reduced membrane content and organization as compared with normal chloroplasts. By contrast, several components of the electron transport system (plastocyanin, cytochrome b_{553}, P700, ferredoxin and ferredoxin-NADP oxido-reductase) and of the Calvin cycle (phosphoribulokinase, phosphoriboisomerase, phosphoglyceric acid kinase, triose phosphate dehydrogenase, triose phosphate isomerase and aldolase) are present in *ac-20* chloroplasts and must therefore be synthesized on cytoplasmic ribosomes [457].

Mutations causing the loss of chloroplast ribosomes have been detected in the higher plants *Pelargonium zonale* [42] and *Gossypium hirsutum* (cotton) [232]. In both species, the mutation is inherited in a non-Mendelian manner and therefore has been assigned to the chloroplast genome. In *Pelargonium*, the primary effect of the mutation is the non-production of chloroplast ribosomal RNA. This provides further evidence that the chloroplast genome codes for chloroplast ribosomal RNA. Like the plastids in the *ac-20* mutants of *Chlamydomonas*, the plastids in mutant individuals of *Pelargonium* have a very much reduced membrane content. Very little else is known about these mutant plastids, but it has been established in both *Pelargonium* and *Gossypium* that the plastids contain DNA and undergo division. The enzymes controlling the synthesis of chloroplast DNA may therefore be presumed to be synthesized on cytoplasmic ribosomes.

4. *Use of specific inhibitors of chloroplast protein synthesis*

Algae or higher plants which are grown in the dark have poorly developed plastids (proplastids or etioplasts) with very little internal organization. Many of the normal chloroplast components are present in the proplastids [403]. Some are present at extremely low levels; others, such as ribosomes, may be present at about half the level of that in fully developed chloroplasts [119]. Illumination of dark-grown plants causes a rapid synthesis of chloroplast components, the establishment of a good deal of internal organization and a marked increase in the size of the plastids as they develop into fully functional chloroplasts. Specific inhibitors of 70*s* ribosome protein synthesis have been widely used in order to distinguish the chloroplast components which are synthesized on 70*s* ribosomes from those which are synthesized on 80*s* ribosomes. The inhibitors are generally supplied to the plants or algal cells before and during the period of illumination. The chloroplasts are then analysed and compared with chloroplasts in cells which had not received inhibitors. Although the use of inhibitors has some limitations, since even the most specific inhibitor may have effects which extend beyond its site of action [359], experiments of this type have provided a good deal of useful information.

The most widely used inhibitors of 70*s* ribosome protein synthesis are D-*threo*-chloramphenicol, lincomycin and spectinomycin. There is close agreement between different investigators as to the effects of these inhibitors on the synthesis of certain chloroplast components. Thus, in algae and higher plants, inhibitors of 70*s* ribosome protein systhesis prevent the synthesis of ribulose diphosphate carboxylase (or the large sub-unit of fraction I protein) and cytochrome b_{559} [45, 118, 119, 403, 457]. These inhibitors also limit the development of chloroplast membranes [118, 345, 403, 457]. In one instance, this has been shown to be caused by the inhibition of the synthesis of four membrane proteins [345]. The synthesis of phosphoribulokinase, ribose phosphate isomerase, triose phosphate isomerase, triose phosphate dehydrogenase, aldolase, P700, plastocyanin, chloroplast RNA polymerase and chloroplast amino-acyl-tRNA synthetases is not prevented by inhibitors of 70s ribosome protein synthesis [45, 118, 119, 185, 403, 457]. These components therefore depend on 80s ribosomes for their synthesis.

Although the results obtained for the components listed above seem unequivocal, there are conflicting reports concerning other chloroplast components. For example, in seedlings of *Pisum sativum*, lincomycin prevents the light-induced increase in the number of chloroplast ribosomes, without preventing the synthesis of chloroplast RNA polymerase or inhibiting incorporation of isotope into chloroplast RNA [118, 119]. Similarly, D-*threo*-chloramphenicol prevents the accumulation of chloroplast ribosomes in the alga *Ochromonas* [435]. These results have been interpreted to suggest that at least one protein of the chloroplast ribosome is made on chloroplast ribosomes. However, in *Chlamydomonas* and *Euglena*, spectinomycin does not prevent the accumulation of chloroplast ribosomes, although in other respects its effects resemble those of other inhibitors of 70s ribosome protein synthesis [275, 457]. This apparent conflict may be resolved by the suggestion that in *Chlamydomonas* and *Euglena*, assembly of chloroplast ribosomes is possible even when one or two ribosomal proteins are absent, whereas in *Ochromonas* and *Pisum* this is not possible. Conflicting evidence has also been obtained for cytochrome b_{553} and for cytochrome f, and at present it is not possible to define their site of synthesis with any degree of certainty.

5. *Synthesis of protein and RNA by isolated chloroplasts*

The most rigorous proof that a particular chloroplast component is synthesized in the chloroplast is the demonstration of its synthesis in isolated chloroplasts. Early experiments on the synthesis of macromolecules in isolated chloroplasts were beset by the problem of bacterial contamination, which led to a number of spurious reports of chloroplast protein synthesis. Even when this problem had been overcome, experiments with isolated chloroplasts were not entirely successful. Although the chloroplast preparations incorporated radioisotopes into protein or RNA, it was not possible to identify any discrete individual products. Thus, in 1967, Spencer and Whitfield reported that the product of *in vitro* RNA synthesis by isolated chloroplasts of *Spinacea oleracea* was very polydisperse, ranging in size from 5s to over 35s [440]. However, as chloroplast isolation techniques have improved, more definitive results have been obtained. It is now possible to isolate large quantities of intact, fully functional chloroplasts in which both protein and RNA synthesis are dependent on light under aerobic conditions. The criterion of light dependence under aerobic conditions readily distinguishes chloroplast protein and RNA synthesis from protein and RNA synthesis in unbroken cells or in bacteria [118].

Light-dependent amino acid incorporation in isolated *Pisum* chloroplasts results in the labelling of one soluble protein (containing 25% of the incorporated radioactivity) and six separate membrane-bound proteins [36, 112]. The soluble protein has been identified as the large subunit of fraction I protein by its mobility on polyacrylamide gels (Fig. 4.11) and by tryptic peptide

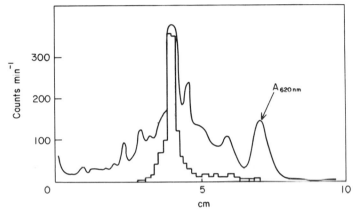

FIG. 4.11. Polyacrylamide gel electrophoresis of soluble chloroplast proteins of *Pisum sativum* (pea) after incorporation of ^{14}C-leucine *in vitro*. The gels were stained and scanned to locate the protein bands (by absorbance at 620 nm) and then sliced and counted to locate the radioactivity. The labelled protein is the large subunit of fraction I protein (from ref. 36).

analysis [36]. The findings that isolated chloroplasts can synthesize the large subunit of fraction I protein is supported by more recent evidence that chloroplasts contain an RNA species which may be translated by a bacterial protein-synthesizing system to give the large subunit of fraction I protein [180]. The messenger RNA for the large subunit of fraction I protein is therefore present in the chloroplast.

The membrane-bound proteins synthesized by isolated chloroplasts have been partially characterized by electrophoresis on polyacrylamide gels following solubilization with digitonin. The proteins have molecular weights ranging from less than 18 000 up to 85 000. The most heavily labelled protein has a molecular weight of 32 000. Attempts to identify these membrane-bound proteins have been unsuccessful, but it has been established that they do not include cytochrome f, or the proteins from the protein–chlorophyll complexes of photosystem I (P 700) and photosystem II [112].

Much of the RNA synthesized by isolated *Spinacea* chloroplasts is polydisperse, as revealed by polyacrylamide gel electrophoresis. The identity of the polydisperse RNA is not known. In addition, four discrete products are synthesized [181]. These have molecular weights of 2·7, 1·2, 1·1 and 0·47 × 10^6 daltons. These molecules have been compared with the chloroplast RNAs which are synthesized *in vivo*: 2·7, 1·2, 1·1, 0·65, 0·56 and 0·47 × 10^6 daltons (Fig. 4.12). The RNA species of molecular weights 1·1 and 0·56 × 10^6

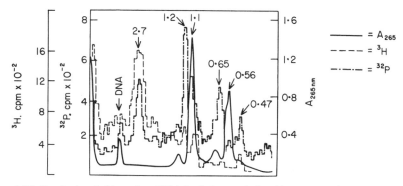

FIG. 4.12. Synthesis of chloroplast RNA in *Spinacea* (spinach) *in vivo* and *in vitro*. Spinach leaves were labelled with ^32P-phosphate. The chloroplasts were extracted and mixed with chloroplasts which had been labelled with ^3H-uridine *in vitro*. RNA was extracted from the mixed chloroplast preparation and fractionated by polyacrylamide gel electrophoresis. The numbers above the peaks are the molecular weights of the RNA molecules (units are 10^6 daltons) (from ref. 181).

are the ribosomal RNA molecules. The RNAs of molecular weights 1·2 and 0·65 × 10^6 have been identified as the immediate precursors of the mature ribosomal RNAs. Their molecular weights are very similar to those of the ribosomal RNA precursors in prokaryotic cells [164]. It has been suggested that the RNA of molecular weight 2·7 × 10^6 daltons is a polycistronic precursor to the chloroplast ribosomal RNAs, similar to the polycistronic precursor to the cytoplasmic ribosomal RNAs [120, 181]. This suggestion has received support from the results of molecular hybridization experiments which have shown that the DNA sequence which is complementary to the 2·7 × 10^6 dalton RNA contains the sequences complementary to the ribosomal RNAs themselves [179]. It is thus likely that the 0·47 × 10^6 dalton RNA is a piece of "excess" RNA, removed from the 2·7 × 10^6 dalton RNA during processing [181]. Although this interpretation of the data is the most straightforward, it still remains to be explained why the 0·56 × 10^6 dalton RNA and its immediate precursor, the 0·65 × 10^6 dalton RNA, do not become labelled in isolated chloroplasts, whereas they do become labelled *in vivo*. It is difficult to envisage any mechanism whereby the $2 \cdot 7 \times 10^6$ dalton RNA can be cleaved to give the immediate precursor to the $1 \cdot 1 \times 10^6$ dalton ribosomal RNA, without also giving the immediate precursor to the $0 \cdot 56 \times 10^6$ dalton ribosomal RNA. Despite this difficulty, it may be concluded that discrete, individual RNA species are synthesized by isolated chloroplasts.

6. Conclusions

The data made available by all these experimental techniques are summarized in Table 4.V. Two features stand out very clearly. Firstly, the large subunit of

TABLE 4.V. Sites of coding and synthesis of chloroplast components

Chloroplast component	Site of coding	Site of synthesis
Cytochrome b_{553}	nucleus	cytoplasm ?
Cytochrome b_{559}	nucleus	chloroplast ?
Cytochrome f		cytoplasm?
Ferredoxin		cytoplasm
NADP ferredoxin oxido-reductase		cytoplasm
Photosystem I (P700)	nucleus	cytoplasm
Photosystem II	nucleus	cytoplasm
Plastocyanin	nucleus	cytoplasm
Unspecified membrane proteins		cytoplasm and chloroplast
Small subunit, fraction I protein	nucleus	cytoplasm
Large subunit, fraction I protein ⎫ Ribulose diphosphate carboxylase ⎭	chloroplast	chloroplast
Aldolase	nucleus	cytoplasm
Triose phosphate isomerase		cytoplasm
Triose phosphate dehydrogenase		cytoplasm
Phosphoribulokinase		cytoplasm
Phosphoriboisomerase		cytoplasm
Phosphoglyceric acid kinase		cytoplasm
Amino-acyl-tRNA synthetases	nucleus ?	cytoplasm
RNA polymerase		cytoplasm
Enzymes of DNA synthesis		cytoplasm
Ribosomal proteins	nucleus and chloroplast	cytoplasm and chloroplast
Ribosomal RNA	chloroplast	chloroplast
Transfer RNA	chloroplast	chloroplast

fraction I protein (ribulose diphosphate carboxylase) and some ribosomal proteins are definitely known to be coded for by the chloroplast genome. The chloroplast genome also codes for chloroplast ribosomal RNA and transfer RNA. This leaves 90% of the genome unaccounted for. Secondly, with one exception, the site of synthesis of a given chloroplast component corresponds to its site of coding. Thus components which are coded for in the nucleus are synthesized in the cytoplasm and subsequently transported into the chloroplast. Components which are coded for in the chloroplast are synthesized in the chloroplast. The exception to this is cytochrome b_{559}. This is coded for in the nucleus, but is thought to be synthesized in the chloroplast, on the basis of results obtained from mutants lacking chloroplast ribosomes and from use of inhibitors of protein synthesis.

A great many aspects of chloroplast development are clearly under nuclear control, and it can be envisaged that the interactions between the chloroplastic system and the nucleocytoplasmic system may well be very complex. This complexity is emphasized by the finding that certain nuclear genes in *Zea mays* and in other higher plants may cause mutations in the chloroplast genome [383, 397]. As yet, there are no known instances of chloroplast genes affecting the nuclear genome. Nevertheless, in considering the roles of the two genomes in chloroplast development, it is reasonable to accept Schiff's suggestion that a good deal of "informational cross-talk" takes place [403].

IV. MITOCHONDRIAL NUCLEIC ACIDS

Much less is known about the molecular biology of mitochondria in green plants than is known about the molecular biology of chloroplasts. This contrasts markedly with the wealth of information available on the mitochondria of lower fungi [43, 397]. However, the improvement in techniques for the isolation of plant mitochondria in high yields has caused an acceleration of progress in this field, and it seems likely that many of the gaps in our knowledge of the molecular biology of mitochondria in green plants will be filled before too long.

DNA was first detected in plant mitochondria by cytological and histological methods [250, 324]. Mitochondria of *Beta vulgaris* and *Lilium* were shown to contain Feulgen-positive, deoxyribonuclease-sensitive fibrils, which were identified as mitochondrial DNA. In electron micrographs, the fibrils resembled the chromosomes of bacteria, thus suggesting that *in vivo*, mitochondrial DNA is not complexed with histones.

In 1966, one year after the discovery of DNA in plant mitochondria was reported, two groups of investigators reported the successful identification of mitochondrial DNA in preparations of DNA from whole cells. Certain mutant strains of *Euglena gracilis* ("bleached" strains) contain no chloroplasts and hence no chloroplast DNA. Centrifugation of DNA preparations from mutant cells in gradients of caesium chloride reveals the presence of a minor DNA species with a buoyant density of $1 \cdot 691$ g cm^{-3} (nuclear DNA in *Euglena* has a buoyant density of $1 \cdot 707$ g cm^{-3}). This DNA species is very much enriched in

DNA preparations made from mitochondria and was therefore identified by Edelman and his colleagues as mitochondrial DNA [116, 117]. This identification has been confirmed subsequently, although there is some doubt about the buoyant density originally assigned to *Euglena* mitochondrial DNA: some investigators have suggested a value of $1 \cdot 689$ g cm^{-3} rather than $1 \cdot 691$ g cm^{-3} [384]. Using similar techniques to those used by Edelman and his colleagues, Suyama and Bonner demonstrated the existence of mitochondrial DNA, with a buoyant density of $1 \cdot 706 - 1 \cdot 707$ g cm^{-3}, making up $0 \cdot 3 - 0 \cdot 5\%$ of the total DNA, in a range of higher plants [459]. The value of $1 \cdot 706 - 1 \cdot 707$ g cm^{-3} for the buoyant density of mitochondrial DNA of higher plants has been confirmed by more recent observations [266, 512] (Fig. 4.13). Thus, like chloroplast DNA, the mitochondrial DNA from higher plants shows a high degree of constancy in buoyant density over a wide range of species. One curious feature concerning mitochondrial DNA is that it has not yet been positively identified in *Chlamydomonas*, despite the great deal of attention focused on cytoplasmic inheritance in that organism.

Like chloroplast DNA, mitochondrial DNA reanneals completely after denaturation (Fig. 4.14). Calculation of the molecular weight of mitochondrial DNA from its reannealing rate gives values of $0 \cdot 70 - 0 \cdot 74 \times 10^8$ daltons in *Pisum, Vicia, Lactuca, Spinacea* and *Nicotiana* [254]. Earlier estimates of $1 \cdot 0 - 1 \cdot 5 \times 10^8$ daltons in the same plants were too high because too high a value had been accepted for the molecular weight of the bacteriophage DNA used as a standard. The kinetic complexity of mitochondrial DNA in higher plants is therefore slightly lower than that of chloroplast DNA, slightly higher than that of yeast mitochondrial DNA, and 10–20 times higher than that of mitochondrial DNA in animals. The amount of DNA in each mitochondrion is of the order of $1 \cdot 0 \times 10^8$ daltons [459], which suggests that each mitochondrion contains one or two copies of the mitochondrial genome.

The mitochondrial DNA in yeast and in a wide variety of animals is circular. Isolation of DNA from plant mitochondria has largely yielded linear molecules. However, Kolodner and Tewari [254] have shown that circular DNA molecules are released by gentle osmotic lysis of *Pisum* mitochondria. The contour length of the molecules corresponds to a molecular weight of $0 \cdot 70 \times 10^8$ daltons. This value agrees very closely with the value derived from the rate of reannealing.

The synthesis of mitochondrial DNA has been studied in bleached strains of *Euglena gracilis*. The synthesis of mitochondrial DNA is limited to one period of the cell cycle, coinciding with the *S*-phase of nuclear DNA synthesis [72]. The mitochondria then divide early in mitosis. There is some doubt as to whether the replication of mitochondrial DNA in *Euglena* is semi-conservative, since density-shift experiments indicate some randomization of the density-label between the two strands [384]. It is only possible to interpret this result in

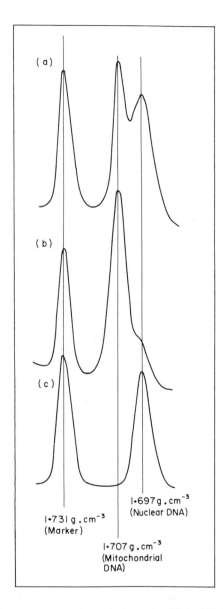

FIG. 4.13. Microdensitometer traces of u.v. photographs obtained from ultracentrifugation of *Brassica rapa* (turnip) DNA. (a) DNA from a crude preparation of mitochondria. (b) DNA from a purified preparation of mitochondria. (c) Total cellular DNA (from ref. 266).

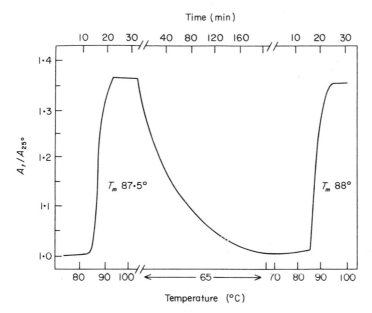

FIG. 4.14. Thermal denaturation and renaturation of mitochondrial DNA from pea (*Pisum sativum*). Thermal denaturation was carried out in the presence of 0·17 M Na⁺ (from ref. 254).

terms of semi-conservative replication if it is assumed that extensive recombination and/or turnover occurs during the replication process. It has in fact been demonstrated that mitochondrial DNA in *Euglena* turns over, with a half-life of 1·8 cell generations, but whether turnover occurs during replication has not been established [384]. It has also been reported that the mitochondrial DNA of yeast (*Saccharomyces cerevisiae*) replicates in an apparently dispersive fashion [518].

The presence of ribosomes in plant mitochondria was first established by electron microscopy[250] and has since been confirmed by the isolation of ribosomes from mitochondria. There has been some controversy over the size of mitochondrial ribosomes in plants. Thus, Wilson and his colleagues estimated the sedimentation coefficient of *Zea mays* mitochondrial ribosomes to be 66s [519], whereas Leaver and Harmey have shown that mitochondrial ribosomes from a range of plants, including *Zea mays*, have sedimentation coefficients of 77–78s, with subunits of 60s and 40s [265, 266]. The size of the major RNA species in the mitochondrial ribosomes (see below) support the results obtained by Leaver and Harmey, and it seems probable that the result obtained by Wilson and his colleagues was an artefact, possibly caused by the presence of plastid ribosomes, rather than mitochondrial ribosomes, in their preparations.

The major RNAs in the mitochondrial ribosome are larger than those in the chloroplast ribosome. The RNA from the large subunit has a sedimentation coefficient of $24s$, while that from the small subunit has a sedimentation coefficient of $18-19s$ [266]. These values are much more similar to the values for cytoplasmic ribosomal RNAs than they are to the values for chloroplast ribosomal RNAs. Considerable confusion has existed as to the molecular

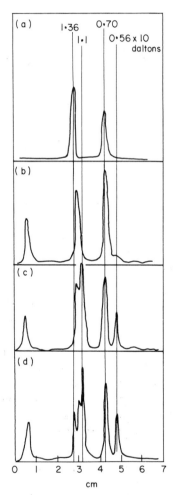

FIG. 4.15. Polyacrylamide gel electrophoresis of nucleic acids from *Brassica rapa* (turnip). (a) Cytoplasmic ribosomal RNA. (b) Mitochondrial DNA and mitochondrial ribosomal RNA. (c) Mitochondrial DNA, mitochondrial ribosomal RNA, and *Escherichia coli* ribosomal RNA. (d) Mitochondrial DNA and cytoplasmic, mitochondrial and *E. coli* ribosomal RNA (from ref. 266).

weights of mitochondrial ribosomal RNAs, as estimated by polyacrylamide gel electrophoresis. The secondary structure of the mitochondrial ribosomal RNAs is very sensitive to change in temperature [266]. Thus, the migration of the RNAs through gel is very much affected by the temperature at which the electrophoresis is carried out. By using conditions most likely to maintain the normal secondary structure, Leaver and Harmey have estimated the molecular weights in various species of higher plants to be $1 \cdot 12 - 1 \cdot 2 \times 10^6$ daltons for the RNA of the large subunit and $0 \cdot 69 - 0 \cdot 78 \times 10^6$ daltons for the RNA of the small subunit of the mitochondrial ribosome (Fig. 4.15). These molecular weights are much more similar to those for RNAs from 80s ribosomes than to those for RNAs of 70s ribosomes (Table 4.VI).

TABLE 4.VI. Properties of higher plant ribosomes

Location	s value	s value of subunits	RNA of subunits	
			s value	Molecular weight (daltons)
Cytoplasm	80	60	25	$1 \cdot 3 \times 10^6$
			5	$3 \cdot 8 \times 10^4$
			5·8	$5 \cdot 0 \times 10^4$
		40	18	$0 \cdot 7 \times 10^6$
Chloroplast	70	50	23	$1 \cdot 1 \times 10^6$
			5	$3 \cdot 9 \times 10^4$
		30	16	$0 \cdot 56 \times 10^6$
Mitochondrion	77/78	60	24	$1 \cdot 12 - 1 \cdot 2 \times 10^6$
			5	$3 \cdot 8 \times 10^4$?
		40	18/19	$0 \cdot 69 - 078 \times 10^6$

Although the ribosomes from plant mitochondria bear some resemblance to the 80s cytoplasmic ribosomes, they are similar in other ways to the 70s ribosomes of the chloroplast (Table 4.VI). The similarities include the instability *in vivo* of the RNA of the large subunit of the ribosome, and the presence of 5s but not of 5·8s RNA [266]. Functionally, too, the mitochondrial ribosomes resemble chloroplast ribosomes. Protein synthesis on ribosomes from plant mitochondria is inhibited by lincomycin and D-*threo*-chloramphenicol, but not by cycloheximide [248, 265, 320]. The mitochondrial ribosomes of higher plants may therefore be considered as ribosomes of the 70s type.

Mitochondria of algae and higher plants contain transfer RNA species which are distinct from those of the cytoplasm and from those of the chloroplast [166, 169, 248, 249, 305]. Mitochondria also contain their own amino-acyl-transfer

RNA synthetases [166, 248, 249]. The specificity of recognition of mitochondrial transfer RNAs by mitochondrial synthetases is not very marked. Thus, a significant level of charging is achieved in heterologous systems, where one component is from the mitochondria and the other from the cytoplasm [166, 249]. There are two species of methionine-accepting transfer RNAs in mitochondria [168, 169]. One of these is capable of formylation by a mitochondrial transformylase and may therefore be designated $tRNA_F^{met}$. The $tRNA_F^{met}$ of the mitochondria of *Phaseolus* is different from that of the chloroplasts, as revealed by analysis of the Tl-ribonuclease hydrolysis products of the two molecules [168]. It is probable that formyl-methionyl-$tRNA_F$ functions as the initiator transfer RNA in mitochondrial protein synthesis, again indicating that the mitochondrial ribosome is of the 70s type rather than the 80s type.

Very little is currently known about the functions of the mitochondrial genome and protein-synthesizing system in green plants. Although the presence of a second extranuclear linkage group (i.e. in addition to that in the chloroplasts) has been established in higher plants, no genes have been mapped in the mitochondrial genome [397]. Further, it is difficult to study the inheritance patterns of mutations affecting mitochondrial function, since most green plants cannot live without functional mitochondria. For the same reason, it is difficult, although not impossible, to study the effects of inhibitors on the synthesis of mitochondrial components. The only result obtained with inhibitors of protein synthesis suggests that none of the proteins of the mitochondrial ribosome are made in the mitochondria, since the accumulation of mitochondrial ribosomes is not inhibited by D-*threo*-chloramphenicol in the alga *Ochromonas* [435]. This result awaits confirmation in other plant species.

One feature which has been established is that isolated mitochondria from algae and from higher plants incorporate amino acids into protein [13, 248, 265, 320]. However, it has not yet been possible to identify any one protein as being the product of mitochondrial protein synthesis, even when tightly coupled mitochondria are used. Nevertheless, it is clear that the protein-synthesizing system in mitochondria is functional, and it seems likely that the identity of some of the proteins synthesized on mitochondrial ribosomes will be established in the near future.

SUGGESTIONS FOR FURTHER READING

ELLIS, R. J. and HARTLEY, M. R. (1974). Nucleic Acids of Chloroplasts. *In* "Nucleic Acids" (K. Burton, ed.). International Review of Science. Vol. 6. Medical and Technical publishing Co., Lancaster.
GOODWIN, T. W. and SMELLIE, R. M. S., eds (1974). "Nitrogen Metabolism in Plants". *Biochem. Soc. Symp.* **38**.

INGLE, J., POSSINGHAM, J. V., WELLS, R., LEAVER, C. J. and LOENING, U. E. (1970). The properties of chloroplast ribosomal RNA. *Symp. Soc. exp. Biol.* **24,** 303–325.

LLOYD, D. (1974). "The Mitochondria of Microorganisms". Academic Press, London and New York.

SAGER, R. (1972). "Cytoplasmic Genes and Organelles". Academic Press, New York and London.

ZALIK, S. and JONES, B. L. (1073). Protein Biosynthesis. *A Rev. Pl. Physiol.* **24,** 47–68.

5. The Cell Cycle

J. A. BRYANT

I. INTRODUCTION

The general observation that the DNA content of non-dividing cells in a given organism remains constant, except when "polyploidy" occurs, has been mentioned previously (Chapter 1). Dividing cells, however, do not show this constancy. Thus, in the cells of growing tissues of the mouse (*Mus*), Swift detected DNA contents varying between 2C (the amount of DNA corresponding to a normal diploid chromosome set) and 4C during the interphase of the mitotic cycle [463]. This suggests that the DNA content doubles before the onset of mitosis. The demonstration that DNA synthesis not only takes place in interphase, but is confined to a restricted period of interphase was made by Walker and Yates in the early 1950s [500]. Using animal cells in culture, they were able to relate the DNA content of a cell to the developmental stage reached by that cell. Their results suggested that following the completion of cell division, there is a period in which the DNA content of a cell remains at the 2C level. This is followed by a period in which there is a more or less linear increase in DNA content from the 2C to the 4C level. The DNA content remains constant at the 4C level until the onset of mitosis, leading to the production of two daughter cells, each with a 2C level of DNA. These observations were followed in 1953 by those of Howard and Pelc, who showed by autoradiographic methods that incorporation of radioisotope into the nuclear DNA of cells in root meristems of *Vicia faba* (broad bean) is confined to a discrete period of interphase [202]. Howard and Pelc introduced the concept of the DNA cycle or cell cycle, divided into four phases: G1 ("gap 1"), the period after daughter cell formation during which the DNA content remains steady at the 2C level, S-phase, the phase of nuclear DNA synthesis, G2 ("gap 2"), the period in which the DNA content remains constant at the 4C level, and M, mitosis or cell division (Fig. 5.1). The cell cycle is therefore a sequence of discrete, temporally separated events which normally follow on one from another in a definite orderly sequence. Study of the cell cycle and of its control is regarded by many investigators as one of the most important areas of modern biology. An understanding of cell division is relevant not only to academic problems, but is also relevant to several extremely important applied problems such as the improvement of yield in crop plants and the control of cancers. This chapter deals with the methods used in study of the cell cycle, with the events that occur in "typical" and modified cell cycles, and with the control of the cell cycle.

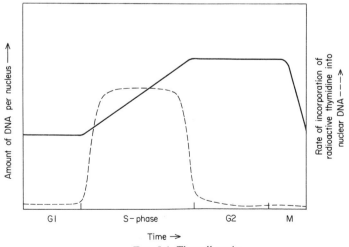

FIG. 5.1. The cell cycle.

II. STUDY OF THE CELL CYCLE IN PLANTS

A. Synchronous populations of cells

In synchronous populations of cells, the cell cycles of the individual cells are all in phase. Detection and measurement of the major phases of the cell cycle is therefore relatively straightforward. The S-phase is the period over which the DNA content of the culture doubles, and during which the cells readily incorporate radioactive thymidine into nuclear DNA. Cell division may be detected by counting mitotic figures and by estimating the total number of cells in the population. Synchronous populations of cells occur naturally only rarely. Usually, therefore, synchronous cell populations must be derived from nonsynchronous populations, using one of two methods, selection or induction. In selection synchrony, cells which are at a given stage in the cell cycle are separated from the total population and grown on as a separate culture. The method relies on the use of a parameter of cell growth which changes through the cell cycle. The most widely used parameters are cell size and cell density; the most widely used method for the separation of cells of different sizes and/or densities is density gradient centrifugation. Cells from a non-synchronous culture are concentrated and layered on to a gradient of sucrose or Ficoll. The gradients are then centrifuged either for a few minutes (velocity sedimentation, resulting in separation of cells by size and density) or to equilibrium (resulting in the separation of cells by buoyant density alone). The smallest and/or least dense cells, which are those which have most recently divided, are removed from the top of the gradient and used to initiate a synchronous culture. This method has been used with a good deal of success with bacteria and with various species of yeast. Synchronous cultures of *Chlorella* have also been obtained with this technique [199], but the technique has not been applied to other green plants. Suspension cultures of higher plant cells are not suitable material for the selection technique, since cells at the same stage of the cell cycle may differ significantly in size or density.

The process of induction of synchrony in non-synchronous populations relies on manipulation of the conditions under which the cells are grown. One of the most commonly employed techniques for induction of synchrony is the use of inhibitors which act at specific points in the cell cycle. The cells are prevented from entering the phase of the cycle on which the inhibitor acts, and a population of cells blocked at the same point in the cycle therefore accumulates. The inhibitor is then removed, and the whole cell population is able to proceed synchronously through the cycle (Fig. 5.2). There are a number of difficulties inherent in this technique. For example, in order to ensure that all cells in a population are synchronized, it is often necessary to use a "double blockade". This is best illustrated by considering the effects of an inhibitor of DNA synthesis. The presence of an inhibitor of DNA synthesis causes the cells to accumulate at the G1–S boundary. However, at the time of addition of the inhibitor, a proportion of the cell population will be in S-phase. Because of the presence of the inhibitor, many of the cells in S-phase will be unable to complete DNA synthesis and will therefore remain "frozen" in S-phase, although not synthesizing DNA. In order to bring these cells into synchrony with the remainder of the population, the inhibitor must be washed out and the cells allowed to grow in the absence of the inhibitor for long enough to allow all the cells (both those held in S-phase and those held at the G1–S boundary) to proceed through S-phase. The inhibitor is then added once more, causing all the cells to be blocked at the G1–S boundary. This double blockade technique has been used very successfully with thymidine, which at high concentrations (25 mM) is an effective inhibitor of DNA synthesis in mammalian cells [360]. However, even at very high concentrations thymidine is not an effective inhibitor of DNA synthesis in plants, and the use of thymidine for induction of synchrony does not give satisfactory results [389]. Other inhibitors of DNA synthesis, such as 5-amino-uracil, 5-fluoro-deoxy-uridine and hydroxyurea, and inhibitors of mitosis, such as colchicine, have been used with both plant and animal cells. However, these compounds are regarded as too toxic to use a double blockade technique and so give only a partial synchrony [125]. Further, even when used as a single blockade, these compounds may have unwanted side-effects such as chromosome damage, and thus their use is open to criticism [243].

Many green algae may be synchronized by repeated cycles of alternating light and dark treatments. In these algae, parts of the cell cycle are light-dependent in autotrophically grown cells and so the effect of an alternating

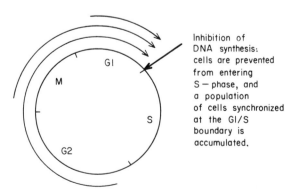

FIG. 5.2. Induction of synchrony by use of an inhibitor.

light—dark regime is similar in some respects to a multiple blockade by an inhibitor, resulting in a high degree of synchrony, but without any of the side-effects of inhibitors. Further, the synchrony may be maintained indefinitely by continuation of the light—dark treatment, since once the cells are synchronized, they are bound to remain in phase with the light—dark regime. Algae which may be synchronized in this way include *Chlorella, Chlamydomonas* and *Euglena* [81, 341, 467]. Alternating warm—cold treatments have also been used to synchronize cultures of *Euglena*. The rationale of the method is similar to that of the light—dark treatment. Division proceeds at temperatures of around 25°C, but is prevented at lower or at higher temperatures. Thus, cycles of 18—25°C and 35—28°C have both been used successfully with *Euglena*. In both instances the first temperature mentioned is restrictive, whereas division may proceed at the second temperature [471].

Perhaps the most widely applicable method for induction of synchrony is the "starvation" method. Cultures of cells are allowed to enter the stationary phase. The cells are then transferred to a fresh medium from which a component essential for DNA synthesis or division has been omitted. The component is then added to the medium, and the cells proceed through the cycle synchronously. For suspension cultures of higher plant cells, the component omitted from the medium is generally kinetin. However, the degree of synchrony achieved by this form of induction is poor in higher plants [228, 389], although satisfactory results are obtained with cells of other organisms.

For synchronizing suspension cultures of higher plant cells, a variant of the starvation method has been developed by Street and his colleagues [243]. Cells are allowed to proceed well into the stationary phase, and are then inoculated at low cell density into a fresh medium. The cells show a high degree of division synchrony, which may persist for six or more cell generations (Fig. 5.3). This technique has been extensively used for synchronizing cultures of *Acer pseudoplatanus* (sycamore) cells. The technique is also suitable for suspension cultures of cells of other higher plants, including *Glycine max* (soya bean) [53]. Suspension cultures of *Daucus carota* (carrot), however, are not completely synchronized by this procedure, but they may be synchronized by incorporating a cold treatment into the procedure [340]. Cells in the stationary phase are inoculated into new medium at 4°C. After several hours at 4°C, the temperature is raised to 27°C, and the cells proceed to synthesize DNA and divide synchronously for at least two generations.

The methods for induction of synchrony discussed so far are mainly applicable to cells grown in suspension culture. However, it is also possible to induce synchrony in explants of higher plant tissue containing large numbers of cells. If explants are excised from dormant or quiescent plant tissues (e.g. the parenchymatous tissue of storage organs such as carrots or potatoes) and cultured in a suitable medium, DNA synthesis and cell division are initiated, and the

J. A. Bryant

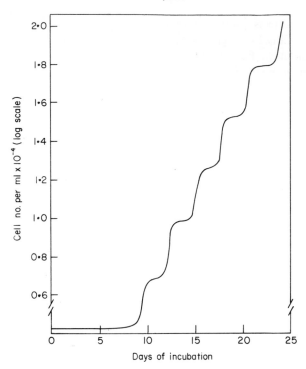

FIG. 5.3. Synchronous divisions in a suspension culture of cells of *Acer pseudoplatanus* (sycamore) (from ref. 243).

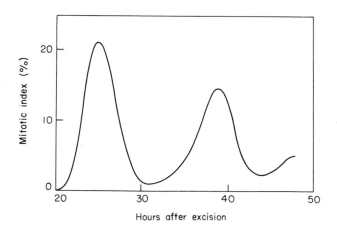

FIG. 5.4. Synchronous divisions in an explant of *Helianthus tuberosus* (artichoke) tuber. About 35% of the cells in the explant divide during the first peak of mitotic activity if the explant is kept in the light (from ref. 536).

explants begin to grow (see section VI). The first two, or even three divisions after excision show an acceptable level of synchrony, particularly in explants of *Helianthus tuberosus* (Jerusalem artichoke) tubers (Fig. 5.4). As with cells in suspension culture synchronized by the starvation technique, the first cell cycle is abnormally long because of the lag phase prior to the onset of DNA replication [175, 536].

Van't Hof and his colleagues have demonstrated that withholding carbohydrate from excised roots of *Vicia faba* or *Pisum sativum* causes the meristematic cells to be held either in late G1 or in late G2 [492]. Carbohydrate starvation therefore leads to the establishment of two separate cell populations in these meristems. The populations may be brought into phase with each other by adding carbohydrate back to the culture medium in the presence of an inhibitor of DNA synthesis, such as 5-fluoro-deoxy-uridine. The cells in late G1 remain there; the cells in late G2 proceed through mitosis and G1 until they too are held in late G1 [256].

There are some instances of populations of plant cells which exhibit a natural synchrony. The first two or three divisions in the root meristem of germinating *Vicia faba* seedlings show synchrony [219], as do the early divisions in the developing endosperms of many seeds [124]. The extended meiotic division in the anthers of many monocotyledonous plants is also synchronous [124].

B. Non-synchronous populations of cells

Synchronous populations of cells are clearly extremely useful. However, synchronous populations of higher plant cells are, with the exceptions mentioned above, essentially unnatural. Study of the cell cycle in natural, non-synchronous populations of cells requires the use of more complex methods. The most popular method for analysis of the cell cycle in natural populations is the "labelled mitosis" method, first developed by Howard and Pelc in the early 1950s [202], and subsequently modified and improved by other investigators [368, 491]. The cell population is supplied with a brief pulse of tritiated thymidine. Samples are taken at intervals, autoradiographs are prepared, and the number of labelled metaphases is recorded, as is the total number of mitoses. As the cells which were in S–phase at the time of the pulse (and hence incorporated radioactivity into DNA) enter mitosis, the number of labelled metaphases rises from zero to a peak; the number then falls as the cells which were in G1 during the pulse enter mitosis (Fig. 5.5). The time between two successive peaks of labelled metaphases in the total cell cycle time (T). The width of the first peak is taken as the length of S–phase. The time from the end of the pulse to the first peak of labelled metaphases is G2 plus $\frac{1}{2}$M (the term $\frac{1}{2}$M arises because metaphase is the second stage of mitosis, and the first stage, prophase, takes about half of the time taken by a complete mitosis). The time

G

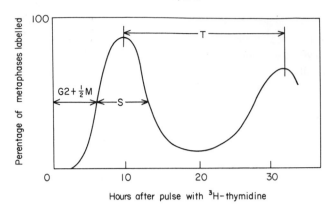

FIG. 5.5. The "labelled metaphase" method for determination of cell cycle time.

taken by the rest of the cycle is calculated by difference:

$$G1 + \tfrac{1}{2}M = T - S - G2 - \tfrac{1}{2}M.$$

The length of M is then established by determining the mitotic index(M.I.). Suppose that at any one moment 10% of a population of dividing cells are actually in mitosis. The mitotic index is then 0·1 (i.e. 10/100). The length of M may be calculated from the equation:

$$\frac{M}{T} = \frac{Log_e\ (M.I. + 1)}{Log_e\ 2} \qquad \text{(see ref. 316)}$$

Having established the length of M, the lengths of G1 and G2 are arrived at by subtraction.

   Although the "labelled mitosis" method is by far the best for estimation of the lengths of all the phases of the cycle, other methods may be used when only total cycle time is needed. Cycle time in plant meristems is measured by giving a short treatment with compounds which interfere with mitosis. The two most often used are colchicine, which prevents the separation of daughter chromosomes, giving rise to tetraploid cells, and caffeine which causes the formation of binucleate cells. The time taken for the tetraploid or binucleate cells to reappear in mitosis is the total cycle time, T [491].

III. THE DURATION OF THE CELL CYCLE

Table 5.I shows the duration of the cell cycle in a number of cell populations. Three interesting features are illustrated by the data. Firstly, in higher plants,

TABLE 5.1. The duration of the cell cycle in higher plant cells.

	Duration of the phases of the cell cycle (h)					Reference
	G1	S	G2	M	Total	
Tradescantia (whole root tips):						
Grown at 30°C	2	10	2	2	16	
Grown at 21°C	6	11	2	2	21	520
Grown at 13°C	15	23	8	5	51	
Zea (root tips):						
Quiescent centre	151	9	11	3	174	
Meristem just behind Q.C.	2	11	7	2	22	85
200 μm behind Q.C.	4	9	6	4	23	
Two populations of *Acer* cells grown synchronously:						
1.	37	15	14	2	68	154
2.	13	15	19	1	48	
Two populations of *Acer* cells grown nonsynchronously:						
1.	4	7	8	2	21	243
2.	26	8	8	3	45	

mitosis is usually the shortest phase of the cycle, occupying between one and four hours in plants grown at temperatures of 20–30°C. Secondly, much of the variation in cycle length between different populations of cells of the same species can be attributed to variation in the length of G1. This is well illustrated by the data on the cell cycle in different zones of the *Zea* root tip, and by the data on cells of *Acer pseudoplatanus* in suspension culture. The variability of G1 as compared with the relative constancy of the other phases has important implications for our understanding of the control of the cell cycle (see section VI). Thirdly, under less favourable conditions, such as low temperatures, not only G1, but all phases of the cycle may be lengthened.

IV. RNA AND PROTEIN SYNTHESIS DURING THE CELL CYCLE

The cell cycle as originally described was defined in terms of gross changes in the amount of the genetic material, DNA. More recently, attention has also been focused on the products for which the genes code, namely RNA and protein. RNA synthesis has been studied in a variety of cell types using labelling techniques, and by measuring total amounts of RNA. Two features emerge very clearly from these experiments. Firstly, RNA is not synthesized during mitosis [98, 490]. The highly condensed state of the chromosomes prevents the occurrence of gene transcription. Secondly, during the preceding three phases of the cycle (G1–S–G2), RNA is synthesized continuously, and at a rate which increases through the cycle. The rate of RNA synthesis at the end of G2 is generally around twice that observed at the beginning of G1. In animal cells, there is some evidence that the doubling of the rate of RNA synthesis coincides with the S–phase [541]. The doubling in the rate of RNA synthesis has therefore been ascribed to a gene dosage effect: a 4C amount of DNA is presumed to be able to support RNA synthesis at twice the rate of a 2C amount. In cells of green algae and higher plants, however, the rate of RNA synthesis and accumulation appears to increase steadily through the cell cycle, with no abrupt doubling during the S–phase [190, 242].

The rate of protein synthesis and accumulation also increases throughout the G1–S–G2 phases of the cycle, the rate generally showing a doubling over this period. The doubling in rate in *Chlorella* and in *Acer* cells is achieved by a gradual exponential increase [242, 404]. In animal cells, the synthesis of protein continues during mitosis, although the rate may be lower than that observed in G2 [366, 469]. The synthesis of protein in mitosis is presumably supported by messenger RNA synthesized in earlier phases of the cycle. There are no data available concerning protein synthesis during mitosis in algae or higher plants.

Although "total" protein is synthesized throughout the cell cycle, the "synthesis" of many (but not all) individual proteins shows a discontinuous pattern (the term "synthesis" in this context means, except where stated, an increase in the extractable activity of an enzyme). The two major types of discontinuous pattern are the "step" and the "peak" patterns (Fig. 5.6). The "step" pattern of activity is generally assumed to represent the discontinuous synthesis of a stable protein, while the "peak" pattern is assumed to represent the discontinuous synthesis of an unstable protein, or of a protein which is degraded by cellular proteases.

Some of the examples of discontinuous synthesis are clearly related to the overall progress of the cell cycle. Many of the enzymes involved in the synthesis of deoxyribonucleoside triphosphates show a peak of activity in late G1 or in S–phase in yeasts, green algae and higher plants [113, 242, 419]. The enzyme histone f1 phosphokinase exhibits a peak of activity in late G2 in the slime-mould *Physarum* [46]. Histone synthesis (as measured by incorporation of radioactive amino acids into histones, and by measurement of the total amounts of histones) is almost completely confined to S–phase [37, 527]. However,

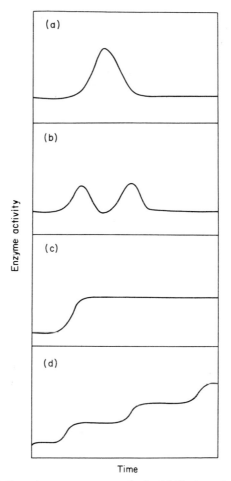

FIG. 5.6. Patterns of discontinuous enzyme synthesis. (a) Single peak. (b) Multiple peak. (c) Single step. (d) Multiple step.

many of the enzymes which show discontinuous synthesis are not closely associated with the gross changes of the cell cycle. Thus, the enzymes listed in Table 5.II as exhibiting discontinuous synthesis include enzymes involved in nucleotide synthesis, respiration, photosynthesis and phosphate metabolism. The period of synthesis of a given "step" or "peak" enzyme is always in the same phase of the cell cycle, even with enzymes which have no obvious involvement with DNA synthesis or mitosis. For example, aspartate transcarbamylase activity always exhibits a peak in G2 in *Acer* cells, even in cell cultures in which G1 is exceptionally long [242].

Two major hypotheses have been developed to explain the phenomenon of discontinuous synthesis. The first of these is the hypothesis of "oscillatory repression", suggested by Masters and his colleagues in 1965 [299] and subsequently taken up by a number of investigators. In essence, this hypothesis suggests that the synthesis of all enzymes that exhibit discontinuous synthesis is subject to feed-back repression, e.g. by end-product. The peak or step in enzyme activity is said to represent a period when enzyme synthesis is derepressed. The increase in enzyme activity causes an increase in the amount of end-product which represses enzyme synthesis. End-product concentration then falls as the end-product is used in metabolism, until the synthesis of the enzyme is derepressed once again. Enzymes which exhibit continuous synthesis are enzymes whose synthesis is permanently derepressed. If this hypothesis is correct, then it should be possible to change a discontinuous pattern of synthesis into a continuous pattern under conditions of permanent derepression. The pattern of synthesis of ribulose diphosphate carboxylase in *Chlorella* may be cited as an example of this. At high growth rates, where synthesis of the enzyme is presumed to be derepressed, enzyme synthesis is continuous; under low growth rates, enzyme synthesis is discontinuous (Table 5.II). However, certain data on the synthesis of inducible enzymes in *Chlorella* are not explicable in terms of oscillatory repression. Isocitrate lyase is induced by the presence of acetate in the growth medium [223]. The induced increase in enzyme activity is accompanied by an increase in the amount of enzyme protein (both enzyme activity and the enzyme protein itself are virtually undetectable in uninduced cells). The induction of the enzyme is inhibited by inhibitors of protein and RNA synthesis, suggesting that the induction depends not only on the *de novo* synthesis of protein but also on RNA synthesis. This suggestion is confirmed by the finding that induction is accompanied by the appearance in the cytoplasm of the messenger RNA for isocitrate lyase (as identified by its ability to direct the synthesis of isocitrate lyase in a protein-synthesizing system from wheat germ) [413a]. According to the hypothesis of oscillatory repression, the enzyme should be inducible at any stage in the cell cycle. In fact, this enzyme is only inducible during the middle part of G 1. The cells' potential for synthesis of this inducible enzyme therefore shows a discontinuous pattern. This problem introduces a related problem, namely the timing of the discontinuous synthesis of the non-adaptive enzymes. As has already been indicated, the step or peak in the activity of a given enzyme always occurs at the same point in the cell cycle. It is obviously possible that major demands for energy and for carbon skeletons occur at certain stages of the cell cycle, and that these demands bring the oscillatory repression into phase with the cycle. However, it would be expected that manipulation of the environment in which cells are grown (e.g. by providing extra carbon sources, or by lowering the concentrations of non-essential nutrients) might bring the timing of synthesis of discontinuous enzymes out of phase with the cycle. In fact, this does not happen, and this suggests that the factors that keep enzyme synthesis in phase with the cycle must be very closely linked to the cell's progress through the cycle.

The second hypothesis is the hypothesis of "linear reading" or "sequential transcription". This hypothesis suggests that the timing of the synthesis of a particular enzyme depends on the position on its chromosome of the gene

TABLE 5.II. Patterns of enzyme "synthesis" in synchronous cultures of *Chlorella* and *Acer pseudoplatanus*.

	Enzyme	*Pattern of Synthesis*	*Reference*
Chlorella sp.	d-TMP kinase	Peak	419
	d-CMP deaminase	Step	419
	Alkaline phosphatase	Step	252
	Acid phosphatase	Step	252
	Ribulose-1, 5-diphosphate carboxylase	Step or continuous	318
	Isocitrate lyase	Inducible for limited period	223
	Nitrate reductase	Inducible for limited period	223
Acer pseudoplatanus	Aspartate transcarbamylase	Peak	
	Thymidine kinase	Peak	
	Glucose-6-phosphate dehydrogenase	Step	242,243
	Succinate dehydrogenase	Two Peaks	

coding for that enzyme. Transcription is envisaged as proceeding from one end of a chromosome to the other. The step or peak in enzyme activity corresponds with the period of availability of the messenger RNA following transcription of the appropriate gene. The timing of the linear transcription may be kept in phase with the cell cycle by the changes which occur in chromatin structure during the cycle (see section VI). This hypothesis also accommodates the observation that certain inducible enzymes are only inducible at a particular phase of the cycle. The phase of the cycle in which the enzyme is inducible corresponds with the period during which the gene is available for transcription. Halvorson and his colleagues have obtained a good deal of evidence, mainly from work on the yeast *Saccharomyces cerevisiae*, in support of this hypothesis [173]. The most compelling piece of evidence is that the order of synthesis of four enzymes (orotidine-5-phosphate decarboxylase, aspartokinase, phosphoribosyl-ATP pyrophosphorylase and threonine deaminase) whose genes map on chromosome V, is exactly consistent with the order expected if linear transcription occurs. Further, inversion of a chromosome segment causes an inversion of the order of synthesis of the enzymes whose genes lie on that segment.

Some of the enzymes that are synthesized discontinuously exhibit two or three peaks or steps of activity (e.g. succinate dehydrogenase in *Acer* cells, α-glucosidase in *Saccharomyces*). This can be explained on the basis of linear reading if it is assumed that the different peaks or steps represent the synthesis of different isozymes, coded for by non-allelic genes well separated from each other on the chromosomes. This situation has in fact been shown to occur with the enzyme α-glucosidase in *Saccharomyces,* where the number of steps in enzyme activity corresponds to the number of different non-allelic genes coding for the enzyme [173].

Although the linear reading hypothesis is attractive, and for yeast cells at least, well founded, there are a number of observations which are not consistent with the hypothesis in its simplest form. However, most of these observations can be accommodated when it is accepted that regulation of protein synthesis may occur at points other than gene transcription (Chapters 2, 3, 6 and 7). The observation that a number of enzymes are synthesized continuously is one such observation. On the basis of the linear reading hypothesis, continuous synthesis may be explained in the following way. The gene for the enzyme is transcribed at the appropriate time, and the mRNA becomes available for translation. If the mRNA is stable, is not subjected to nuclease action and remains available for translation there will be continuous synthesis of the particular enzyme through the cell cycle. This interpretation in fact moves the emphasis away from the process of transcription. The important events in the control of protein synthesis may be concerned with post-transcriptional events, as well as with the synthesis of mRNA. According to this view, an enzyme showing a peak or a step in activity does so not only because the gene is available for transcription for a limited period, but also because the particular messenger RNA has only a limited working life. The importance of post-transcriptional control is emphasized by observations on nitrate reductase in *Chlorella* [223]. This enzyme is induced in response to nitrate, but only at a particular stage of the cell cycle. The situation therefore seems similar to that observed with isocitrate lyase. However, nitrate reductase can be induced in the presence of 6-methyl-purine, which reduces the rate of RNA synthesis in *Chlorella* to a negligible level. Thus, it appears that the induced synthesis of nitrate reductase depends on the presence of mRNA synthesized earlier in the cycle. It is possible that the mRNA for nitrate reductase is synthesized under non-inducing conditions, but is only used under inducing conditions. The synthesis of nitrate reductase and the timing of that synthesis in the cell cycle therefore seem to be under post-transcriptional control. Further, the synthesis of isocitrate lyase, which appears to be controlled transcriptionally under most circumstances may also be subject to post-transcriptional control: cells grown in the presence of the inducer (acetate) but under limiting light regimes, accumulate the mRNA for isocitrate lyase, but it is not translated [413a].

Of the two hypotheses discussed, it is clear that for eukaryotic cells in culture, the linear reading hypothesis is better able to explain all the available data. However, it is difficult to apply the linear reading hypothesis in its present form to non-dividing cells. The importance of this difficulty is emphasized by the fact that a majority of cells in a higher plant are non-dividing. Reference to the enzyme aspartate transcarbamylase (the first enzyme in the pathway for synthesis of the purine nucleotides) in *Acer* cells illustrates the difficulty very well. In *Acer* cells in synchronous culture, the extractable activity of aspartate transcarbamylase rises from a low level at the end of S–phase to reach a marked peak in mid-G2, and then falls back to a low level before the onset of mitosis [243]. The peak in aspartate transcarbamylase activity is consistent with both the oscillatory repression theory and the linear reading theory. According to the oscillatory repression theory, the nucleotide pool is at a low level after S–phase. Synthesis of aspartate transcarbamylase is then derepressed; the resultant burst of enzyme activity allows the synthesis of enough purine nucleotides to satisfy the demands of the two daughter cells up to the beginning of G2. According to the linear reading hypothesis, the gene for aspartate transcarbamylase is transcribed late in the sequence; again, the enzyme activity present in G2 must allow for synthesis of enough purine nucleotides to enable the two daughter cells to complete G1 and S–phase. However, *Acer* cells in natural populations (i.e. in the plant) do not go on dividing. Further, the majority of non-dividing cells are held in G1 (some exceptions to this are mentioned in section V). If protein synthesis relies on either oscillatory repression which is entrained with the division cycle, or on linear reading which is rigidly in phase with the cycle, then enzymes such as aspartate transcarbamylase which are normally synthesized in G2 will not be synthesized. It is extremely unlikely, for a number of reasons, that these non-dividing cells depend on the peak of enzyme activity which occurs in the G2 phase of the last complete cell cycle before division ceases, Firstly, the cells which cease dividing maintain a high metabolic activity as they elongate and differentiate. Secondly, many of the differentiated cells have a very long working life. These cells therefore have a high demand for metabolites. In order to include non-dividing cells within the compass of linear reading or oscillatory repression, it must either be assumed that these processes do not remain in phase with the division cycle in non-dividing cells, or that non-dividing cells rely on export from dividing cells for many essential metabolites. The difficulties involved in applying these hypotheses to non-dividing cells strongly suggest that the mechanisms exerting coarse control over protein synthesis in differentiating or differentiated non dividing cells are different from those in undifferentiated dividing cells. Further, it should be pointed out that the hypotheses of oscillatory repression and linear reading were both developed for unicellular organisms. There is as yet no concrete evidence that either hypothesis is applicable to cells of higher organisms, even when those cells are grown in suspension culture.

V. CELL DIVISION AND DIFFERENTIATION

In higher plants, cell division takes place mainly in specialized regions known as meristems. Two types of meristem may be distinguished: primary meristems such as those located at the apices of roots and shoots, and secondary meristems, e.g. vascular cambium, which are located in the mature differentiated tissues of the plant. Although the cells in one meristem may differ markedly from those in another, all meristematic cells have two features in common: they do not have a single, large vacuole and they possess thin, highly plastic cell walls.

The relationship between cell division and differentiation is well illustrated in the root apex. A cell in the apical meristem divides to give two daughter cells. The daughter cells in their turn progress through a further cell cycle, each dividing to give two more daughter cells, and so on. Commonly, in root meristems, cell lines go through up to six division cycles [19], but after this, most of the cells reach a position in the meristem where they cease dividing and go on to differentiate. Three major processes occur during differentiation of the cells: (i) elongation and vacuolation, (ii) an increase in the thickness of and a decrease in the plasticity of the cell wall, and (iii) the assumption of characteristics related to the specialized functions of the cell. Once this process is complete, the ability to divide is very much reduced and the cell does not normally divide again. There are, however, two exceptions to this. Firstly, vascular cambium is a tissue whose specialized function is division; cambium cells, although elongated, retain their thin plastic walls and do not develop a single, large, vacuole. Secondly, differentiated, vacuolated, thick-walled cells located in mature tissues have occasionally been observed to divide [128, 431].

Although division is only rarely observed in fully differentiated cells, those which are undergoing differentiation may divide. Thus, it is not unusual to observe cells that are identifiable as vascular elements, albeit in a very early stage of differentiation, undergoing division. It seems likely that the ability to divide is not lost suddenly, but gradually. As the cell becomes more committed to differentiation, it becomes less committed to mitotic activity. This idea, which has been developed by a number of investigators [59, 536], is supported by observations that during differentiation plant cells may undergo modified or truncated cell cycles.

In order to understand the modified cycles which may occur in differentiating cells, it is first necessary to discuss the events of the normal cell cycle in more detail (see Fig. 5.7). The cycle starts with G 1, during which time the total DNA content of the cell does not change (although some DNA turnover may occur: Chapter 1). The cell then enters S-phase. During this time, the DNA is replicated, and the DNA content therefore rises from the 2C level to the 4C level. The replication of DNA is accompanied by the synthesis of histones [37, 527]. Analysis by autoradiography of the chromosomes during S–phase has

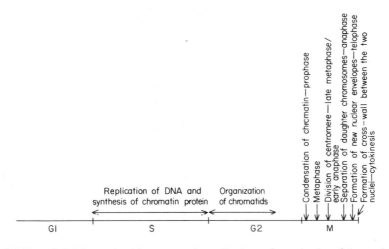

FIG. 5.7. The cell division cycle with respect to the replication and organization of chromatin.

shown that different regions of a given chromosome are replicated at different times during S–phase [71]. The growing points for DNA replication are initiated in a controlled and ordered sequence, but the sequence is not necessarily related to the position of the initiation sites along the chromosome. In both plant and animal cells, the heterochromatin is known to be replicated late in S–phase [279]; this observation is supported by the finding that reiterated DNA, much of which is located in heterochromatin (Chapter 1) is replicated later than the unique sequences [138]. Following S–phase, the two daughter DNA molecules become fully integrated with the chromatin proteins, and separate from each other for much of their length. The two DNA molecules have thus become two chromatids, held together by the centromere. This process of chromatid organization is completed during G2. In the first phase of mitosis, prophase, the chromosomes become condensed, and hence visible. The nuclear envelope also begins to disintegrate. In metaphase, the chromosomes align themselves on the cell equator, and the centromeres then divide. Each chromatid is now a daughter chromosome. In anaphase the daughter chromosomes separate, one of each pair moving to one pole and one to the other pole. This movement is controlled by the spindle microtubules. Telophase, in which the new nuclear envelopes are formed, follows, giving rise to two daughter nuclei. Strictly, the process of mitosis finishes at the end of telophase. The term "mitosis" therefore refers to nuclear division. However, when the term "mitosis" is used in the context of the cell cycle, the occurrence of cytokinesis (i.e. the division of the cell in which mitosis has occurred, by the formation of a cell wall between the two nuclei) is also often implied. Mostly, this usage of the term "mitosis" presents no problems, since cell wall formation usually accompanies telophase, or follows very soon after it. There are some exceptions to this, and in such instances care must be taken to distinguish between mitosis and cytokinesis.

The various types of modified cell cycle which occur in differentiating cells are listed in Table 5.III. The nature of the factors which cause the cell cycle to become blocked at particular points is discussed in section VI. The interesting feature which is relevant to the current discussion is that cells have the ability to by-pass both the "block" and all phases that would normally take place after the "block". Thus, truncated cell cycles can occur. For example, cells in which centromere division has failed to occur, and which therefore contain diplo-chromosomes, with a 4C amount of DNA, can by-pass the remainder of the cycle to enter G1 and then S-phase again. The second cycle again proceeds as far as chromatid formation, giving rise to cells with "quadro-chromosomes", containing an 8C amount of DNA [97]. The repeated cycles of DNA synthesis which occur in cells where the organization of the chromatids is inhibited can lead to even more spectacular results. In the suspensor cells of *Phaseolus coccineus* (runner bean) for example, there may be up to 13 complete S—phases, giving rise to sets of enormous polytene chromosomes, with 16 384C amounts of DNA [48]!

One of the most interesting modified cell cycles is that which occurs during differentiation of the protocorm of the orchid *Cymbidium* [322]. This has already been discussed in detail in Chapter 1, but will be summarized again here. Many of the cells in the protocorm exhibit cell cycles blocked at the end of S—phase. They undergo between two and four successive rounds of DNA synthesis, giving rise to sets of polytene chromosomes containing 8, 16 or 32C amounts of DNA. As differentiation proceeds, this endo-reduplication of the total DNA ceases. However, in some of the cells, the synthesis of the DNA in the heterochromatin (presumed to be reiterated DNA) continues for a further one to five cycles, giving rise to cells in which the heterochromatin is replicated to levels of between 64C and 1024C. Whether synthesis of reiterated DNA normally occurs late in S—phase in these cells, as it does in other plant cells, is not known. Nevertheless, it is certain that this cell cycle, consisting essentially of a very short G1 plus one part of S—phase, is the most truncated cell cycle so far described in plants.

TABLE 5.III. Modified cell cycles

Phase of cell cycle at which "block" occurs	Consequence	Example	Reference
1. During S-phase	"Over-replication" of part of genome	Protocorm of *Cymbidium*	322
2. At end of S-phase	No organization of chromatids. Leads to formation of polytene chromosomes	Suspensor cells of *Phaseolus*	48
		Cymbidium protocorm	322
		Differentiated cells in roots of higher plants	97,490
3. Between end of S–phase and metaphase	Centromeres fail to divide leading to formation of diplochromosomes		
4. Between end of metaphase and end of anaphase	Nuclear division fails to occur; polyploid nuclei are formed	Cells in tapetum of *Lycopersicum*	59
5. Between end of telophase and beginning of cytokinesis	Bi-ucleate cells	Tapetum of *Lycopersicum*	59

The term "block" does not necessarily imply the presence of an inhibitor. It simply refers to the point at which progress through the cell cycle is halted.

The term "polyploid" is used loosely by many authors to include cells with polytene chromosomes and cells with diplochromosomes. Cells which are truly polyploid are those which possess more sets of completely separated chromosomes than the usual two.

Truncated cell cycles in which S–phase is omitted also occur. Infection of plants of *Vicia faba* with *Agrobacterium rubi* leads to the formation of tumours. In some of the tumour cells repeated rounds of DNA synthesis take place, giving cells with up to 16C levels of DNA. At certain phases of tumour growth, cells with 4C–16C amounts of DNA have been shown to undergo successive cell divisions (i.e. mitosis plus cytokinesis) without the occurrence of S–phase, eventually leading to the formation of cells with the "normal" 2C amount of DNA [371].

VI. CONTROL OF THE CELL CYCLE

A. Control points in the cell cycle

The events of the cell division cycle form a controlled sequence in which the completion of one event is normally followed by the initiation of the next. Once a cell has entered S–phase, it may normally be expected to complete the division cycle. With the exceptions of the cells discussed in the previous section, DNA replication only takes place in preparation for mitosis. Similarly, mitosis does not usually occur unless it has been preceded by DNA synthesis.

The linkage between DNA synthesis and mitosis which normally exists has led to the idea that the major control point for the cell cycle is at the end of G1. Once cells have passed this point, they are, under most circumstances, committed to complete the remainder of the cycle. The restriction point represents, in effect, a "point of no return". Support for this idea comes from estimation of the lengths of different phases of the cell cycle in different populations of cells of the same species, e.g. cells in different parts of the meristem in *Zea mays* or different cultures of *Acer pseudoplatanus* (sycamore) cells. As was noted earlier, the length of G1 is the major source of intraspecific variation in the length of the cell cycle. Under most conditions, the time occupied by S + G2 + M is more or less constant. This is clearly consistent with the existence of a major control point at the end of G1 [154].

However, truncated cell cycles do occur, and in some plant tissues, cells which have undergone a truncated cell cycle form a significant proportion of the population. The commonest of the truncated cell cycles is a cycle in which DNA synthesis and chromatid organization occur, leading to the formation of diplochromosomes. The diplochromosomes remain uncondensed, mitosis does not take place and the cell is thus maintained in G2. On the basis of this evidence it may be argued that the existence of a major control point in late G2 is likely. The hypothesis that there are two major control points in the cell cycle has received some support from the data obtained by Van't Hof and his colleagues [492]. If excised roots of *Vicia faba* or *Pisum sativum* are starved of carbohydrate, the cells in the meristem stop dividing. The cessation of cell division is not haphazard. Cells are held in either G1 or G2. No cells are held in S–phase or mitosis. This suggests that once S–phase and mitosis are initiated, they are normally completed, even when the cell is deprived of an exogenous source of carbohydrate. On renewing the supply of carbohydrate to the excised roots, the cells held in G1 enter S phase, and the cells held in G2 enter mitosis, both events being preceded by a lag phase (Fig. 5.8). Further, a given cell can be held in G1 by carbohydrate starvation, then allowed to proceed through S–phase by the provision of carbohydrate, and finally held in G2 by withdrawing carbohydrate once more. Thus, each cell in the meristem is subject to control both in G1 and G2

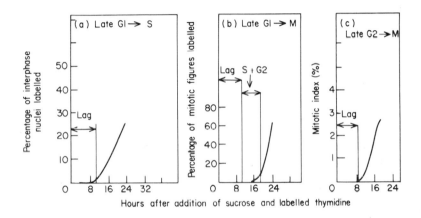

FIG. 5.8. The resumption of progress through the cell cycle in cells in the root tip of *Pisum sativum* (pea). Root tips were cultured for 72h in the absence of sucrose. Sucrose and ³H-thymidine were added at time zero. Samples of root tips were analysed by autoradiography at intervals after time zero (from ref. 492).

In addition to the major control points in G1 and G2, it is obvious that other control points must also exist. The existence of truncated cycles which are blocked at points between S–phase and the end of mitosis is clear evidence for the existence of such "minor" control points in mitosis, in G2, and possibly in S–phase; it seems very likely that "minor" control points also exist in G1. The remainder of this chapter is devoted to a consideration of the main lines of evidence concerning the nature of the control points and of the factors acting on those control points. Much of the evidence comes from work on animal and fungal (i.e. yeast) cells; this will be considered together with the evidence from cells of green algae and higher plants.

B. The structure of chromatin during the cell cycle

The major phases of the cell cycle take their names from the more obvious facets of the behaviour of chromatin during the cycle. In addition to the obvious and gross changes in chromatin structure, it is now known that more subtle changes in chromatin structure also occur. The first of these takes place in late G1/early S–phase. During this part of the cycle in HeLa cells (human cancer cells in culture) the actinomycin D-binding capacity of the chromatin rises to a sharp peak [355]. Since actinomycin D is a drug which binds specifically to DNA, this suggests that a change in chromatin structure allowing greater access of the drug to the DNA occurs in late G1. At the same time, the DNA in the chromatin becomes more vulnerable to DNA endonuclease, again indicating a change in the structure of chromatin making the DNA more accessible. This has led to the proposal that S–phase is started by a change in chromatin structure which allows a specific DNA endonuclease to introduce single-stranded nicks in the DNA at the initiation points for DNA synthesis. This idea is supported by observations that the ability of isolated nuclei from rat liver cells to synthesize DNA is directly correlated with the activity of a calcium– and magnesium–dependent DNA endonuclease [69]. The hypothesis is consistent with what is known about the replication of nuclear DNA (Chapter 1) and also suggests a possible identity for the major control point at the G1–S boundary.

Another change in chromatin structure which occurs at a specific stage of the cell cycle is the phosphorylation of histone f1. The serine residues of the histones, particularly of histones $f2a_1$ and f2b are readily phosphorylated by histone phosphokinase(s). The process is reversible, and the extent of phosphorylation at any one time depends, at least partly, on the availability of phosphate. The phosphorylation of histone f1, however, shows a pattern which is dependent on the cell's passage through the cell cycle [46, 297]. The level of phosphorylation of histone f1, both in HeLa cells and in the slime mould *Physarum*, remains low throughout G1 and early S–phase. Late in S–phase, or early G2, the level of phsophorylation rises, reaching a peak at the G2–mitosis boundary. The level of phosphorylation then falls again (Fig. 5.9). The rise and fall in the level of phosphorylation is paralleled by the ability of cell-free extracts to phosphorylate histone f1 *in vitro*. Further, in both types of cell the onset of mitosis may be accelerated by the addition of extracts from cells in late G2. Bradbury suggests that the phosphorylation of histone f1 causes a change in the charge distribution in the chromatin complex which rapidly leads to the condensation of the chromatin [46]. Other changes must then occur to maintain the chromatin in a condensed state, since histone f1 is rapidly dephosphorylated as mitosis proceeds. This hypothesis is an attractive one in that it suggests a role for histone f1 (which, unlike the other histones, seems to play no part in maintaining the structural integrity of chromatin); the hypothesis also suggests an identity for the major control point at the G2–mitosis boundary. Evidence to support this hypothesis has recently been obtained from a higher plant system, in that induction of mitotic activity in explants of *Helianthus tuberosus* (Jerusalem artichoke) is accompanied by an increased level of phosphorylation of histone f1 [451].

The two hypotheses discussed above relate the major control points at the end of G1 and G2 to changes in chromatin structure. In both instances a highly specific enzyme, working in the nucleus, is involved. The specific endonuclease

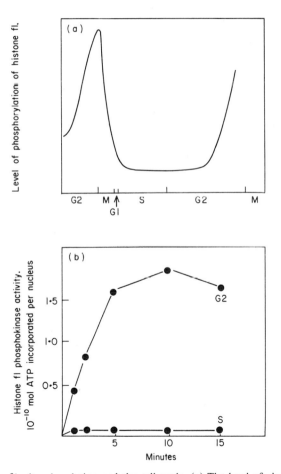

FIG. 5.9. Histone f1 phosphorylation and the cell cycle. (a) The level of phosphorylation of histone f1 though the cell cycle of the slime-mould *Physarum polycephalum* (b) The activity of histone f1 phosphokinase in nuclei isolated from cells in S–phase and from cells in G2 (from ref. 46).

is able to act after the change in chromatin structure at the end of G1. The action of the specific histone f1 phosphokinase brings about the change in chromatin structure at the end of G2. The involvement of these enzymes introduces the possibility of "minor" control points relating to the synthesis or activation of the enzymes, or to their transport into the nucleus.

A facet of chromatin structure that markedly affects the cell cycle is the amount of DNA in the chromatin. The great interspecific variation in nuclear DNA content in higher plants has already been discussed in Chapter 1. The variation in DNA content is correlated with variation in the total cell cycle time (Fig. 5.10); cells containing high levels of DNA require more time to complete the cell cycle than cells with lower levels of DNA. This interspecific variation in

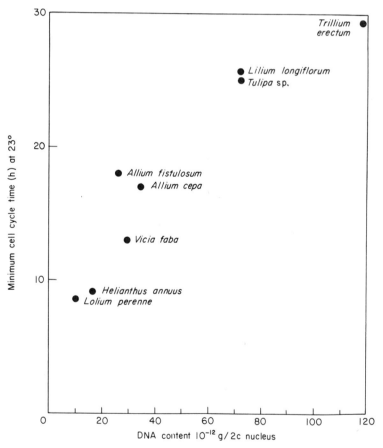

FIG. 5.10. The relationship between nuclear DNA content and minimum total cell cycle-time. (Diagram compiled from data in refs 29 and 380).

cycle time is due to variation in the length of S–phase (this contrasts with the intraspecific variation in cycle time, which, as noted earlier, is largely due to variation in the length of G1). The cycle time is not in fact completely proportional to DNA content, particularly at high DNA levels. There is evidence to suggest that the cells with high DNA levels may exhibit higher rates of DNA synthesis, thus ameliorating to some extent the effect of the high DNA levels [29].

In some higher plant species, the correlation between DNA content and cell cycle time breaks down. In the genera *Anacyclus*, *Anthemis* and *Artemisia*, species which contain proportionately more heterochromatin than usual have a cell cycle time which is shorter than predicted from their nuclear DNA levels [323]. Thus the presence of "extra" heterochromatin increases the speed of passage through the cycle. Nagl suggests that the basis of this phenomenon is a greater rate of DNA synthesis in heterochromatin than in euchromatin. It is also possible that the presence of a greater proportion of heterochromatin shortens the time necessary for chromosome condensation in prophase, since heterochromatin is in a permanently condensed state.

C. Cell fusion and transplant experiments

A very large number of experiments have been carried out with mammalian, protozoan and slime-mould cells in which cells in one phase of the cycle have been fused with cells in another phase of the cycle. The results obtained with these different types of cell are all essentially similar. One example, taken from the work of Rao and Johnson, is therefore sufficient to indicate the type of result obtained. When HeLa cells in G1 are fused with cells in S–phase, the G1 cells are stimulated to enter S–phase early [370]. The effect of S–phase cells is additive: if two S–phase cells are fused with one G1 cell, the cell in G1 enters S–phase even earlier. This suggests that S–phase cells contain a factor which is readily transported into G1 nuclei, and which is necessary for the initiation of DNA synthesis. The factor may not be absolutely necessary for the completion of DNA synthesis, since cells in S–phase which are fused with cells in other phases (and which therefore suffer dilution of the factor) complete S–phase normally.

When S–phase cells are fused with G2 cells, the G2 cells do not synthesize DNA. This means that under normal circumstances, some control factor must be operating to prevent DNA synthesis once one round of replication has occurred. It seems unlikely that G2 cells contain a soluble inhibitor of DNA synthesis, since S–phase cells which are fused with G2 cells complete DNA replication normally. It is possible, however, that G2 cells contain an inhibitor which is tightly bound in the cells, for example in chromatin. It would be very interesting to carry out this experiment with plant cells, since plant cells with

8C or even higher levels of DNA are relatively common (at least in comparison with animals). This suggests that the control which operates to prevent DNA synthesis in G2 may not be as rigorous in plants as in animals.

The results obtained with S–G2 fusions provide evidence for a soluble factor necessary for initiation of mitosis. The G2 cells in such a fusion suffer a delayed entry into mitosis. The S–phase cells show an accelerated entry into mitosis, G2 being very much shortened. The effect of this is that both nuclei in an S–G2 fusion enter mitosis together. This is consistent with the idea that a soluble factor is accumulated in G2, and eventually reaches a level at which mitosis is initiated. Fusion of a G2 cell with an S–phase cell causes, for the G2 cell, a decrease in the amount of the factor, but for the S–phase cell, an increase in the amount of the factor. Mitosis is initiated in both nuclei when the concentration of the factor in the double cell reaches the requisite level.

Experiments in which nuclei from non-dividing cells are transplanted into the cytoplasm of dividing cells have also indicated that a soluble factor is involved in the initiation of S–phase [170]. These experiments have also provided evidence that the factor is cytoplasmic in origin, and is very unspecific (in terms of the types of cell which respond to it). Thus, nuclei from cells in adult mouse liver are stimulated to synthesize DNA by transplanting into the enucleated cytoplasm of frog's eggs.

D. Soluble factors isolated from dividing cells

The properties of the various different factors which have been isolated from dividing cells of animals and plants are shown in Table 5.IV. Most of the factors are associated with the initiation of S–phase. These factors which stimulate DNA synthesis have nearly all been shown to be able to pass through dialysis membrane, and have therefore been termed "low molecular weight". However, the term "low molecular weight" is a relative one. Substances of molecular weights as high as 10 000–12 000 can pass through a dialysis membrane, albeit slowly. Since none of the investigators has stated for how long dialysis was carried out, the molecular weights of these factors could be as high as 10 000. At least two of the factors, the "wedge" from the tumour fluid of rats carrying Erhlich ascites tumour (factor 1 in Table 5.IV) and the factor from the cytoplasm of infant rat brain cells (factor 4 in Table 5.IV) have been shown to possess properties characteristic of protein. Most of the factors are not inactivated by heat, or by protein denaturants (this is even true of the two factors that have been shown to possess properties characteristic of proteins).

The transplantation experiments reviewed in section VIC privided evidence for a very non-specific cytoplasmic factor, able to stimulate DNA synthesis in non-dividing cells and in cells in G1 of a normal division cycle. Of the factors listed in Table 5.IV, only factor 1 (the "wedge") has been shown to be non-

TABLE 5.IV. Soluble factors which promote DNA synthesis and mitosis

Source of factor cycle	Phase of cell stimulated	Type of cell receptive to factor	Properties of factor	Suggested identity of factor	Reference
1. Ehrlich ascites fluid from rats	S–phase	Rat liver nuclei Protozoa, HeLa cells, *Nicotiana* pith	Passes through dialysis membrane. Resistant to heat and protein denaturants. Positive to Folins reagent and to ninhydrin	"The wedge," a modifier of DNA polymerase, causing initiation of DNA replication	123,426
2. Medium containing fast growing mouse L-cells	S–phase	BHK/21 hamster cells	Passes through dialysis membrane. Resistant to heat and freeze–thaw treatment		425
3. Serum from adult rat brain (before cessation of cell division)	S–phase	Isolated nuclei from infant or adult rat brain cells	Passes through dialysis membrane. Heat-labile		422
4. Cytoplasm of infant rat brain	S–phase	Isolated nuclei from infant brain cells only	Passes through dialysis membrane. Resistant to heat. Destroyed by protease		422
5. Cytoplasm of HeLa cells in S–phase	S–phase	Late G1 nuclei of HeLa cells only	Heat labile. Not destroyed by RNase		257
6. Cytoplasm of rapidly dividing mammalian cells	S–phase	Mammalian cells	Resistant to heat		475
7. Medium containing fast growing L-cells	Mitosis	BHK/21 hamster cells	Does not pass through dialysis membrane. Heat labile		425
8. Crown-gall tumour of *Vinca*	Mitosis?	*Vinca* cell cultures	Low molecular weight, heat stable	3, 7-dialkyl-2-alkylthio-6-purinone, an inhibitor of cyclic AMP phosph-diesterase	526

specific, being capable of stimulating DNA synthesis in animal and plant cells, although it is also possible that factors 2 and 3 are non-specific. In fact, some of the factors show a very high degree of specificity, either for the cell type on which they act (factor 4) or for the time in the cell cycle at which they act (factor 5). This observation raises the possibility that there are two types of factor, one general and one specific, involved in the stimulation of DNA synthesis. The isolation of two different factors (factors 3 and 4) from rat brain cells supports this idea.

The nature of these soluble factors remains essentially unknown, but on the basis of their properties listed in Table 5.IV, it is likely that they are small proteins of very simple secondary and tertiary structure. Erhan and his colleagues have suggested that the "wedge" (factor 1) is a small protein which combines with DNA polymerase, increasing its ability to use native DNA as a template for DNA synthesis [123]. Unfortunately, the evidence cited to support this view is not entirely convincing. However, the observations that the onset of S–phase coincides with an increased accessibility of the DNA in chromatin (section V1B), suggest that these factors may be involved in modification of chromatin structure.

The two mitosis-promoting factors (factors 7 and 8 in Table 5.IV) are very different from each other. The factor isolated from animal cells has several characteristics that suggest that it may be an enzyme. Its molecular weight is greater than 10 000, and its mitosis-promoting activity is destroyed by heat and by freezing and thawing. Nothing is known about its mode of action, but in view of the hypothesis that mitosis is initiated by a high level of phosphorylation of histone f1, it is interesting to speculate that this factor may be involved in histone phosphorylation.

The factor isolated from plant cells is a thermostable compound, of low molecular weight, and has been identified by Braun and his associates as a substituted purinone (a purinone is a purine base linked to a glucose molecule via a glycosidic linkage). Substituted purinones, particularly the 3, 7-dialkyl-2-alkylthio-6-purinone isolated from *Vinca* (periwinkle) crown-gall tumour, are powerful inhibitors of adenosine $3':5'$-cyclic monophosphate (i.e. cyclic AMP) phosphodiesterase. Braun has suggested the following scheme for the control of mitosis [526]. Cytokinins (growth substances essential for the maintenance of mitotic activity) promote the synthesis of substituted purinones (to which Braun refers as "cytokinesins"). These inhibit cyclic AMP phosphodiesterase, causing a rise in the level of cyclic AMP, which in turn promotes mitotic activity. Braun does not suggest any mechanism for the promotion of mitosis by cyclic AMP, but by comparison with its action in animal cells, it is possible that it operates by stimulating histone f1 phosphokinase activity. There is some controversy as to whether or not "cytokinesins" may in fact be regarded as a separate class of compounds from the cytokinins. The suggested structure of the "cytokinesin" from *Vinca* (a substituted purine glycoside) is somewhat similar to the general structure of cytokinins (substituted purine ribosides). Indeed, Miller has suggested that the "cytokinesin" extracted from *Vinca* is, in reality, ribosyl-zeatin, a natural occurring cytokinin [310]. Further, many compounds that are classed as cytokinins are also inhibitors of cyclic AMP phosphodiesterase activity [184]. However, whether cytokinins inhibit cyclic AMP phosphodiesterase activity directly or indirectly (by promoting synthesis of a "cytokinesin"), it is known that cyclic AMP and cyclic AMP phosphodiesterase are widely distributed in plants [53, 150, 329]. In some instances, mitosis is correlated with high levels of cyclic AMP. For example, Brewin and Northcote have shown that kinetin causes a rise in the level of cyclic AMP in cells of *Glycine max* (soya bean) synchronized by sub-culturing from a stationary phase culture [53]. There is a 20 h lag before the level of cyclic AMP starts to rise. This lag may be associated with the formation of an intermediate (a "cytokinesin"?). The rise in the level of cyclic AMP precedes the first mitosis in the synchronous culture. Further, 8-bromoadenosine 3′:5′ cyclic monophosphate promotes division in excised *Nicotiana* parenchyma tissue [525]. There is thus a good deal of circumstantial evidence to implicate cyclic AMP in the control of mitosis in plants.

E. Poly-(ADP-Ribose) and NAD

In the mid-1960s a somewhat bizarre polynucleotide, poly-(ADP-ribose), was discovered in the nuclei of rat liver cells [74]. Poly-(ADP-ribose) is a polymer of ADP units, each of which is linked to an "extra" ribose molecule (Fig. 5.11). The substrate for polymerization is NAD [532]. Each NAD molecule is cleaved to yield nicotinamide plus an ADP-ribose unit which is immediately linked to the growing end of the poly-(ADP-ribose) chain. The synthesis of poly-(ADP-ribose) takes place in the nucleus (as does that of NAD), is absolutely dependent on DNA or on artificial poly-deoxyribonucleotides, and is inhibited by nicotinamide [532]. *In vivo*, the polymer appears to be linked to one or several of the chromatin proteins [69]. Although the relevant experimental evidence is not entirely clear, it seems likely that poly-(ADP-ribose) is an inhibitor of DNA synthesis, either because its presence in chromatin prevents access of endonuclease to the initiation sites, or because the poly-(ADP-ribose) is bound to the endonuclease itself, preventing it from "nicking" the DNA [69]. Consistent with the view that poly-(ADP-ribose) inhibits DNA synthesis are the findings that in rat liver nuclei, high concentrations of NAD promote synthesis of poly-(ADP-ribose) and inhibit DNA synthesis [69], and that in the slime mould *Physarum polycephalum*, the activity of the enzyme NAD pyrophosphorylase shows a marked peak in S–phase [437]. Since this enzyme breaks down NAD to give ATP and nicotinamide mononucleotide, it

Fɪɢ. 5.11. Diagram of part of a poly-(ADP-ribose) molecule.

may be acting to prevent accumulation of NAD in the nucleus, thus preventing synthesis of poly-(ADP-ribose) by removing the substrate for poly-(ADP-ribose) polymerase. Further, in *Physarum,* the activity of poly-(ADP-ribose) polymerase is high at all stages of the cell cycle except S–phase, during which the activity falls to a very marked trough [55]. This suggests that an inability to synthesize poly-(ADP-ribose) is correlated with the ability to synthesize DNA. These findings thus raise the possibility of the occurrence of negative control of the normal cell cycle, mediated by inhibition of DNA synthesis, in addition to positive controls, acting to stimulate or promote DNA synthesis and mitosis.

F. Protein synthesis, DNA synthesis and mitosis

The synthesis of protein is an integral part of the growth of a cell, and is therefore essential for the normal functioning of the cell cycle. In addition to this general involvement, protein synthesis is also involved with control of specific phases of the cycle. This is illustrated by the results obtained by Van't Hof [256, 492]. As has been indicated previously, depriving excised roots of *Vicia* or *Pisum* of carbohydrate causes meristematic cells to be held in G1 or in G2. When the supply of carbohydrate is restored, there is a lag before the cells enter S–phase or mitosis (Fig. 5.9). The length of the lag phase is dependent on the length of the starvation period. Inhibition of protein synthesis during the lag phase prevents cells from entering S–phase or mitosis, suggesting that the lag phase represents a time when proteins necessary for S–phase and mitosis are synthesized. The necessity for protein synthesis after carbohydrate starvation in cells apparently held in late G1 or late G2 is discussed below.

The proteins necessary for S–phase are of two types. Firstly, there are the enzymes involved in the synthesis of deoxyribonucleotides, the initiation of replication and the synthesis of DNA. Many of these enzymes show a peak of activity which coincides with S–phase. If these peaks of activity represent *de novo* synthesis, then inhibition of protein synthesis in G1 should lead to an inhibition of DNA synthesis. This in fact has been observed in green algae and higher plants [37, 501, 502] as well as in a very wide variety of other cell types.

Secondly, in many eukaryotic organisms, proteins which are synthesized during S–phase itself are essential for the completion of S–phase. Thus, in *Chlorella,* the progress of S–phase is halted by cycloheximide at concentrations which abolish protein synthesis [501, 502]. This suggests either that the enzymes involved in DNA synthesis turn over very rapidly (i.e. with a half-life of a few minutes) such that inhibition of protein synthesis leads to a rapid decrease in the amount of enzyme, or that synthesis of other (non-enzyme) proteins is essential for DNA synthesis. The second hypothesis has received the most support. The proteins thought to be essential for DNA synthesis are the histones [508]. It is envisaged that DNA synthesis and the synthesis of chromosomal proteins are normally linked together so that the newly synthesized DNA is rapidly packaged into chromatin and DNA replication therefore cannot take place in the absence of histone synthesis. Evidence to support this hypothesis is largely circumstantial. Firstly, plant proteins in general do not turn over very fast, even under starvation conditions [485]. Therefore, unless the enzymes involved in DNA synthesis are atypical, it is unlikely that the inhibition of DNA synthesis by inhibitors of protein synthesis is mediated via enzyme decay. Secondly, in prokaryotic organisms (which do not possess histones), once DNA synthesis has been initiated, it may be completed even when protein synthesis is inhibited.

The completion of S–phase in the absence of protein synthesis has in fact been observed in one lower eukaryote, the yeast *Saccharomyces cerevisiae* [189]. In this organism, the chromosomes contain 10^2– 10^3 times less DNA than the chromosomes of higher eukaryotes. It is therefore perhaps less essential that the DNA in *Saccharomyces* is packed immediately into a stable structure, and thus DNA synthesis is possible even when histone synthesis is inhibited. The timing of the response to inhibitors of protein synthesis in *Saccharomyces* is particularly interesting. DNA synthesis is inhibited by inhibitors of protein synthesis at any point up to the end of G1. At any point after this (i.e. after the initiation of DNA synthesis) inhibition of protein synthesis has no effect on DNA synthesis. This suggests that during G1, synthesis and accumulation of proteins occurs until the level of a particular protein or group of proteins reaches a threshold for the initiation of S–phase. This observation is relevant to the results obtained by Van't Hof with excised roots (see section VIA). In cells blocked in late G1 under conditions in which protein turnover may occur (such as carbohydrate starvation), the level of the proteins necessary for initiation of S–phase will fall below threshold level during the blockade. On restoring the carbohydrate supply, there is a necessity for protein synthesis in order to bring the level of the particular protein(s) back up to the level necessary for the initiation of S–phase. There is thus a lag between the restoration of the carbohydrate supply and the passage of the G1 cells into S–phase.

The relationship between protein synthesis and mitosis is similar to that between protein synthesis and DNA synthesis. Inhibition of protein synthesis in G2 prevents the onset of mitosis in a wide variety of eukaryotic cells (see ref. 316). Inhibition of protein synthesis after the initiation of mitosis generally does not prevent completion of mitosis. This is again evidence for a major control point caused by the need for a particular protein (or proteins) to reach a threshold level before mitosis is initiated. At present there is no knowledge of the identity of the "division protein(s)". On the basis of Bradbury's model for control of the initiation of mitosis via the level of phosphorylation of histone f1, the enzyme histone f1 phosphokinase may be one such division protein. It is also possible that proteins involved in spindle formation (microtubule proteins) may be among the proteins which must be synthesized before mitosis can occur. However, it should be emphasized that there is no direct experimental evidence to support either of these suggestions.

Figure 5.12 sums up the observations that have been discussed in sections VIA–F. The diagram incorporates results obtained with protozoan, fungal,

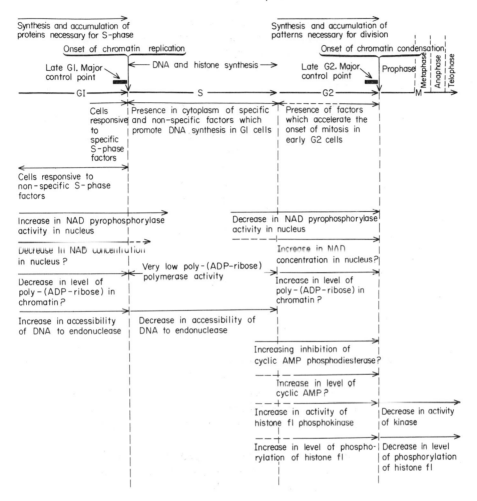

Fig. 5.12. Regulation of the cell cycle. This diagram summarizes the data discussed in sections VIA–F.

slime-mould and mammalian cells in addition to those obtained with cells of green plants. There is some justification for this, since many of the control factors discussed operate in a number of widely different types of cell. Even when results obtained with several types of cell are drawn together, it is clear that our knowledge of the internal regulation of the cell cycle is very fragmentary, and in fact mainly descriptive. Much work remains to be done before any ideas on the significance of different events, or on the linkage (if any exists), between one event and another, can be formulated. The problem of internal regulation of the cell cycle is still largely a problem for the future.

G. The induction of division in differentiated plant cells

In sections VIA–F the control of the cell cycle in cells undergoing a series of successive divisions has been discussed. No mention has yet been made of the control of division in non-dividing, differentiated cells, nor of the roles of neighbouring cells and extracellular factors in the control of cell division. In attempts to study these particular problems, much attention has been focused on the induction of division in differentiated plant tissues. If small explants are excised from mature storage organs of higher plants and incubated under suitable conditions, mitotic activity is induced. The conditions necessary for the induction of division in various different species are listed in Table 5.V. There is a lag period between excision of the tissue and the resumption of DNA synthesis and mitosis. This lag period is characterized by a very marked increase in the metabolic activity of the tissue. The increased metabolic activity includes an increase in the rate of respiration, a breakdown of storage compounds, an increased rate of synthesis and accumulation of RNA and protein, and an enhanced ability to take up ions and other nutrients. The overall process has been described by many authors as "de-differentiation". This should not be taken to mean that the cells become like meristematic cells. The changes that occur are in fact a reversal of the events that occur during the onset of quiescence, albeit telescoped into a much shorter time period than the onset of quiescence [60]. The change in metabolic activity following excision is therefore a change from the metabolism typical of mature, quiescent, storage parenchyma to the metabolism typical of rapidly elongating cells which have recently ceased dividing. Excision does not lead to a loss of the major structural features associated with the differentiated state, such as the vacuole, and the non-plastic cell wall. The de-differentiation which follows excision is therefore only partial.

Three major hypotheses have been developed concerning the basis of the induction of division in excised plant tissue. These are (i) that the induction of division is caused by loss of inhibitors of division from the tissue, (ii) that the induction of division relies on a supply of plant growth substances in the tissue, and (iii) that the induction of division is caused by an increased osmotic potential in the tissue. These three hypotheses are discussed separately here, but they are in fact not mutually exclusive. It is quite possible that all three types of factors invoked by the three hypotheses may be involved in the induction of division.

Several investigators have suggested that differentiation is accompanied by a progressive inhibition of mitosis and DNA synthesis [59, 536]. The truncated cell cycles that occur in some differentiating cells (section V) are interpreted as cycles that have been "frozen" by the presence of an inhibitor which is synthesized during differentiation. The process of partial de-differentiation is seen as an essential pre-requisite to the resumption of mitotic activity, and is

TABLE 5.V. Induction of division in explants of storage tissue

Tissue	Type of explant	Incubation medium	Events observed:				Reference
			Increased metabolic activity	Increase in DNA polymerase activity	DNA synthesis	Mitosis	
Tuber of Solanum tuberosum	Discs, 10mm × 1mm	Simple buffer solution	√		√	√	57
Root of Daucus carota	Discs, 10mm × 1 mm	Simple buffer solution	√		√	?	60
Root of Beta vulgaris	Discs, 10mm × 1mm	Simple buffer solution	√	√	×	×	109,495
Tuber of Helianthus tuberosus	Discs 10mm × 1mm	Water	√		×	×	416
		Water plus CaCl₂ Auxin, Cytokinin	√		√	√	
Tuber of Helianthus tuberosus	Cylinders 2mm × 2mm	Nutrient Solution	√	×	×	×	175
		Nutrient solution plus auxin	√	√	√	√	

assumed to be accompanied by the loss of the inhibitor. Yeoman has suggested that the high degree of synchrony shown during the first induced division in explants of *Helianthus tuberosus* (Jerusalem artichoke) is evidence that an inhibitor is lost during the lag period [536]. The synchrony is likened to that induced in cell cultures by treating the cells with an inhibitor of DNA synthesis and then removing the inhibitor. However, synchrony can be induced by means other than the use of inhibitors (section II), and so the division synchrony in *Helianthus* explants is not, in fact, evidence for loss of an inhibitor. Further, no inhibitors that are specifically associated with the differentiated state have been isolated from plants or animals. The compound poly-(ADP-ribose), mentioned in section VIE, may well be an inhibitor of DNA synthesis (and hence of cell division) but there is as yet no evidence that it is present in higher concentrations in differentiated cells than in undifferentiated cells, nor indeed that it is present in plant cells at all.

The plant species listed in Table 5.V show a good deal of variation with respect to their dependence on growth substances during the induction of division. In explants of *Solanum tuberosum* (potato) tubers, DNA synthesis and mitosis occur in the absence of any added growth substances or nutrients. By contrast, explants of *Helianthus tuberosus* tubers do not undergo cell division, do not synthesize DNA and do not show any increase in DNA polymerase activity in the absence of added growth substances. The other species listed fall between these two extremes with respect to their dependence on growth substances for the induction of division. This varying dependence on growth substances introduces the second hypothesis, namely that division in differentiated cells is induced by plant growth substances. The two growth substances classically associated with DNA synthesis and cell division are the auxins and cytokinins (Chapter 7). Both are known to be produced in cells whose contents are breaking down [336, 418]. It is therefore probable that both auxin(s) and cytokinin(s) diffuse or are transported into tissue explants from the wounded cells at the surface of the explant. Indeed, indol-3yl-acetic acid (IAA) is known to be produced by explants of *Solanum tuberosum* tubers [187]. A different type of growth substance, gibberellic acid, is synthesized in discs of *Helianthus tuberosus* tuber, where it is involved in the induction of synthesis of some of the hydrolytic enzymes that contribute to the increase in metabolic activity [47]. This is clearly comparable to the induction of α-amylase activity by gibberellic acid in germinating cereal seeds (Chapter 7). The induction of division may therefore be regarded as a series of events that are dependent to an increasing extent on growth substances which are either synthesized in response to wounding or which occur in the breakdown products of wounded cells. The interspecific variation with respect to dependence on exogenous growth substances may be due to different tissues producing different amounts of growth substances.

Whilst there is evidence that growth substances are involved in the induction of division, there is also evidence that their effects may be modified by other factors. Discs (10 mm diameter × 1 mm thick) of *Helianthus tuberosus* tubers incubated in water do not synthesize DNA and do not exhibit mitotic activity (Fig. 5.13). The highest rates of DNA synthesis and division are achieved only

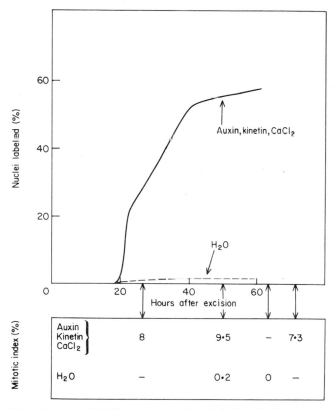

FIG. 5.13. The induction of DNA synthesis and cell division in discs cut from tubers of *Helianthus tuberosus* (artichoke). Discs were incubated in water or in water containing auxin, kinetin and calcium chloride (from ref. 416).

when auxin, cytokinin and $CaCl_2$ are added to the water, although the addition of either auxin or cytokinin alone does induce a low level of DNA synthesis and division. In small cylinders (2 mm diameter × 2mm thick) of *H. tuberosus* tuber, incubated in a complex nutrient medium, the highest rates of DNA synthesis and division are achieved simply by the addition of auxin to the nutrient medium [175, 536]. Addition of a cytokinin together with the auxin has no effect on DNA synthesis or cell division. This suggests that the nutrient status of the tissue and/or the geometry of the explants (cylinders, 2 mm × 2 mm, have a greater surface:volume ratio than discs, 10 mm × 1 mm) may modify the tissue's requirement for an exogenous supply of growth substances.

The possible involvement of the geometry of the explant is also suggested by the localization of the mitotic activity in the more peripheral regions of the explant. This effect is very marked in explants of *Solanum tuberosum* tubers, where DNA synthesis and mitosis are confined to cell layers 2, 3 and 4 (counting from the surface) whereas most other facets of the increased metabolic activity are exhibited by all the cells in the explant [57]. The localization is less marked in discs (10 mm × 1 mm) cut from roots of *Daucus carota* (carrot) in which the increase in DNA content is consistent with the occurrence of DNA replication in about 60% of the cells [60], and in explants of *H. tuberosus* tubers incubated in the dark, in which up to 50% of the cells may divide [536]. The tendency for mitotic activity to be localized towards the periphery of the explants is consistent with the idea that concentration gradients exist between the centre and the surface of the explant. Such gradients could arise by the diffusion or transport of growth substances and/or nutrients into the explants, either from the external medium, or from the surface layer of wounded cells. Gradients may also be established by the movement of inhibitors from the explant into the external medium.

The hypothesis that mitotic activity in tissue explants may be induced by a sudden increase in osmotic potential has been developed by Kahl [229]. An increase in osmotic potential certainly occurs in explants from *Solanum* (potato) tubers [229]; this is at least partly related to the great increase in the ability of the explant tissue to accumulate nutrients. Further, conditions which cause increases in osmotic potential also induce division in buds of *Solanum*, leaves of *Echeveria* [260] and cotyledons of *Raphanus* (radish) [247]. As is pointed out in Chapter 7, growth substances are known to affect membrane properties. It obviously is possible that in doing so they stimulate an uptake of nutrients leading to an increased osmotic potential. However, in view of the complexity both of the action of growth substances and of the cell cycle itself, it seems unlikely that increased osmotic potential is the sole trigger for the onset of mitotic activity in tissue explants.

Although the different types of explant show differing requirements for growth substances with regard to the induction of division, all the species are dependent on growth substances and nutrients for the continuation of division: mitotic activity ceases after one to three division cycles unless growth substances and nutrients are present. In the presence of a complex nutrient medium containing mineral salts, vitamins, a suitable carbon source and auxin and cytokinin, cell division continues, leading to the formation of a callus. The daughter cells produced by successive divisions become more like true meristematic cells, small in size, thin walled, and with much smaller vacuoles than the parenchyma cells from which they were derived [536]. Eventually, the rate of division in the callus decreases markedly, and differentiation is re-initiated, leading to the formation of meristematic regions and regions of vascular tissue within the callus [536]. However, callus cells can be maintained in a dividing state either by taking small explants of the callus and incubating them on the full growth medium (this will cause the explant from the callus to go through the same cycle of division and differentiation as the parent callus) or by culturing small callus explants in shaking suspension culture. The latter treatment leads to separation of the cells from each other. The cells then go on dividing until the level of essential nutrients in the medium falls to a limiting value, at which time the cells stop dividing and the culture enters stationary phase [243]. Cells in stationary phase in suspension culture do not differentiate, but may be induced to divide again by inoculation into fresh medium (as has been described in section II). Thus, the balance between differentiation and division depends not only on nutrients and growth substances, but also on intercellular interactions. This again indicates the importance of gradient formation.

The maintenance of mitotic activity in explants and in single cells requires a high nutrient status, with particular concentrations of auxin and cytokinin, and it is probable that similar requirements exist in normal meristematic cells. The existence of concentration gradients of growth substances and nutrients in plant organs may well be involved in the cessation of mitotic activity and the onset of differentiation. Indeed, it has been widely suggested that the trigger which initiates differentiation in young, undifferentiated cells may be a decreased nutrient status leading to an inability to continue dividing. However, it seems likely that the situation in reality is more complex. The existence of inhibitors of DNA synthesis and of cell division in differentiated cells, although not yet demonstrated, has become a real possibility with the discovery of poly-(ADP-ribose). The prevention of division in differentiated cells is almost certainly mediated by a variety of factors, including a decreased nutrient status, changes in the concentration of growth substances and possibly by a decrease in osmotic potential and an increase in the concentration of inhibitor(s). Elucidation of the interaction of these factors will lead to an increased understanding of the control of both division and differentiation.

SUGGESTIONS FOR FURTHER READING

BROWN, R. and DYER, A. F. (1972). Cell division in higher plants. *In* "Plant Physiology, a Treatise" (F. C. Steward, ed.), Vol. VIC. Academic Press, New York and London.
CLOWES, F. A. L. and JUNIPER, B. E. (1968). "Plant Cells". Blackwell Scientific Publications, Oxford.

H

HALVORSON, H. O., CARTER, B. L. A. and TAURO, P. (1971). Synthesis of enzymes during the cell cycle. *In* "Advances in Microbial Physiology" (A. H. Rose, ed.), Vol. 6, pp. 47–106. Academic Press, London and New York.

MITCHISON, J. M. (1971). "The Biology of the Cell Cycle". Cambridge University Press, London.

STREET, H. E., ed. (1973. "Plant Tissue and Cell Culture". Blackwell Scientific Publications, Oxford.

VAN'T HOF, J. and KOVACS, C. J. (1972). Mitotic cycle regulation in the meristem of cultured roots: the principal control point hypothesis. *Adv. exp. Med. Biol.* **18,** 15–32.

YEOMAN, M. M., ed. (1976). "Cell Division in Higher Plants". Academic Press, London and New York.

6. Molecular Aspects of Differentiation

J. A. BRYANT

I. INTRODUCTION

Morphogenesis, the development of an adult plant with several different types of organ, each with different types of tissue, involves two processes: cell division and cell differentiation. Cell division leads to an increase in cell numbers and thus provides the units from which the plant is built up. Cell differentiation is the formation from "unspecialized" embryonic cells of specialized cells with specific functions. In plants, this process of differentiation is an on-going one. Mature plants, unlike mature animals, contain embryonic regions (meristems). As long as mitotic activity continues in these meristems, the process of differentiation continues.

Directionality or polarity is an integral facet of cell division and development: division in only one plane gives rise to a filament of cells joined end to end; division in two dimensions gives rise to a sheet of cells, one cell thick; division in three planes at right-angles to one another gives rise to a three-dimensional body of cells. In fact, the number of planes of division possible is much greater than three, as new cell walls are not usually formed exactly at right-angles to existing ones. Variation in the incidence of division occurring in any one of the possible planes of division can therefore lead to the attainment of a great variety of shapes in plant organs. The process of cell enlargement (which

217

is the first phase of differentiation) is of particular importance in contributing to plant form. During cell enlargement, the volume of an individual cell may increase to up to 100 times its original value. Cell enlargement is therefore largely responsible for the size of a plant. The directions in which a cell expands determine the final shape of a cell. This is obviously an important part of differentiation, since part of the specialization of a cell is the attainment of a shape and size suited to the cell's functions. The direction and extent of cell enlargement are also relevant to the final shape attained by plant organs. For example, the basic planar structure of a leaf may be modified by control of the size and shape attained by the individual cells in the leaf (as is seen in the differences between "sun" and "shade" leaves on an individual plant). Thus, morphogenesis relies on the control and integration of the processes of cell division and cell differentiation.

The overall control of morphogenesis is largely genetic. Each individual of a species undergoes a similar (although not necessarily identical) pattern of morphogenesis to every other individual, and thus attains a particular set of characteristics by which a plant species may be recognized. However, it is clear also that the environment can play a large part in morphogenesis, modifying the basic patterns imposed on the plant by its genome. The environmental control of morphogenesis lies outside the scope of this monograph, as does morphogenesis as seen at the level of the plant organ. This chapter is concerned with one aspect of morphogenesis, namely the differentiation of specialized types of cell from "unspecialized" cells, particularly as seen at the molecular level. Although we are still very far from understanding the process of differentiation and its regulation, a vast literature exists on this topic. It has therefore been necessary to be very selective. This chapter is not an exhaustive review of all the known facts concerning differentiation, but contains discussion of a number of examples that clearly illustrate general principles and current hypotheses.

II. DIFFERENTIATION OF VASCULAR TISSUE

A. Formation of vascular tissue *in vivo*

Plants are particularly useful organisms in which to study differentiation of particular types of cell. As has already been pointed out, differentiation is a process that continues throughout the life of a plant: the meristems continue to produce new cells which then undergo differentiation. Plant cells are enclosed in relatively rigid walls which prevent the cells from moving in relation to each other. Thus it is possible to trace a developmental sequence by taking samples of cells at increasing distances from a meristematic region. This technique has been used by Northcote and his colleagues in their very extensive studies of the differentiation of vascular elements in higher plants [333, 335, 336].

Much of Northcote's work has been concerned with the changing composition of the cell wall during differentiation. The growth of the cell wall is initiated at cytokinesis (see Chapter 5). There is a proliferation of Golgi bodies in the region where the new cell plate will form. Vesicles are budded off from the Golgi cisternae and are guided by microtubules to the region of the cell plate (Fig. 6.1). These vesicles fuse and eventually traverse the whole distance between the two old cell walls. The vesicles budded off from the Golgi bodies

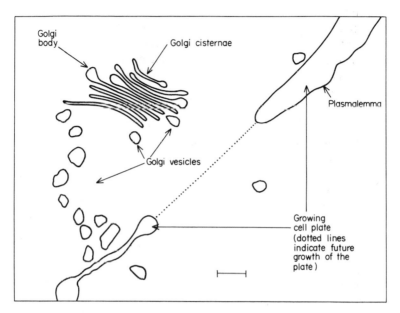

FIG. 6.1. Association of Golgi bodies with cell plate. Scale line = 200 nm (Drawn from electron micrographs in ref. 335).

evidently contain the material from which the cell plate is formed. Analysis of the walls of cells that have recently divided shows that acidic pectins (mainly heteropolymers of galacturonic acid and rhamnose, known as galacturonorhamnans) are present in high concentrations, and it is likely that the cell plate consists very largely of galacturonorhamnans [333, 335].

Following the formation of the cell plate (which becomes the middle lamella of the mature wall), the primary wall is built up. The period of deposition of the primary wall corresponds to the period of rapid cell enlargement, although in some cells, further enlargement may occur after the completion of primary wall deposition and in others, cell enlargement ceases before the completion of primary wall deposition. The composition of the primary wall differs from that of the middle lamella in four ways (Fig. 6.2). Firstly, the pectins laid down in the primary wall are very much less acidic than the galacturonorhamnans of the cell plate. Analysis of the pectins of the primary wall indicates that while galacturonorhamnans continue to be deposited, less acidic pectins are also deposited. These less acidic pectins consist mainly of arabinans (polymers of L-arabinose), galactans (polymers of galactose) and arabinogalactans (hetero-polymers). The ratio of neutral pectins to acidic pectins increases as deposition of the primary wall progresses [333, 335]. Secondly, another group of poly-mers, the inappropriately named "hemicelluloses" are laid down [177, 333]. The major components of the hemicellulose fraction are xylans, polymers of the pentose sugar, xylose. The different types of xylan are classified according to the monosaccharides that make up the short side-chains. The most commonly occurring xylans are glucoxylans (xylans with glucose side-chains) and glucuronoxylans (xylans with glucoronic acid side-chains). In some plants, polymers of mannose with short glucose side-chains (glucomannans) also occur in the hemicellulose fraction [335]. The third feature of primary wall growth is the deposition of the cellulose elementary fibrils, which are laid down in increasing amounts as the primary wall becomes older [333]. Fourthly, as the primary wall approaches maturity and the rate of cell enlargement de-creases, increasing amounts of a protein which is extremely rich in hydro-xyproline are laid down in the wall [234, 259]. Hydroxyproline occurs only rarely in most proteins, but in this structural protein of the plant cell wall (known, perhaps rather misleadingly, as "extensin") hydroxyproline accounts for 20–30% of the amino acid residues. The function of the protein in cell wall structure is not known, but its presence in the cell wall in appreciable quantities seems to be associated with cessation of rapid extension growth. By analogy with collagen, a structural protein from animals which contains large amounts of proline and hydroxyproline [369], it seems likely that extensin is a stiffening element, by virtue of the particular secondary and tertiary structure imposed on the protein by its unusual amino acid composition.

The different components of the primary wall must not be thought of as being discrete and separate entities within the wall. This is illustrated by the finding, by Northcote and his associates, that the very acidic galacturonorhamnans laid down early in the growth of the primary wall apparently become less acidic as the wall grows older [335, 336]. The major reason for this is that the more neutral arabinogalactan molecules become covalently linked to galacturonor-hamnans. The covalent attachment of one cell wall component to another is in fact a general feature of cell wall structure. Albersheim and his colleagues [234] have carried out a detailed analysis of mature primary walls in a number of plants, using specific enzymes and chemical hydrolytic procedures combined with sensitive fractionation and assay techniques. They showed that each component of the primary wall is in covalent linkage with at least one other

(a) General composition of the cell wall of
     a xylem element

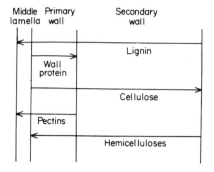

The arrows indicate
directions of increasing
relative concentration
of the components within
the general composition
of the cell wall

(b) The pectins

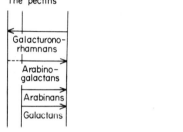

The arrows indicate
directions of increasing
relative concentration
of the components within
the pectin fraction

(c) The hemicelluloses

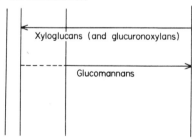

The arrows indicate
directions of increasing
relative concentration
of the components
within the hemicellulose
fraction

FIG. 6.2. The composition of the plant cell wall (from refs 333 and 335).

component, with the exception of cellulose which is hydrogen-bonded to the xyloglucan (Fig. 6.3). The hydrogen-bonding between cellulose and xyloglucan provides a labile linkage which may be readily broken during extension growth [376, 377] (see also Chapter 7).

The budding-off of Golgi vesicles to form the cell plate has already been mentioned. The Golgi bodies are again active throughout the growth of the primary wall. The Golgi vesicles fuse with the plasmalemma by a process of reverse pinocytosis, and discharge their contents into or onto the growing cell wall. It is likely, therefore, that the Golgi bodies may be a major site of synthesis of cell wall polysaccharides. This suggestion has been confirmed by results obtained from a series of elegant experiments performed by Northcote and his associates [337]. Roots of *Triticum* (wheat) were supplied with radioactive glucose for a short period, and then transferred to non-radioactive glucose.

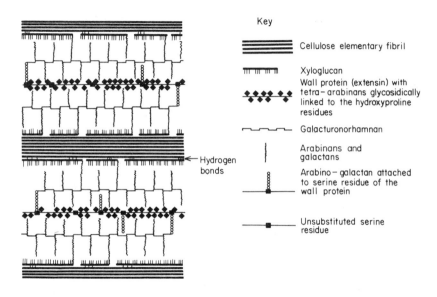

Key

Cellulose elementary fibril

Xyloglucan

Wall protein (extensin) with tetra–arabinans glycosidically linked to the hydroxyproline residues

Galacturonorhamnan

Arabinans and galactans

Arabino–galactan attached to serine residue of the wall protein

Unsubstituted serine residue

←Hydrogen bonds

FIG. 6.3. Model for structure of the primary cell wall (from ref. 234).

Samples of cells undergoing cell plate and cell wall deposition were taken at intervals and examined by high resolution autoradiography. (During preparation of samples for autoradiography, freely soluble substances are lost from the cells; the autoradiography therefore reveals the presence of radioactivity in insoluble compounds, such as structural carbohydrates.) After a few minutes in the presence of radioactive glucose, radioactivity was detected in the cisternae of the Golgi bodies. During the cold chase, the radioactive compounds were detected in Golgi vesicles; the vesicles were transferred across the cytoplasm, and the radioactive compounds were eventually discharged into the cell wall (Fig. 6.4). This is clear evidence that the Golgi bodies are responsible for synthesis of cell wall polysaccharides. Analysis of the polysaccharides labelled in these experiments indicated that the Golgi bodies synthesize exclusively the polysaccharides of the pectin and hemicellulose fractions. Further, Golgi

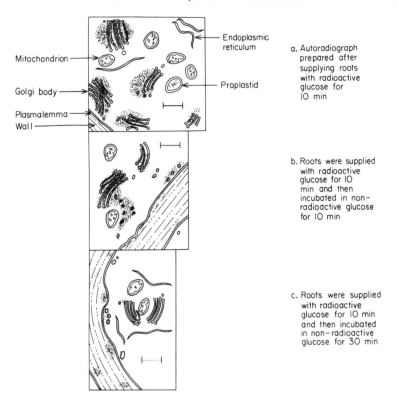

Mitochondrion

Golgi body

Plasmalemma
Wall

Endoplasmic
reticulum

Proplastid

a. Autoradiograph
   prepared after
   supplying roots
   with radioactive
   glucose for
   10 min

b. Roots were supplied
   with radioactive
   glucose for 10
   min and then
   incubated in non-
   radioactive glucose
   for 10 min

c. Roots were supplied
   with radioactive
   glucose for 10 min
   and then incubated
   in non-radioactive
   glucose for 30 min

FIG. 6.4. Involvement of Golgi vesicles in synthesis of cell wall polysaccharides. Roots of *Triticum* (wheat) were incubated in radioactive glucose for 10 min and then transferred to non-radioactive glucose. Samples were harvested at intervals and examined by high-resolution autoradiography. The diagrams (drawn from autoradiographs in ref. 337) show the passage of radioactive polymers from the Golgi cisternae to the cell wall. The blackened areas in the original autoradiograph (indicating the presence of radioactivity) are represented by shading in this figure. Scale line = 1·0 μm.

bodies isolated from pea roots incorporate radioactivity from labelled glucose into polysaccharides of the pectin and hemicellulose fractions *in vitro* [178]. The enzymes responsible for interconversion of monosaccharides, in addition to those catalysing their polymerization, are therefore located in Golgi bodies. There is no evidence either from *in vivo* labelling experiments or from experiments with isolated Golgi bodies, that the Golgi bodies are responsible for synthesis of cellulose [336, 528]. Indeed, the current evidence suggests that the site of cellulose synthesis is the outer surface of the plasmalemma (see below).

The synthesis and transport of cell wall polysaccharides requires the activity of a number of subcellular components, including Golgi bodies, microtubules and the plasmalemma. At least one further organelle, the ribosome, is involved in the synthesis of the primary wall. Despite being a somewhat unusual protein, the cell wall structural protein is synthesized in the usual manner on 80s ribosomes. Hydroxylation of the proline to form hydroxyproline takes place after synthesis of the protein [82]. Before its incorporation nto the cell wall, the protein is taken up into the Golgi bodies where it is combined with linear tetra-arabinose units [64, 148]. The tetra-arabinose units are linked to the protein via the hydroxyproline molecules. The protein–arabinose complex is then transported to the cell wall in Golgi vesicles as has previously been described for the polysaccharides of the pectin and hemicellulose fractions.

The growth of first the cell plate and then the primary wall depends on the co-ordinated synthesis and deposition of a number of different components, the relative proportions of which change during the growth of the wall. The synthesis and deposition of these components relies in turn on the co-ordinated activity of a number of different organelles and also on the varying activities of a large number of different enzymes. This interplay of different enzymes is also a feature of secondary wall deposition.

Cell wall deposition does not necessarily cease after the completion of cell enlargement. In many cells, the material laid down after the cell has attained maximum size is similar in composition to the material laid down during cell enlargement, and is thus regarded as primary wall. The extent of this "late" primary wall deposition is usually limited. This is true of a large proportion of the parenchyma cells in the ground tissue of the plant (e.g. pith and cortex), and of most of the photosynthetic cells in leaves. However, many other types of cell go on to deposit a thick secondary wall on the inner surface of the primary wall. Although the transition from primary wall to secondary wall is normally gradual, five major differences may be distinguished between the composition of primary and secondary walls (Fig. 6.2). Firstly, no pectins are deposited during secondary wall growth. Secondly, the proportion of hemicelluloses decreases progressively during secondary wall growth. The composition of the hemicellulose fraction also changes, with larger amounts of glucomannans and lesser amounts of xyloglucans in the secondary wall than in the primary wall. Thirdly, the amount of cellulose deposited per unit weight of wall increases progressively, so that the most recently deposited layers of a mature secondary wall consist almost entirely of cellulose. Further, the orientation of the elementary fibrils in the secondary walls is parallel with the long axis of the cell, contrasting with the more transverse orientation in the primary wall. In some plants (e.g. *Gossypium*—cotton) there is also an increase in the size of the individual cellulose polymers: cellulose molecules in the primary wall have a molecular weight of $3–4 \times 10^5$, whereas those in the secondary wall have a molecular weight of around 2×10^6 [298]. These changes in cellulose orientation and structure are associated with the increased rigidity of the secondary wall. The fourth major difference between primary and secondary walls is that secondary walls do not contain the hydroxyproline-rich structural protein. A further feature of some secondary cell walls is that they contain lignin, although it must be pointed out that this feature is not diagnostic, since primary walls may also become lignified.

The changes in composition of the wall during secondary growth are paralleled by changes in the activity of the Golgi bodies. Radioactivity from ^{14}C-glucose does not label galacturonic acid, galactose or arabinose during secondary wall formation, whereas xylose and glucuronic acid continue to be labelled [336]. The cessation of pectin synthesis is therefore reflected in the inability of the cell to synthesize the monomers from which pectins are synthesized. The most probable reason for this is a decline in activity of the epimerase enzymes which are responsible for the conversion of the glucose series of monosaccharides to the galactose series [335, 540] (Fig. 6.5). In addition to the cessation of pectin deposition at the end of the period of primary wall deposition, the deposition of hemicelluloses declines throughout the period of secondary wall deposition. The synthesis of polysaccharides in the Golgi bodies, and the transport of the polysaccharides in Golgi vesicles similarly

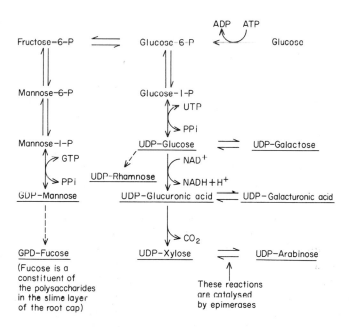

FIG. 6.5. Pathways of biosynthesis of the monosaccharide units from which cell wall polysaccharides are synthesized. The compounds directly involved in cell wall biosynthesis are underlined. Dotted lines indicate multi-state conversions.

decline. By contrast, the cellulose-synthesizing activity of the plasmalemma increases markedly during the growth of the secondary wall, as would be expected from the increased contribution made by cellulose to the bulk of the wall. The changing roles of the Golgi bodies and plasmalemma in polysaccharide biosynthesis are well illustrated by results of autoradiographic experiments carried out by Northcote's research group. If radioactive glucose is fed to young cells in root tips, the major sites of synthesis of radioactive polymers are the Golgi bodies [336]. In older cells, for example in stems of *Acer pseudoplatanus* (sycamore) seedlings, the surface of the plasmalemma is the major site of polymer synthesis, and the particular polymer synthesized in these cells is a $\beta 1$–4 glucan, namely cellulose [528].

The pattern of cell wall deposition described so far is characteristic of a number of relatively unspecialized cells, such as those occurring in xylem or phloem parenchyma. However, the more specialized cells in vascular tissue exhibit further structural modifications which are associated with the very specific functions performed by these cells. Two such cell types are the xylem vessel element and the phloem sieve element. Xylem vessel elements are very elongated cells, with thick secondary walls. The secondary thickening in xylem vessel elements is very often not complete, in the sense that parts of the wall remain as primary wall, with no deposition of secondary material. Some of the many patterns of secondary wall formation in xylem vessel members are shown in Fig. 6.6. The microtubules and the membrane system of the endoplasmic reticulum are involved in this localized deposition of secondary wall material. During primary wall growth, microtubules lie close to the plasmalemma at all parts of the cell boundary [335, 336]. This is in accordance with their role in the transport of Golgi vesicles to the plasmalemma. At the end of primary wall growth, in the cells which are to become tracheids or vessel elements, the microtubules become localized in groups. In the regions of the cell boundary where there are no microtubules, there is a proliferation of endoplasmic reticulum, giving the appearance, in micrographs, that the endoplasmic reticulum is closely pressed against the plasmalemma. As secondary wall deposition proceeds, those regions of the wall directly across the plasmalemma from the endoplasmic reticulum remain unthickened, whereas the rest of the wall is thickened by deposition of a secondary wall (Fig. 6.7). This suggests that the endoplasmic reticulum prevents the deposition of hemicelluloses at the plasmalemma, and also prevents the synthesis of cellulose by the enzymes on the plasmalemma surface [334, 336]. In addition to the unthickened regions on the side walls of xylem vessel elements, the end walls also remain unthickened. This is again associated with the presence of endoplasmic reticulum close to the plasmalemma during the period of secondary wall deposition.

Many of the cells making up the xylem become lignified. Lignification may occur as part of secondary wall formation, as in wood parenchyma and

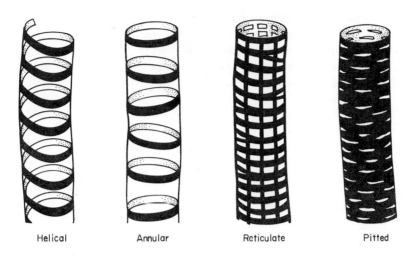

Helical   Annular   Reticulate   Pitted

FIG. 6.6. Patterns of wall thickening in xylem vessel elements. Scale line = 20 μm.

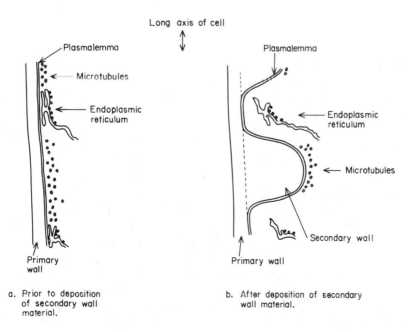

FIG. 6.7. Diagram illustrating the relationship between the endoplasmic reticulum and the pattern of secondary wall deposition in xylem vessel elements.

sclerenchyma, or, as in xylem vessel elements and tracheids, may occur when secondary wall formation is essentially complete. Lignin is a very complex polymer, built up from phenylpropanoid alcohols (Fig. 6.8). The precursor for these phenylpropanoid alcohols is phenylalanine, and entry of the phenylpropanoid unit of phenylalanine into the pathway for biosynthesis of the alcohols is

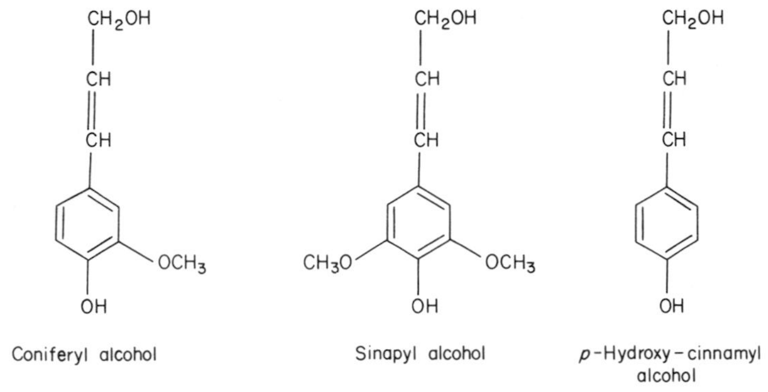

Coniferyl alcohol   Sinapyl alcohol   p-Hydroxy–cinnamyl alcohol

FIG. 6.8. The phenylpropanoid alcohols, precursors for lignin biosynthesis.

mediated by phenylalanine-ammonia-lyase (Fig. 6.9). In tissues which have low basal levels of phenolic compounds, the onset of lignification is preceded by a very marked increase in the activity of phenylalanine-ammonia-lyase. In explants of *Glycine max* (soya bean) and *Coleus,* the presence of phenylalanine-ammonia-lyase activity may be used as a specific marker for xylem vessel element differentiation [394]. In *Apium* (celery) petioles, phenylalanine-ammonia-lyase activity is almost completely confined to the vascular bundles [395]. In stems of *Larix* (larch), the synthesis of the phenylpropanoid alcohols occurs only in differentiating xylem, and "laccase", the enzyme complex responsible for polymerizing the alcohols, is confined to cells undergoing lignification [145]. The mechanism of deposition of lignin into the wall has not been elucidated. However, three features have been established. Firstly, the Golgi bodies are not involved [528]. Secondly, in most xylem vessel elements, lignin is mainly deposited in areas of the wall that have previously undergone secondary thickening, although in secondary xylem (derived from the vascular cambium, rather than from the apical meristem), lignification may also occur in the regions of primary wall that have not been thickened by deposition of a secondary wall. Thirdly, the lignin is deposited throughout the entire thickness of the wall, despite its being a large, complex and very insoluble polymer (Fig. 6.2).

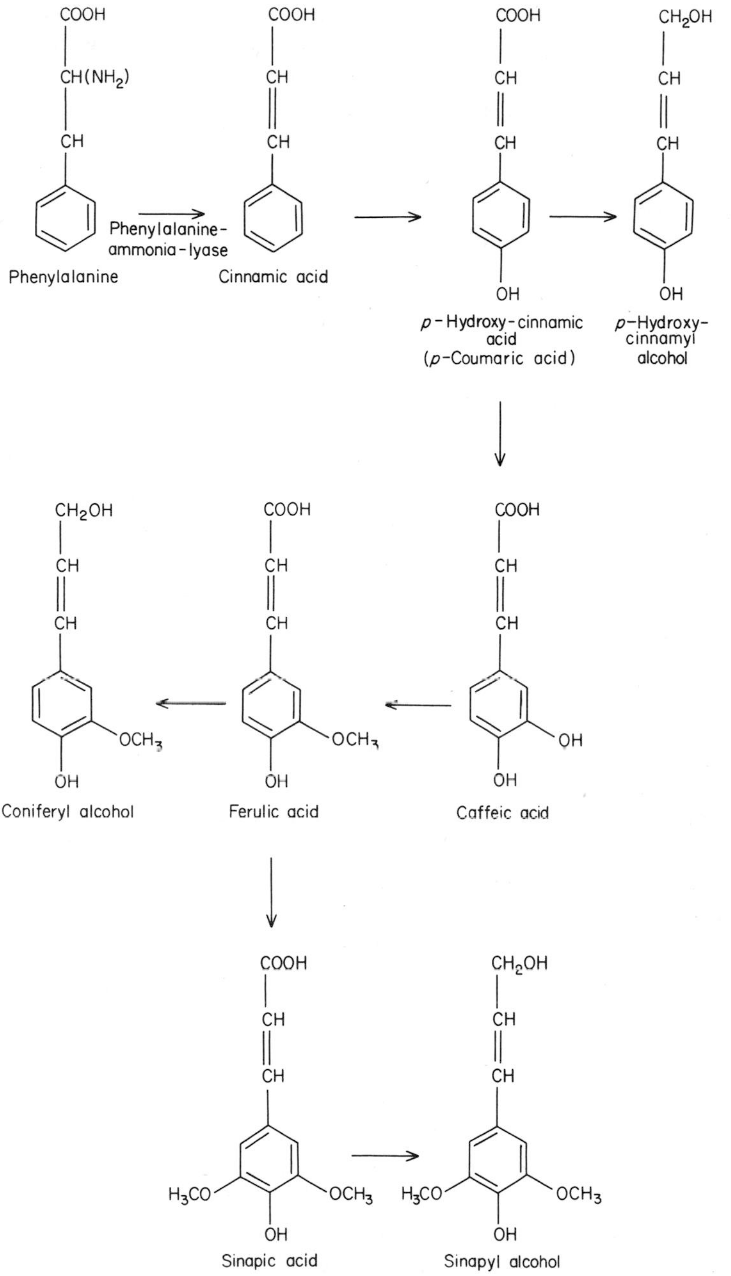

FIG. 6.9. The synthesis of phenylpropanoid alcohols from phenylalanine.

As lignification approaches completion, the unthickened end walls of the vessel element are broken down and the cytoplasm begins to degenerate. The breakdown of the end walls is presumed to be mediated by the action of polysaccharide hydrolases. Such enzymes may well be synthesized on ribosomes bound to the endoplasmic reticulum in the region of the end wall. The monosaccharide units from which cell wall polysaccharides are built up, including glucose, galactose, xylose, arabinose and uronic acids, have been detected in the sap of newly differentiated xylem vessels (418). This is consistent with the view that end-wall breakdown is achieved by hydrolysis of the constituent polysaccharides. The xylem sap also contains hydrolytic enzymes which are presumed to be involved in the breakdown of the cell contents: nucleases, proteases and acid phosphatases [418]. It is not yet clear whether the presence of these enzymes is a result of *de novo* synthesis of enzyme protein, or of release of previously inactive enzymes from lysosomes [147]. The breakdown of the end walls and the disintegration of the cell contents results in the formation of a vessel, a long tube consisting of a number of dead cells which are joined end to end. The programme of differentiation of a xylem vessel element is therefore a programme leading to the formation of a dead cell which has a specific function in the living plant body. The developmental sequence described here is summarized in Fig. 6.10.

The programme of differentiation of the phloem sieve element, the other major conducting element of the plant, is somewhat different from that of the xylem vessel element (Fig. 6.10). The pattern of secondary wall deposition does not show the range of variation observed in xylem vessel elements. Secondary wall material is deposited over the entire primary wall, although the secondary wall adjacent to the companion cell is usually thicker than the other walls of the cell. Sieve elements differentiated from vascular cambium (secondary phloem) often have thicker walls than sieve elements differentiated from apical meristems (primary phloem). The composition of the secondary wall in a phloem sieve element is different in certain important respects from that in a xylem vessel element, having a much lower proportion of hemicelluloses and no lignin [333].

The secondary growth of the wall in the regions which eventually become sieve plates or sieve areas is particularly interesting (Fig. 6.11). The sieve plates are formed in parts of the wall where there are numerous plasmodesmata. During growth of the secondary wall, sheets of "rough" endoplasmic reticulum (endoplasmic reticulum to which ribosomes are bound) become closely associated with the plasmalemma in and around the plasmodesmata. As is observed in xylem cells, the presence of the endoplasmic reticulum effectively prevents the deposition of a secondary wall. However, the wall in this region does not remain unthickened. Instead of normal secondary wall growth, callose (a polysaccharide made up of glucose units joined by β 1–3 linkages) is deposited

Elongation ——————→

Unequal division

— Smaller cell: Growth of primary wall | Growth of secondary wall. Some callose deposition around plasmodesmata which connect with sieve cell | Cell retains organelles and exhibits high metabolic activity → COMPANION CELL

— Larger cell: Elongation and vacuolation* | Growth of primary wall | Growth of secondary wall. Callose deposition around plasmodesmata in end walls | Digestion of callose in end walls. Synthesis of P protein | Sieve plate formation complete. Reduction in numbers of organelles | Distinction between vacuole and cytoplasm lost. P-protein reorganised in strands passing through sieve pores → SIEVE CELL

"Normal" division: Elongation and vacuolation* | Growth of primary wall | Growth of secondary wall (often localized). No secondary material deposited on end walls / Lignification mainly of secondarily thickened areas of wall | End walls digested | Lysis of cell contents → VESSEL ELEMENT

* The term vacuolation refers to the formation of a large, single vacuole. The companion cell, which is often termed "non-vacuolate", in fact contains, in common with other "non-vacuolate" cells, many very small vacuoles.

FIG. 6.10. Developmental sequences involved in the differentiation of xylem vessel elements, phloem sieve cells and phloem companion cells.

into and onto the primary wall around the plasmodesmata. The callose is synthesized by a UDPG→β1–3 glucan glycosyl transferase ("callose synthetase"). This enzyme is tightly bound to a membrane component of the cell [140]. The identity of this membrane component has not been established, but it is thought to be either the endoplasmic reticulum or the plasmalemma [91, 140]. If the transferase is in fact associated with the endoplasmic reticulum or with the plasmalemma adjacent to which the endoplasmic reticulum is located, it is likely that the enzyme protein is synthesized on the ribosomes which are bound to the endoplasmic reticulum [335]. During the phase of callose deposition, a structural protein, P protein, is synthesized in the cytoplasm. The molecules of protein are organized to form fibrils and these in turn aggregate to form "protein bodies" which are often located in cytoplasmic strands that traverse the vacuole [95].

The later stages of differentiation of the phloem sieve element are not yet clearly understood, even at the purely descriptive level. The overall process is clearly complex, involving a number of spatially separated but temporally integrated events. The plasmodesmata between adjacent sieve elements are enlarged. The enlargement is caused by a degradation of the cell wall, including the newly deposited callose, starting in the middle and working outwards (Fig. 6.11). This leads to the formation of an area of cell wall which is traversed by pores. The pores are lined with callose and are usually 0·1–0·5 μm in diameter. This area of the cell wall is known as the sieve plate. As with the breakdown of the end walls of xylem vessel elements, the breakdown of cell wall material is presumed to be mediated by the activity of hydrolytic enzymes. In phloem sieve cells the hydrolysis is obviously closely regulated since it is confined to the regions of wall which are impregnated with callose. This suggests that the major hydrolytic activity involved is a β1–3 glucanase [126]. This enzyme is unable to hydrolyse cellulose, which is the major component of the "normal" secondary wall in phloem sieve cells, but is able to hydrolyse callose.

As the process of pore enlargement proceeds, the subcellular organelles begin to degenerate. The nucleus disappears completely, and the plastids and mitochondria lose most of their internal membranes (lamellae and cristae respectively). Much of the endoplasmic reticulum is lost, including that passing through the pores, and there is a great reduction in the number of ribosomes. It must be emphasized, however, that phloem sieve cells do not degenerate completely. They retain at least rudiments of all their organelles, with the exception of the nucleus, and they exhibit metabolic activity. Further, the plasmalemma remains intact and apparently functional. The degeneration of cell contents, which in xylem vessel elements proceeds to completion, is only a partial degeneration in phloem sieve cells, and is clearly a regulated process. During the final stages of this specialization of phloem sieve cells, the distinction between cytoplasm and vacuole is lost; it is not clear whether this is

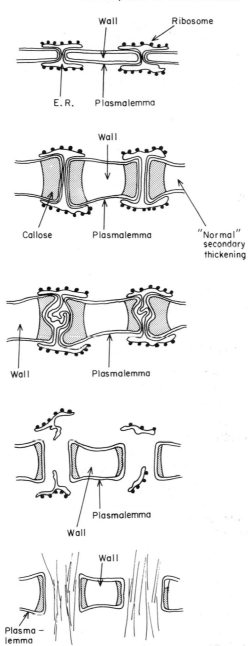

FIG. 6.11. Diagram illustrating the formation of a phloem sieve plate.

caused by a breakdown of the tonoplast, or by a resorption of the vacuole with a concomitant increase in the volume of the cytoplasm. As the vacuole disappears, the P protein in the protein bodies becomes reorganized. The protein bodies disperse and the protein fibrils become orientated in a longitudinal direction [95]. Some of the protein fibrils pass through the pores in the sieve plate, thus forming strands which connect adjacent sieve elements [95]. A very large number of questions remained to be answered, particularly in relation to the structure and function of P protein. The mechanism which brings about the change of orientation of the protein fibrils during the transition from protein bodies to strands is not understood. The relationships between molecular structure and the structure of the fibrils, and between the structure of the fibrils and the structure of the strands has not been elucidated. The function of P protein is still a matter for conjecture, although it is assumed by many investigators (but not all) to have a role in translocation. Despite these unanswered questions, it is clear that the differentiation of phloem sieve elements is characterized by four features in particular: the synthesis of a specific protein (P protein), the synthesis of a specific polysaccharide (callose), a closely regulated partial breakdown of the cell wall (to form sieve areas), and a marked specialization of the cell contents.

B. Formation of vascular tissue *in vitro*

The differentiation of vascular elements, described in detail above, is particularly interesting since it is possible to study the process under controlled conditions *in vitro*. Formation of vascular elements may be induced in undifferentiated masses of callus tissue by the application of auxin and sucrose. This has been achieved with callus derived from *Syringa vulgaris* (lilac) by Wetmore and Rier [514] and with callus derived from *Phaseolus vulgaris* (French bean) by Northcote and his associates [221, 222]. Northcote and his colleagues carry out their experiments in the following manner. Blocks of callus tissue are placed on a maintenance medium containing agar, mineral salts, coconut milk and low concentrations of indol-3yl-acetic acid (IAA) and sucrose. A wedge-shaped notch is cut in the top of each block of callus tissue, and a wedge-shaped block of agar, containing IAA ($0 \cdot 1$ mg 1^{-1}) and sucrose (20 g 1^{-1}) is inserted into the notch (Fig. 6.12). Use of ^{14}C-labelled sucrose and IAA shows that the sucrose and auxin pass from the agar into the block of callus tissue by diffusion, thereby creating concentration gradients of the two substances within the callus tissue. This leads to the formation of nodules of vascular tissue in a band $0 \cdot 05 - 0 \cdot 5$ mm below the base of the induction wedge (Fig. 6.12). Each nodule contains a band of cambium, with phloem on one side (towards the induction wedge) and xylem on the other. The auxin and sucrose therefore cause the formation of organized regions of dividing cells (cambium) giving rise to phloem and xylem in an arrangement very similar to that observed in a normal vascular bundle. The cell walls in the nodules of vascular tissue have a higher xylose: arabinose ratio than the undifferentiated callus cell. This is consistent with the changes in cell wall composition that occur during secondary thickening. The synthesis of lignin and callose is also detected during induction of vascular tissue. Further, the induction is accompanied by marked increases in the activities of phenyla-

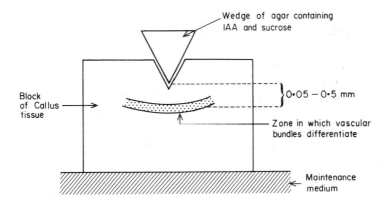

FIG. 6.12. Induction of vascular tissue formation in callus tissue of *Phaseolus vulgaris* (from ref. 222).

lanine-ammonia-lyase and $\beta 1$–3 glucan (i.e. callose) synthetase [171]. Thus, the vascular elements induced in the blocks of callus tissue are characterized by the presence of the specialized polymers that occur in "normal" vascular elements, and by the presence of enzymes involved in the synthesis of those polymers.

Jeffs and Northcote used sugars other than sucrose in combination with IAA. All were ineffective in inducing the formation of vascular tissue, with the exception of maltose and trehalose, which, like sucrose are α-glycosyl disaccharides. However, even maltose and trehalose were not as effective as sucrose. The role of sucrose in controlling the induction of vascular tissue is not therefore to provide an energy source, but to act as a specific regulatory agent, presumably recognized in cells by virtue of a specific feature of its structure. Jeffs and Northcote also varied the relative proportions of auxin and sucrose, showing that increasing amounts of IAA cause the preferential induction of xylem, and increasing amounts of sucrose cause preferential induction of phloem. Since xylem differentiation involves the breakdown of cell contents, leading to the presence in newly differentiated xylem of IAA (derived from tryptophan released from peptide linkage by proteolysis), and since the function of phloem is the transport of sucrose, Northcote has suggested that the differentiation of vascular tissue may be self-perpetuating [221, 222, 334, 336]. This is an attractive hypothesis, but it does raise the problem of the origin of the auxin and sucrose necessary for induction of the formation of the first vascular elements in the plant embryo.

III. GENE EXPRESSION AND DIFFERENTIATION

The examples of cell differentiation discussed in section II show that differentiation is a very complex process, even when viewed at the level of a single cell within the plant body. However, it is clear that the developmental programme of a cell involves, among other things, the appearance and disappearance at specific times of specific proteins (as detected by changes in enzyme activities, and by changes in the structure and content of organelles and membrane components). Further, different types of cell are characterized by different types of enzymes and structural proteins. This illustrates, perhaps somewhat simplistically, that differentiation involves differential gene expression: different genes are expressed in different types of cell. Even though specialized cells are characterized by the presence of specific gene products, many differentiated cells are totipotent, in that they contain all the genetic information necessary for growth of a whole plant. Thus, xylem vessels differentiate from pith parenchyma cells in *Coleus* stems in which a vascular bundle has been severed [432]. More spectacularly, whole plants can be grown from single cells in culture. This was first achieved with callus cells derived from the phloem parenchyma of *Daucus carota* (carrot) roots [448], and it is now known that a wide variety of specialized cells, including cells from leaves, stems and ovary tissue, can give rise to plantlets either directly, or via the formation of callus [205]. Some specialized cells, however, are not totipotent. The most obvious examples are xylem vessel elements and tracheids, which are dead cells, and phloem sieve cells, which are enucleate and therefore contain no primary genetic information.

It must be emphasized that the term "gene expression" has a much wider meaning than merely "gene transcription". Gene expression means the appearance in the cell of the gene product in an active form. During the years that immediately followed the publication of Jacob and Monod's famous paper on the induction of β-galactosidase activity in *Escherichia coli* [214], it was widely assumed that differential gene expression is eukaryotes was effected by some sort of genetic operator system, with the major sites of control being the structural genes or regulatory genes associated with the structural genes. Indeed, there is a good deal of general evidence that developmental changes may be accompanied by changes in gene transcription. The availability of the DNA in chromatin to act as a template for RNA synthesis increases markedly in a number of situations, including the breakage of seed dormancy [220] and the induction, by indol-3yl-acetic acid, of root formation in shoot explants (see Chapter 7). The chromatin from pea (*Pisum sativum*) cotyledons has a much lower template activity than the chromatin from pea shoot apices (Chapter 1). These changes in the template activity of chromatin are associated with changes in the composition of the RNA synthesized *in vivo*, as detected by competitive molecular hybridization (Chapter 7). There has been one report that these general changes in template activity and RNA synthesis actually reflect changes in the transcription of specific genes. Bonner used chromatin from pea cotyledons and from shoot apices as templates for *in vitro* RNA synthesis. The RNA was then used as a message for *in vitro* protein synthesis. Bonner's results suggested that the RNA transcribed from cotyledon chromatin was able to support the synthesis of globulin, the major storage protein of the cotyledons, whereas RNA transcribed from shoot apex chromatin was not able to do so [40]. It must be pointed out that although these experiments were carried out in the early 1960s, they have not been repeated successfully in other laboratories. However, if the results are correct, they strongly suggest that the expression of at least one plant gene, that coding for globulin, may be regulated by controlling the availability of the gene for transcription. Although there is clearly little data relating to this topic in plants, there is good evidence from experiments with cells from chicken that the chromatin from reticulocytes supports the synthesis of globin mRNA whereas the chromatin from liver does not. In this instance, the non-histone proteins have been identified as the specific regulatory agents (see Chapter 1).

Regulator genes, which are a feature of the genetic operator model, have been discovered in higher plants. For example, the synthesis of zein, a storage protein in *Zea mays,* is very much reduced in "floury-2" and "opaque-2" mutants. The genes concerned do not code for zein, but clearly affect the expression of the zein gene. These genes are therefore regulatory genes [326]. Their mode of action is not known, and it must not be assumed that they act in the same way as the regulatory genes in *Escherichia coli.*

Although examples of transcriptional control are known, it is now realized that control of gene expression in eukaryotic cells is not necessarily at the transcriptional level. This is clearly illustrated by results of experiments with the unicellular alga, *Acetabularia*. The giant cell develops from a zygote that elongates to form a stalk which may be up to five centimetres in length. Rhizoids grow out from the base of the stalk, and the nucleus becomes located in one of the rhizoids. Several weeks later, a cap is formed (Fig. 6.13). The single cell of *Acetabularia* is therefore more complex than most other eukaryotic cells. The formation of the cap may be regarded as a process of differentiation. It involves the sequential appearance and disappearance of a number of enzymes [176, 486], the synthesis of protein [174], and the synthesis of

238 J. A. Bryant

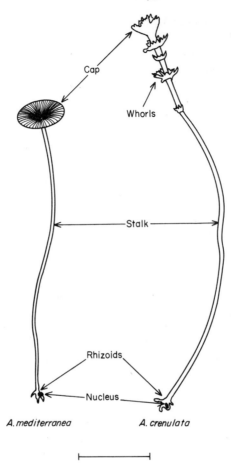

Cap

Whorls

Stalk

Rhizoids

Nucleus

A. mediterranea   *A. crenulata*

FIG. 6.13. Cap morphology in *Acetabularia*. Scale line = 1 cm.

specific polysaccharides [540]. The process is thus analogous to the developmental sequences outlined in section II.

Removal of the nucleus from an *Acetabularia* cell is a relatively straightforward procedure. Cells that have been enucleated form a cap which is identical in appearance to the cap formed by a normal cell. Further, the normal patterns of increased protein synthesis, the appearance and disappearance of specific enzymes and the synthesis of specific polysaccharides all occur in enucleate cells. This means either that the genes controlling cap formation are cytoplasmic (e.g. in the plastids) or that cap formation can be controlled post-transcriptionally. The former idea is ruled out by the finding that cap morphology, and some of the enzymes involved in cap formation, are specified by nuclear genes [157, 174]. It therefore seems probable that in *Acetabularia,* post-transcriptional events may regulate the expression not just of one gene, but of all the genes involved in a developmental sequence. Further, enucleation may be performed several weeks before the initiation of cap formation, and cap formation still proceeds normally [174]. This indicates that the messenger RNA molecules may be very long-lived, or that the enzymes involved in cap

TABLE 6.I. Possible points of control of gene expression.

Transcriptional control
    Availability of the DNA template (i.e. gene) for transcription

    Initiation of transcription (recognition of the template and binding of RNA polymerase to the template)

    Rate of transcription (number of and activity of RNA polymerase molecules)

    Termination of transcription and release of messenger RNA

Translational control
    Processing and "maturation" of messenger RNA versus degradation (i.e. availability of messenger RNA for translation)

    Transport of messenger RNA

    Availability of transfer RNA and amino acyl-transfer RNA synthetases

    Availability of ribosome subunits

    Formation of initiation complex

    Initiation

    Peptide bond formation and translocation

    Termination of protein synthesis and release of protein

    Continued availability of messenger RNA versus degradation

Post-translational control
    Modification of protein structure (primary, secondary, tertiary or quaternary)

    Activation or inactivation of protein

    Rate of turnover (synthesis versus degradation)

formation may be stored in inactive forms in the cytoplasm for long periods. The necessity for protein synthesis before and during the process of cap formation suggests (but does not prove) that the former explanation (long-lived mRNA) is more likely to be correct.

Although the process of cap formation in *Acetabularia* provides evidence for the existence of post-transcriptional control, it must not be assumed that all gene expression in eukaryotes is controlled post-transcriptionally. Indeed, the examples discussed earlier in this section provide evidence that transcriptional control also occurs. In fact, three general levels of control are now recognized: transcriptional control (relating to the synthesis of messenger RNA), translational control (relating to the utilization of messenger RNA) and post-translational control (relating to the modification of protein structure after protein synthesis). Translational and post-translational control are often dealt with together as "post-transcriptional" control (as in the discussion of cap formation in *Acetabularia*). Each of these levels of control incorporates a large number of points at which gene expression may be regulated (Table 6.I). Very little is known about which of these specific control points are the most important. This is perhaps understandable, since only for a very small number of genes has the general level of control been established.

The absolute criteria for establishing whether the appearance of a particular enzyme or structural protein, at a particular developmental stage, is controlled transcriptionally, translationally or post-translationally, are very straightforward. Transcriptional control means that the appearance of the protein must be preceded by the synthesis of a species of messenger RNA that had not previously been synthesized. Translational control involves the synthesis of a protein that had not previously been synthesized (i.e. *de novo* synthesis) without the synthesis of any "new" species of messenger RNA. Post-translational control involves the appearance of the protein in an active form without the need for the synthesis of any "new" species of protein. However, it is much easier to state these criteria than to apply them in a given situation. For example, only four messenger RNAs from higher plants, those coding for globulin [40], the large subunit of Fraction I protein [180], leghaemoglobin (the haemoglobin that occurs in the root nodules of legumes) [498] and cellulase [497a], have been detected by virtue of being translated in *in vitro* protein-synthesizing systems. In addition, the mRNA for isocitrate lyase has been detected in the alga *Chlorella* [413a]. Of these, only the messenger RNA for leghaemoglobin has been purified (see Chapter 2). Therefore, it is not possible to establish at present whether the synthesis of particular species of messenger RNA is initiated at particular stages in development, or in particular types of cell. Even were it possible to demonstrate that a particular messenger RNA is only detectable in cells synthesizing the protein for which the messenger RNA codes, this does not constitute completely unequivocal proof that the control of the synthesis of that protein is at the transcriptional level. The absence of a particular messenger RNA may be caused by its rapid degradation immediately after synthesis. Although this possibility may seem very remote, it must be pointed out that a large proportion of the RNA synthesized in the nucleus of a eukaryotic cell is degraded without ever leaving the nucleus [176]. Whether or not the degraded RNA represents, at least in part, "unwanted" messenger RNA species has not been established.

Establishing that the presence of a particular enzyme has been mediated via *de novo* synthesis is less difficult. It is possible to measure the amount of enzyme protein, as has been carried out for isocitrate lyase in *Chlorella*: the induction of the enzyme is accompanied by a very dramatic increase in the amount of enzyme protein [223]. It is also possible to detect synthesis of a protein by use of radioisotopes. During the gibberellic acid-induced increase in α-amylase activity in *Hordeum* (barley) seeds, the α-amylase enzyme protein becomes radioactive if the seeds are incubated in ^{14}C-labelled amino acids [493]. An alternative to the use of radioisotopes is the use of deuterium oxide ("heavy water") or $H_2^{18}O$. The deuterium or ^{18}O is incorporated into protein molecules during peptide bond formation, causing the buoyant density of the protein to increase. Isopycnic centrifugation of a crude extract in gradients of caesium chloride, followed by assay of enzyme activity along the gradient, is used to measure the buoyant density of the enzyme from plants incubated in D_2O or $H_2^{18}O$ and from control plants. This technique obviates the need for purification of the enzyme protein (which is a necessary part of the procedure for studying ^{14}C incorporation into enzymes), and hence is becoming increasingly popular.

Although measurements of the amount of enzyme protein, or study of the incorporation of ^{14}C, ^{18}O and deuterium may indicate that an increase in enzyme activity is accompanied by synthesis of the enzyme, they do not necessarily indicate that *de novo* synthesis (i.e. synthesis of a protein not previously synthesized, or a very marked increase in rate of synthesis of a protein previously synthesized at a very low rate) is involved. For example, it is possible that an increase in enzyme activity is mediated via a greatly reduced rate of degradation of an enzyme which is continuously turning over. Under such circumstances, the enzyme will become labelled with ^{14}C, ^{18}O or deuterium, and an increase in the amount of enzyme protein will be detectable, but clearly there is no *de novo* synthesis of the enzyme. Thus, any attempt at detecting *de novo* synthesis of an enzyme should incorporate estimates of the rate of turnover of that enzyme (see Chapter 7 for a fuller discussion of this topic). When this criterion is applied to the detection of enzyme synthesis by density labelling (with ^{18}O or deuterium), the published data fall into three classes. Firstly, some enzymes show a dramatic increase in activity without any incorporation of the density label. Provided it can be shown that other proteins in the same tissue do incorporate density label (and hence that the density label is available in the tissue), this is a clear indication that the increase in the activity of the enzymes is not mediated by *de novo* synthesis. The increase in activity is therefore an example of post-translational control. The increase in acid ribonuclease activity in the cotyledons of germinating *Pisum arvense* (field pea) seedlings is an example of this type [18]. Secondly, a very large number of enzymes have been shown to increase in buoyant density in tissues incubated in density label during the rise in enzyme activity. Unfortunately, in the majority of the experiments performed, no attempts were made at estimating the rate of enzyme turnover. The conclusion to be drawn from such data is that the increase in enzyme activity is accompanied by protein synthesis, but the data do not permit a distinction between *de novo* synthesis and changes in the rate of turnover. Examples are the increases in the activities of isocitrate lyase and malate synthetase in the cotyledons of *Arachis hypogea* (peanut) and of acid phosphatase in the cotyledons of *Pisum sativum* [224, 288]. It is perhaps improbable that increases in the activities of a large number of enzymes are mediated by decreases in the rate of degradation. It is thus likely that, despite the lack of unequivocal proof, many of the enzymes in this second group are in fact synthesized *de novo*. Support for this suggestion comes from the third class of data. In a very small number of instances, investigators carrying out density-labelling experiments have made estimates of enzyme turnover rate. In all such experiments reported so far, an increase in buoyant density of the enzyme has been shown to arise by an increased rate of synthesis (i.e. *de novo* synthesis), and not by a decreased rate of degradation. In these instances the increase in enzyme activity is clearly not a post-translational phenomenon. An example of this is the phytochrome-mediated increase in acid ribonuclease activity in the hypocotyls of *Lupinus albus* [2].

The technique of enucleation has already been discussed in connection with the control of cap formation in *Acetabularia*. Unfortunately, it is not yet possible to carry out enucleation experiments with differentiating cells of higher plants. However, during the course of differentiation, phloem sieve cells lose their nuclei and thus become enucleate cells. These cells continue to synthesize protein for several months after the loss of their nuclei. This is another indication of the longevity of messenger RNA in eukaryotic cells. Unfortunately, there is no information to indicate whether or not there are any dramatic changes in the activities of particular enzymes in sieve cells after enucleation. We therefore do not know whether developmental sequences can be controlled completely post-transcriptionally in these cells.

The most widely used tools in the study of the control of gene expression are inhibitors of protein synthesis (such as puromycin and cycloheximide) and of RNA synthesis (e.g. actinomycin D). The use of inhibitors has been widely criticized, since many of them may have unwanted side-effects. However, if used carefully, inhibitors can give general information on the control of gene expression. This is particularly true of negative results obtained with inhibitors. For example, actinomycin D, at concentrations that completely inhibit seedling growth, does not prevent the increase in the activities of isocitrate lyase and malate synthetase in the cotyledons of *Citrullus vulgaris* (water melon) [196]. Similarly, 6-methyl-purine, at concentrations which abolish RNA synthesis, fails to prevent induction of nitrate reductase in *Chlorella* [223]. It is thus likely that the increases in activities of these enzymes are not dependent on the synthesis of messenger RNA, and control is therefore post-transcriptional. Further, the increases in isocitrate lyase and malate synthetase activity in *Citrullus* and in nitrate reductase activity in *Chlorella* are all inhibited by cycloheximide. In the absence of information on the possible side-effects of cycloheximide in these systems, this result merely suggests, but does not prove, that the increases in enzyme activity are dependent on protein synthesis. In the absence of additional data, dependence on protein synthesis cannot necessarily be interpreted in terms of *de novo* synthesis of enzyme protein. By contrast, the increase in acid ribonuclease activity in cotyledons of *Pisum sativum* is not prevented by cycloheximide, even at concentrations that are highly toxic [61]. This increase in enzyme activity therefore does not depend on the synthesis of protein, and is mediated by a post-translational control mechanism.

Even from this brief discussion, it is clear that obtaining information on the level of control of gene expression in plants is difficult. The number of instances where the level of control of the synthesis of a particular protein has been unequivocally established is very small. Nevertheless, a number of conclusions may be drawn. Firstly, although it is widely assumed that much gene expression is regulated at the transcriptional level, there is very little evidence to support this assumption. Secondly, it is apparent that the control of the expression of a number of genes is at the translational and post-translational levels. However, it is not yet clear whether translational and post-translational control are more important than transcriptional control, as has been suggested by Harris [176]. One problem here is that a negative result obtained with an inhibitor (e.g. failing to prevent an increased enzyme activity with concentrations of cycloheximide that abolish protein synthesis) is easier to interpret than a positive result. A further problem is that it is at present much easier to demonstrate that an increase in enzyme activity depends on *de novo* protein synthesis than it is to demonstrate a requirement for *de novo* synthesis of the messenger RNA concerned. A hypothetical example serves to illustrate these two problems. Let it be supposed that three enzymes, A, B, and C, show dramatic increases in activity at a particular stage of development. The increases in the activities of A and B are inhibited by cycloheximide, whereas the increase in the activity of C is not. The increases in the activities of A and B are shown by density labelling to be accompanied by the synthesis of enzyme protein, in the absence of any change in the rate of enzyme degradation. Density label is not incorporated into enzyme C. The increase in the activity of enzyme A is prevented by actinomycin D, whereas the increases in the activities of enzyme B and C are not. It is clear that the increase in the activity of enzyme C is controlled post-translationally (not inhibited by cycloheximide; no incorporation of density label), whereas enzymes A and B are synthesized *de novo*. It is also clear that the synthesis of enzyme B does not depend on RNA synthesis (actinomycin does not prevent the synthesis of enzyme B). Control of the synthesis of enzyme B is therefore at the translational level. However, no more can be concluded about the synthesis of enzyme A unless two assumptions are made. These are: (a) that actinomycin D specifically inhibits RNA synthesis (this assumption has been challenged by a large number of investigators, and it is clear that in some types of mammalian and avian cells at least, actinomycin D may also cause degradation of RNA and protein, and membrane breakdown: see refs. 176 and 347); and (b) that no change in the rate of degradation of the messenger RNA coding for enzyme A occurs. If these assumptions are made, then the results suggest that synthesis of A depends on *de novo* synthesis of mRNA, and hence that control is probably transcriptional. At present, there is very little information relating to the synthesis and turnover of individual messenger RNA species, or to the availability of individual genes for transcription. Until such information is available, it is extremely difficult to obtain evidence for the transcriptional control of the expression of individual genes. The current emphasis on translational and post-translational control may therefore be a reflection of the availability of techniques rather than of the relative importance of these processes in cell differentiation.

So far, discussion has centred on the regulation of the appearance of particular proteins at particular developmental stages, or in particular cells. Attention must also be given to the regulation of the disappearance of proteins, since this too is an integral part of any developmental sequence (see section II). In a situation in which the messenger RNA coding for an enzyme has a short half-life, and in which the enzyme itself is subject to dilution by cell division, the regulation of the disappearance of that enzyme may be relatively straightforward. Under these conditions, a cessation of transcription of the appropriate gene will lead to a fall in the concentration of the enzyme in each cell. However, plant cells undergoing differentiation do not undergo repeated division cycles. Furthermore, the available evidence indicates that messenger RNA molecules in the cytoplasm of plant cells are long lived. A rapid decline in the activity of a particular enzyme must therefore be achieved by one or more of the following methods:

(a) *Increasing the rate of degradation of the messenger RNA.*

(b) *"Inactivation" of the messenger RNA.* No examples are known of the degradation or "inactivation" of specific messenger RNAs; this is merely a reflection of our lack of knowledge of the metabolism of messenger RNA in eukaryotic cells.

(c) *Increasing the rate of degradation of the enzyme.* The enzyme nitrate reductase, which is inducible by nitrate in a number of plants, turns over even under steady-state conditions in the presence of a constant supply of nitrate [542]. The steady state is thus achieved by a balance of synthesis and degradation. When cells are removed from inducing conditions, the activity of nitrate reductase falls rapidly. This is brought about by an increased rate of degradation [11, 186]. The degradation system itself may involve synthesis and degradation of a specific proteolytic enzyme, since the reduction in nitrate reductase activity under non-inducing conditions is prevented by cycloheximide [480]. Similar reductions in the rate of decline of activity in the presence of inhibitors of protein synthesis have been reported for a number of other enzymes. These results illustrate once again that protein degradation as well as protein synthesis must be taken into account when considering the control of gene expression.

(d) *Inactivation of the enzyme.* The decline in invertase activity that occurs when tubers of *Solanum tuberosum* (potato) are transferred from a low temperature (0–5°C) to a higher temperature (10–15°C) is caused by an increase in the amount of a specific inhibitor of invertase [367]. The inhibitor is also a protein and, like the invertase itself, is subject to turnover [367]. Control of the level of invertase activity in potato tubers could take place completely at the post-translational level, by variations in the degradation rate of either the enzyme or the inhibitor. The existence of specific inhibitors or inactivators has been reported for a number of other enzymes, including phenylalanine-ammonia-lyase in *Cucumis sativus* (gherkin) and deoxyribonuclease in *Chlorella* [144, 405].

IV. FACTORS AFFECTING GENE EXPRESSION

The differentiation of a number of types of cell from apparently identical embryonic cells requires the expression of different genes in the different types of cell. The different cell types occur in specific positions within the plant body, giving rise to an external form and an internal anatomy which is more or less constant for a given species of plant. This suggests that one element involved in the control of gene expression is a spatial element: different cells develop in different ways because they are in different positions in the plant body. This further implies that cells recognize their position with regard to other cells, or that there is available to each cell some form of positional information. For plant cells, which may contain up to four different photoreceptors, the amount of light reaching a cell is an obvious facet of positional information, and light is known to be an important factor in plant morphogenesis. The involvement of light in the control of differentiation illustrates that factors external to the plant may be important elements in the spatial information that is available to the cell. However, internal factors are also important, and it is becoming clear that patterns and/or gradients of concentration of various substances are involved in the differentiation of different cell types from undifferentiated cells. Knowledge of the substances involved in pattern formation is fragmentary, but the limited information available implicates both plant growth substances and simple nutrients such as inorganic ions and sucrose.

Attention has already been drawn to the role of gradients of auxin and sucrose in the control of vascular tissue formation (section IIB). Sucrose may also be involved in the control of plastid development. When callus cells of certain strains of *Daucus carota* (carrot), are exposed to light, the proplastids develop into fully functional chloroplasts and the cells become photosynthetic. If sucrose is included in the medium on which the cells are grown, the development of the plastids is almost completely inhibited [114, 115]. As with the induction of vascular tissue in blocks of callus, this is a specific effect of sucrose. Glucose is not an effective inhibitor of plastid development. In cell lines which are characterized by high activities of acid invertase in the cell wall free space, plastid development is not inhibited by sucrose. The breakdown of sucrose to the constituent monosaccharides therefore removes the inhibitory effect.

The involvement of sucrose in processes as different as phloem differentiation and the regulation of plastid development illustrates that the effect of a particular compound may vary according to the type of cell in which the compound is acting (see also Chapter 7). The effects of compounds such as sucrose and auxin are clearly affected by other factors. One such factor is the cellular environment. Single cells grown in suspension culture do not differentiate [452]. Addition of auxin and sucrose to the growth medium never induces the formation of xylem or phloem cells in a population of single cells, whereas, as mentioned earlier, these substances cause the induction of very natural-looking vascular bundles in blocks of callus tissue. Contact with other cells is therefore a prerequisite for differentiation. This is another indication that cells are sensitive not only to particular concentrations of growth substances and nutrients, but also to gradients of concentration of these substances. A single cell suspended in a growth medium is surrounded by that medium. All the components of the medium are at the same concentrations on every side of the cell. A cell in a plant body, or in a block of callus, in which substances may diffuse or be transported from a single source, is subjected to different concentrations of those substances on different sides. This emphasizes once more the importance of positional information.

Another factor, perhaps associated with cellular interaction, is the previous developmental history of the cell. As stressed in Chapter 7, any one growth substance can have a wide variety of effects, the actual effect being dependent on the type of cell. The same is true of the nutrients, such as sucrose, which affect patterns of differentiation (see above). The effects of growth substances and nutrients on gene expression are thus modified by previous gene expression.

The involvement of cellular interactions and the developmental backgrounds of cells in the regulation of differentiation illustrates the complexity of the process of differentiation. Two examples illustrate this point further. The first example is concerned with the formation of a number of different types of cell within one type of tissue. The hypothesis proposed by Wetmore and developed by Northcote (see section IIB) that the differentiation of vascular tissue may be controlled by gradients of concentration of auxin and sucrose, with the previously differentiated xylem and phloem being sources of these compounds, has received wide support. Indeed, when examined in the context of types of tissue (e.g. xylem or phloem), the hypothesis is supported by a large number of experimental results, including those discussed earlier. However, when examined in terms of individual cells within a tissue, it is obvious that factors other than simple concentration gradients of auxin and sucrose are also involved. For example, phloem contains a number of very different types of cell. The extremes in phloem development are the sieve element and the companion cell. Despite being so different from each other, the sieve element and its attendant companion cell are daughter cells arising from the same cell division. The cell division that gives rise to these cells is an unequal one. The larger cell becomes the sieve element, undergoing the marked specialization of cell contents outlined earlier. The smaller cell becomes the companion cell, retaining the nucleus and all other organelles, and exhibiting a high metabolic activity [127, 529]. This raises a number of questions. For example, do the two cells develop differently simply because of the difference in their sizes? If so, what causes the asymmetric cell division which gives rise to unequally sized cells? It seems likely that there are "local" factors interacting with the overall concentration gradients of sucrose and auxin, perhaps giving rise to complex patterns of concentrations of a number of relatively simple substances. Such local pattern formation must be closely regulated, since the position of the different types of cell within the phloem is extremely important. It is essential for phloem function, for example, that the sieve cell develops exactly in alignment with the previously differentiated sieve cell. The same is of course true of the alignment of vessel members within the xylem.

The second example concerns the formation of the xylem vessel elements. During the differentiation of these cells, complex subcellular patterns develop. Thus, secondary thickening is laid down on some regions of the primary wall, but not on others. The pattern of deposition is controlled by the distribution of endoplasmic reticulum and microtubules adjacent to the plasmalemma (see

I

section IIA). The factors controlling the movement of cell components are unknown. However, it is known that they are under genetic control. A mutation of a single gene in *Lycopersicum esculentum* (tomato) causes a condition known as "wilty-dwarf" where the vessel elements are anomalous: instead of losing their end-walls, they retain them [3]. The end-walls are retained, because, unlike those in normal vessel elements, they become thickened by the deposition of a secondary wall and are then not vulnerable to breakdown by hydrolytic enzymes. The deposition of secondary wall material on the end-walls of these anomalous vessel elements is associated with a failure of the endoplasmic reticulum to become localized adjacent to the plasmalemma in the region of the end-wall. This is therefore an example of a gene which apparently controls the movement of cell organelles, thereby contributing to the formation of a particular type of highly specialized cell. The primary lesion in "wilty-dwarf" mutants is not known, neither is it known whether the gene is a structural gene, coding for an enzyme, or a regulatory gene. Indeed a number of mutations that affect differentiation have been detected in a variety of plants, and in most, if not all instances the primary lesion has not beeen identified. It is clear, however, that the process of differentiation is itself under genetic control.

SUGGESTIONS FOR FURTHER READING

CLOWES, F. A. L. and JUNIPER, B. E. (1968). "Plant Cells". Blackwell Scientific Publications, Oxford.

DAVIES, D. D. and BALLS, M., eds (1971). Control mechanisms of growth and differentiation. *Symp. Soc. exp. Biol.* **25.**

GRAHAM, C. F. and WAREING, P. F., eds (1976). "Textbook of Developmental Biology". Blackwell Scientific Publications, Oxford.

HARRIS, H. (1974). "Nucleus and Cytoplasm" (3rd edition). Clarendon Press, Oxford.

SMITH, H. (1975). "Phytochrome and Photomorphogenesis". McGraw-Hill, Maidenhead, Berks.

7. Plant Growth Substances

A. J. TREWAVAS

I. INTRODUCTION

All organisms containing more than a few cells co-ordinate the activities of the various parts of the whole, either to achieve balanced growth and differentiation, or to maintain a form of *status quo* against a changing environment. The means of co-ordination was first recognized in mammals and indeeed exists in its most complex form in these organisms. The co-ordinating substances are called "hormones", a term coined by their discoverers to refer to substances produced in ductless glands and initiating action at a distance. It is unfortunate that in the past the term "hormone" has also been used by plant physiologists to describe the somewhat ill-defined group of substances which regulate plant growth and differentiation. These substances are not produced in ductless glands, and at least one of them (ethylene) can affect the cells in which it is produced. For these reasons, the terms "plant growth substance" or "plant growth regulator" are now preferred, and the term "plant hormone" is going out of general usage.

Animal hormones and plant growth substances have one feature in common, that of co-ordination, but there is very little further similarity. The major steps in mammalian differentiation occur during embryogenesis. Subsequent growth occurs by an increase in the number of cells in each of the differentiated parts. The chemicals regulating cellular differentiation and organogenesis in the developing mammalian embryo have still not been identified. The majority of known animal hormones are concerned with the maintenance of the adult organism and its adaptation to a changing environment. This applies even to the sex hormones, which, although they are concerned with the differentiation of secondary sexual characteristics, are also concerned with the maintenance of some of these characteristics in the adult mammal.

In contrast to most animals, plants grow by enlargement and differentiation of cells produced by cell division in embryonic areas (meristems) which are maintained even in the adult form (see Chapters 5 and 6). Plant growth substances are concerned with regulating both the cell division in these embryonic areas and the subsequent differentiation of the new cells, rather than with the control of function in the true adult form. It is thus clear that plant growth substances and mammalian hormones satisfy totally different requirements in their respective organisms. One therefore would not expect plant growth substances and mammalian hormones necessarily to have molecular modes of action that were in any way similar. Indeed, it is likely that the whole pattern of regulation in plants is entirely different from that in mammals.

One feature clearly related to general regulatory mechanisms is the extraordinary plasticity exhibited by plant organs and tissues. This contrasts with the much more limited plasticity of mammalian organs and tissues, which is mainly confined to phenomena such as wound healing. Plant meristems in particular exhibit many facets of this plasticity. Meristems may be rendered inactive (dormant) for many years as a result of treatment with growth substances or by severe dehydration. They may be re-activated by supplying water or by removing the inhibitory agent. Shoot meristems may be mechanically divided, causing the formation of autonomous daughter meristems. Meristems can be generated from relatively mature, differentiated cells, as is seen in the rooting of cuttings. Thus, even when a mature organ has been produced, it retains the capacity to produce another organ, or even a whole plant, vegetatively. In mosses, the reverse has been observed: buds on protonemal filaments can degenerate if the inducing substance is removed.

The lability of plant tissues is also shown very strikingly by the phenomenon of callus formation. Virtually every mature plant tissue is capable of this undifferentiated and apparently disorganized proliferation, which generally occurs in response to wounding (see Chapter 5). Callus tissue may be grown in culture, and the ordered form or polarity of the plant, lost in callus formation, can be restored by applying a directional source of growth substances in an appropriate balance. Thus, root and shoot meristems, or whole plant organs, or even whole plants, can be generated from callus tissue. It is thus likely that polarity, like other aspects of plant development, is malleable; this view is supported by observations on regeneration in some root and shoot cuttings in which the polarity may be reversed by simply turning the cutting upside-down [325, 430]. The potential of callus, derived from mature, differentiated plant tissue, for regeneration of whole plants suggests that the mature plant tissues contain cells which are totipotent. This suggestion is borne out by the findings that whole plants can be regenerated from single cells in culture (see Chapter 6).

The plant's response to growth substance is a further example of plasticity, since one growth substance may regulate many different phases of plant growth and differentiation. For example, gibberellins have been implicated in the regulation of enzyme secretion from the aleurone in cereal grains, of embryo growth, of shoot and internode length, of leaf development, of tropic responses and of flowering. Growth substances also exhibit an extraordinarily large range of concentration over which they act. Concentrations as high as 10^3–10^5 times the minimum effective dose may still elicit the same response (as with auxin control of cell extension in stem tissue or gibberellin control of α-amylase secretion). Animal hormones, by contrast, work over a much narrower range of concentrations (10–20-fold, as opposed to 10^3–10^5-fold). The very wide range of doses to which plants respond suggests that they do not regulate their levels of growth substances as accurately as animals regulate their hormone levels. Plants must therefore be able to tolerate a certain level of "error".

The phenomena described above show clearly that in comparison with mammals, plants exhibit a low degree of tissue specialization combined with a high degree of plasticity. A vascular plant has about one-fifth the number of cell types possessed by a mammal, and the high degree of intracellular cytoplasmic differentiation which is a feature of, for example, muscle, nerve, gland or digestive cells, is almost totally lacking in plants (but see the discussion concerning phloem sieve cells in Chapter 6). Regeneration (with the exception of wound healing) and totipotency are common properties of only the most primitive multicellular animals, whilst dormancy by dehydration occurs widely only in protozoans. These phenomena are relatively widespread in plants, as has been mentioned above. It remains one of the most exciting of challenges to discover and understand the molecular basis of this plasticity of plant cells and tissues. Growth substances, concerned as they are with cellular differentiation and organization, emerge as one of the basic keys with which our understanding will be unlocked.

II. REGULATION OF ENZYME ACTIVITIES BY GROWTH SUBSTANCES

It is a tacit assumption that the regulation of plant growth and differentiation has its origin partly in changes in the pattern of enzyme activities. Even if this is admitted to be true, it does not greatly advance our understanding, since the questions which are then raised reveal the extensive ignorance that prevails. For example, is the whole pattern of enzymes changed, or just a related group, or only one? If many enzymes change, are all the changes of equal significance, or is only one enzymic event crucial to the physiological response? Is there a required temporal order in the enzyme changes, or may the interlinking of metabolic pathways account for apparent sequences of events? Which enzymes are necessary for the production of a xylem or phloem cell? The information summarized in Table 7.I is an attempt to provide tentative answers in a few specific cases to some of these questions. The table shows some of the enzyme changes which have been measured; the selection has been limited to those to which direct physiological significance can be attributed. Some of these enzyme changes are discussed in detail later in the chapter. Meanwhile, two general points can be made from the information presented in the table. Firstly, every growth substance can apparently, under appropriate conditions,

TABLE 7.I. Examples of enzymes which have been shown to respond to growth substance treatment. Increase in activity = +. Decrease in activity = −. * denotes effect on activity inferred from chemical measurements on RNA and DNA levels.

Growth substance	Enzyme	Plant, tissue or organ	Increase or decrease	Reference
Auxin	ATPase (in Plas-malemma)	Oat coleoptile	+	172
	Glucan synthetase	Oat coleoptile, pea stem	+	375
	Cellulase	Pea stem	+	84
	RNA Polymerase	Soya bean hypocotyl	+	339
	DNA Polymerase	Soya bean hypocotyl	+	269
Gibberellin	α-Amylase	Barley aleurone	+	537
	Protease	Barley aleurone	+	537
	β1–3 Glucanase	Barley aleurone	+	537
	RNase	Barley aleurone	+	537
	Phytase	Barley aleurone	+	537
	Phosphoryl choline–cytidyl transferase	Barley aleurone	+	537
	Phosphoryl choline–glyceride transferase	Barley aleurone	+	537
	RNA polymerase	Pea stem	+	225
	Invertase	Artichoke tuber discs	+	264
Ethylene	Cellulase	Leaf abscission zone and fruit	+	1
	Amylase	Various fruits	+	1
	Malic decarboxy-lase	Various fruits	+	1
	Pectinase	Various fruits	+	1
	Chlorophyllase	Various fruits	+	1
	Phenylalanine-ammonia-lyase	Fruit peel (various)	+	1
	RNA polymerase	Soya bean hypocotyl	+	1
	RNase	*Ipomoea* petal	+	1
	DNA polymerase	Pea seedlings	−	10
Abscisic acid	RNA polymerase	Radish hypocotyl	−	308
	DNA polymerase *	*Lemna minor*	−	449
	Protease	Cotton cotyledon	−	207
	Isocitratase	Cotton cotyledon	−	207
Cytokinins	Nitrate reductase	*Agrostemma* embryo	+	237
	Protease	Various leaves	−	237
	RNase	Various leaves	−	237, 444
	RNA polymerase *	Various calluses	+	237
	DNA polymerase *	Various calluses	+	237

regulate the synthesis of RNA (and probably also the synthesis of DNA, and in turn cell division, although effects on DNA polymerase have only been demonstrated for two growth substances). Secondly, every growth substance can regulate the activity of one or more hydrolytic enzymes. Of these, an interesting group are the cell wall polysaccharidases, since these have to pass through the plasmalemma in order to reach their substrates. The possible role of growth substances in the process of secretion is put forward later in the chapter as an important simplifying hypothesis around which to group a unified theory of molecular action (section IX).

Determination of which enzymes change in activity after treatment with growth substances is relatively straightforward. Determining how their activity is regulated is a more difficult and, to many investigators, a more interesting problem. Three general levels of control may be recognized: transcriptional, translational and post-translational. The criteria for distinguishing between these levels of control are discussed in detail in Chapter 6. The following sections in this chapter show how these criteria may be applied to enzyme changes induced by growth substances.

III. DETECTION OF SELECTIVE ALTERATIONS IN THE SYNTHESIS OF MACROMOLECULES AND THE ROLE OF DEGRADATION

It is a frequent finding that after treatment of plant tissue with a growth substance, the activities of different enzymes change by different degrees. An excellent example of this is seen in the effects of high concentrations of auxin on etiolated pea (*Pisum sativum*) stem tissue, as reported by Datko and McLachlan [99]. Three days after auxin treatment, the measured activity of cellulase had increased 30-fold, pectinase 7-fold and β 1–3 glucanase 3–4-fold. The total protein content of the tissue also increased by 3–4-fold. Two possible interpretations of these data may be considered. Firstly, it can be supposed that, following auxin treatment, very specific alterations occur in the synthesis of cellulase and perhaps of pectinase, but not in the synthesis of β 1–3 glucanase (since the increase in activity of this latter enzyme merely paralleled the increase in protein content). Alternatively, it can be supposed that the synthesis of all three enzymes is increased to the same extent, and that differential effects which are observed originate from the enzymes having different rates of degradation.

In general, the level of any enzyme in a cell is the result of a balance between synthesis and degradation. In the steady state, the rate of synthesis equals the rate of degradation. When the steady state is upset by, for example, an increase in the rate of synthesis of an enzyme, then the variation of the enzyme level with time is described by the simple equation:

$$\frac{\mathrm{d}E}{\mathrm{d}t} = V_s - V_D \qquad\qquad \text{(i)}$$

where V_s = rate of synthesis, V_D = rate of degradation and E = amount of enzyme. In mammals, and probably also in plants, protein synthesis obeys zero order kinetics (i.e. is independent of substrate concentration) while protein degradation obeys first order kinetics (i.e. is dependent on the concentration of a single substrate) [30]. Thus, eqn (i) can be re-expressed as

$$\frac{\mathrm{d}E}{\mathrm{d}t} = K_s - K_D. \qquad\qquad \text{(ii)}$$

where K_s = rate constant of synthesis and K_D = rate constant of degradation. In the steady state, $\mathrm{d}E/\mathrm{d}t = 0$, and thus $K_s = K_D.E$.

Now consider the situation where the rate of synthesis of the enzyme is increased to a new and higher constant, K_s^1. The variation of E with time can be found by integrating eqn (ii), as described by Berlin and Schimke [30]:

$$\frac{E}{E_0} = \frac{K_s^1}{K_s} - \left(\frac{K_s^1}{K_s} - 1\right)e^{-K_D t} \qquad\qquad \text{(iii)}$$

Where E_0 is the amount of enzyme at $t=0$. The equation shows that eventually the level of enzyme reaches a new steady state. At this time

$$\left(\frac{K_s^1}{K_s} - 1\right)e^{-K_D t}$$

becomes very small, and thus the magnitude of the increase in enzyme activity (expressed as a multiple of the "basal" activity) becomes equal to the ratio of the "new" rate of synthesis to the "old" rate of synthesis of the enzyme. The shape of the curve of E/E_0 against time is determined by $e^{-K_D t}$ which is, of course, related to the degradation of the enzyme and not to its synthesis. The importance of this is illustrated by Fig. 7.1a. In this figure, E/E_0 has been plotted using eqn (iii) for three imaginary enzymes with half-lives of 0·5 days, 9 days and 18 days. (Half-life, $t_{\frac{1}{2}}=\ln 2/K_D$.) For each enzyme it has been assumed that the rate of synthesis has increased 30-fold, i.e. $K_s/K_s=30$. It can be seen that after 3 days, the enzyme with a half-life of 0·5 days has increased 30-fold in activity, the enzyme with a half-life of 9 days has increased 7-fold in activity, and the enzyme with a half-life of 18 days has increased 3·5-fold in activity. The figures agree very well with the observed auxin-induced increases in the activities of cellulase, pectinase and $\beta1-3$ glucanase quoted earlier! Thus, apparently selective increases in enzyme activities could be brought about by a non-selective and general increase in the rate of synthesis of many enzymes, the

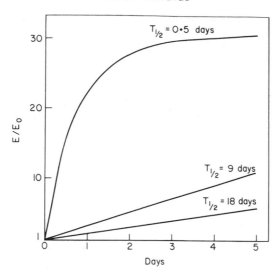

FIG. 7.1 (a). A plot of E/E_0 (i.e. factor by which enzyme activity increases) against time for three enzymes with half-lives of $0·5$ days, 9 days and 18 days. In each case, it has been assumed that the rate of synthesis of each enzyme has been increased 30-fold at time zero ($K'_s/K_s = 30$) and the values of E/E_0 were calculated as described in the text.

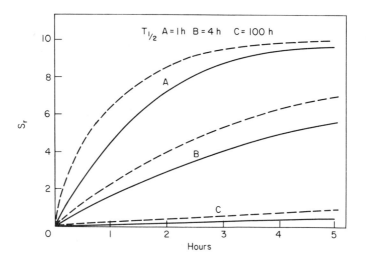

FIG. 7.1 (b). A plot of the specific activity (S_r) against time for three RNA species with half-lives of 100 h, 4 h, 1 h. S_r has been calculated as described in the text. The precursor specific activity (S_p) has been given an arbitrary value of 10. Continuous lines show the progress curves for the specific activities of the three RNA species in control tissue and the dotted lines show the progress curves if the rate of synthesis of each RNA species has been increased by 50% at zero time.

differences arising from the different rates of degradation of the different enzymes. It must also be noted that given sufficient time in the presence of the inducing stimulus, the enzymes with half-lives of 9 and 18 days would reach the same plateau levels (i.e. $E/E_0=30$) as the enzyme with a half-life of 0.5 days, but they would take 50–100 days to do this. It is extremely unlikely that any plant tissue would be exposed to an inducing stimulus for so long, since most growth substances are degraded or transported away from the site of action by the plant. The effects of growth substances on the activities of enzymes with low rates of degradation may therefore never be fully expressed. The little information which is available concerning degradation rates of enzymes in plants suggests that in plants, as in mammals, different enzymes have different half-life times. On the basis of the turnover hypothesis, it is thus possible that selective changes in enzyme activities may be initiated by a relatively non-selective stimulus. In the context of this hypothesis, it should be noted that Datko and McLachlan concluded that cellulase (which, in their experiments, showed the most dramatic increase in activity) exhibits a more rapid degradation rate than other cellular proteins [99].

It can be objected, of course, that this model is much too simple, and that assumptions have been made about unknown turnover rates of enzymes and about apparently instantaneous changes in rates of synthesis. These objections have some validity, but they do not detract from the obvious importance of degradation in causing selective alterations in enzyme activity, nor from the necessity to determine rates of synthesis and degradation of enzymes during events such as treatment with growth substances.

Degradation of macromolecules must also be taken into account in interpretation of data from experiments in which radioisotopes are used. This is well illustrated by an example taken from the work of Key and Ingle [241]. In soya bean (*Glycine max*) hypocotyls, nucleic acids were labelled by supplying the plant tissue with ^{32}P-orthophosphate for 2 h. One sample was treated with the synthetic auxin, 2, 4-dichlorophenoxy-acetic acid (2, 4-D), at a growth-promoting concentration, while the other sample served as a control. After the 2 h incubation the nucleic acids were extracted and then separated by chromatography on columns of methylated albumin-kieselguhr (MAK). The increases in specific activity caused by auxin treatment were measured. It was found that 2, 4-D caused the specific activity of ribosomal RNA to be increased by 47%, of "D–RNA" (DNA-like RNA) by 38%, and of "TB–RNA" (RNA which was *tightly bound* to the MAK column) by 24%. The obvious interpretation of these data is that 2, 4-D causes a selective increase in the synthesis of rRNA as compared with D–RNA and TB–RNA. Further experiments seemed to confirm this interpretation, showing that the differential between the specific activities of rRNA and D–RNA became greater with longer incubation times. However, an alternative interpretation is also possible:

that the incorporation of labelled precursors into the three species of RNA is increased to the same extent by auxin treatment, the apparent differences in the size of the increase arising from different turnover rates.

The stabilities of rRNA, D–RNA and TB–RNA are very different. The half-life of rRNA is of the order of 4 days, that of D–RNA 4 h, and that of TB–RNA 1 h or less. The metabolically active pool of inorganic phosphate is small and has a very high turnover rate. Provided the uptake of ^{32}P-orthophosphate is constant, as is apparently the case for stem tissue, the metabolic pool of inorganic phosphate reaches a constant specific activity in very much less than 1 h. The specific activity of RNA molecules drawing their phosphate from this pool of constant specific activity is described by the equation

$$S_R = S_P \ \ 1 - e^{-K_s t} \tag{iv}$$

where S_R = specific activity of RNA, S_P = specific activity of orthophosphate and K_s = turnover constant of synthesis. This constant is defined as V_s/R, where V_s = rate of synthesis and R = amount of nucleic acid (see reference 423 for the derivation).

In Fig. 7.1b curves have been drawn using eqn (iv) for the variations with time of the specific activities of three macromolecules, A, B and C, with half-lives of 1 h, 4 h and 100 h. The dotted lines show the change in specific activities with time if the rate of synthesis of each macromolecule is increased by 50%. It is instructive to examine the increase in the specific activity of the three macromolecules after 2 h at the increased rate of synthesis. The increase in the specific activity of A is 17%, of B 38% and of C 50%. These figures are somewhat similar to those quoted earlier: TB–RNA, 24%; D–RNA, 38%; rRNA, 47%. Similar agreement is seen when longer incubation times are considered.

The model proposed here is easily criticized mainly on the basis of its simplicity. Possible lag periods in the effects of auxin have been ignored. It has been assumed that the specific activity of the precursor phosphate becomes constant in an infinitely short period of time and then remains constant. Inorganic phosphate has been taken as the actual precursor. Possible effects of auxin on the uptake of ^{32}P-orthophosphate, and on the size of the inorganic phosphate pool have been ignored, and the assumption has been made that RNA levels in the control tissue are in a steady state. However, none of these criticisms detracts from the obvious importance of taking the turnover rates of macromolecules into account when claiming selective enhancement of the synthesis of one type of molecule or another. Figures 7.1a and b also raise three other important points. Firstly, they illustrate the necessity for a kinetic approach to alterations in the catalytic activities or labelling of macromolecules. Secondly, they point to a largely unknown area of control at the level of degradation. The level of any macromolecule can be increased by an enhanced rate of synthesis or a reduced rate of degradation. In both cases, the increase requires synthesis. Thirdly, in a system of macromolecules of varying stability, any effect, general or specific, on the rate of synthesis will initiate specific alterations in the levels of the macromolecules.

The next five sections are devoted to some specific molecular aspects of the physiological activities of auxins, gibberellins, abscisic acid and ethylene. The topics are arranged according to the level at which it is believed the particular growth substance operates. Section IV deals with post-translational control, sections V and VI with translational or post-transcriptional control and sections VII and VIII with transcriptional control.

IV. AUXIN AND CELL EXTENSION

A. Introduction

In the "classical" experimental materials, the coleoptile and the etiolated stem, endogenous auxins probably regulate the rate of cell extension. In other organs which exhibit cell extension, such as non-etiolated stems, leaves or fruits, growth may also be regulated by gibberellins and cytokinins. It is therefore clear that in dealing with auxin and extension growth, we are dealing with only one aspect of a wide-ranging topic. Reference to the oat coleoptile places the role of auxin firmly in perspective. This organ, which can extend to 40–50 mm before senescence intervenes, acquires its maximum sensitivity to auxin when it has reached about one-third of its final length. The effect of auxin is to make what are already elongated cells somewhat more elongated, and then to accelerate the rate of senescence and the approach of death.

B. The lag period

A number of very sensitive optical and electronic methods have been developed in recent years which have enabled plant physiologists to measure the extension of whole coleoptiles, or even single cells in coleoptiles, over extremely short periods of time. A typical time course for extension growth after the addition of indol-3-yl acetic acid (IAA) is shown in Fig. 7.2. There is a lag period of 10–15 min after IAA addition before increased cell extension can be detected. From a biochemical point of view, the molecular events occurring in this lag period may be regarded as precedents of the growth response. The properties of the lag period give insights into the molecular control of cell extension.

The lag period has the following characteristics:

(i) It is unaffected by changes in auxin concentration within the range 10^{-5}–10^{-3}M and is only marginally lengthened at 10^{-7}M IAA.

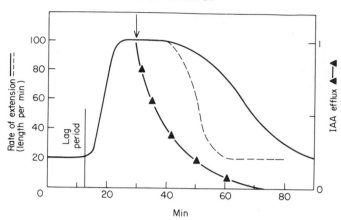

FIG. 7.2. Rate of extension of corn coleoptile sections against time. Batches of corn coleoptile sections were treated at time 0 with 10^{-6}M IAA. The lengths of the sections were measured continuously and rates of extension (length/unit time) determined. At time 30 min either the IAA was removed (—————) or cycloheximide added together with IAA (- - - - -). In a separate experiment, coleoptiles were labelled with radioactive IAA and then transferred to non-radioactive medium at time 30 min. The labelled IAA remaining in the coleoptiles was then determined (▲ ▲ ▲ ▲). (Data calculated from ref. 130).

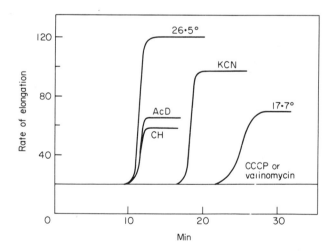

FIG. 7.3. The effects of temperature and various inhibitors on the lag period of auxin-induced extension growth of oat coleoptiles. At time 0, IAA was added to batches of oat coleoptile sections together with actinomycin D (Ac.D), cycloheximide (CH), KCN, valinomycin (VMN), or CCCP. One batch of sections was incubated with IAA alone at 17·7°C, whilst all the others were incubated at 26·5°C. Lengths of sections were measured continuously and rates of extension (length/unit time) calculated. (Data calculated from ref. 131.)

(ii) Lowering the incubation temperature by half results in an approximate doubling of the length of the lag period (Fig. 7.3). The lag period thus has a Q_{10} of about 2, suggesting that it probably has a chemical basis.

(iii) The lag period is lengthened by a prior treatment with potassium cyanide (Fig. 7.3).

(iv) The lag period is not affected by a 60 min pretreatment with actinomycin D, although the final rate of elongation is reduced to 50% of the control rate (Fig. 7.3). On the other hand, the incorporation of labelled orotic acid into RNA is inhibited by more than 90% under the same conditions.

(v) The lag period is unaffected by a 10 min pretreatment with cycloheximide (Fig. 7.3). Amino acid incorporation into protein is inhibited by over 90% in 3–5 min under these conditions. However, cycloheximide does have at least one effect on extension. As Fig. 7.2 shows, if cycloheximide is added after the end of the lag period, the auxin-induced rate of extension continues for about 10 min, and then over the next 20 min falls to the original control (i.e. no auxin) level. Continued protein synthesis may therefore be necessary for continued extension. The lag period is also unaffected by a 3 h pretreatment with puromycin (another inhibitor of protein synthesis), although the final rate of extension is only one-third of that induced by auxin in the absence of puromycin.

(vi) The lag period can be considerably shortened by use of the methyl ester of IAA, rather than IAA itself. Since the methyl ester is more readily taken up than unmodified IAA, this suggests that one component of the lag period may be the time taken for uptake of IAA into the cells of the coleoptiles. Since IAA transport is an active process, this would partly explain the relationship between the length of the lag period and temperature.

Further information may be gained from study of the kinetics of cell extension after removal of IAA from the incubation medium. This is shown in Fig. 7.2, together with a curve showing the efflux of IAA from the coleoptiles following the removal of the auxin from the medium. Increased extension does not continue for very long after the loss of IAA from the coleoptile sections. It thus appears that IAA does not promote any permanent change in metabolism in the promotion of extension growth.

What then may be inferred from these data concerning the nature of the lag period? Part of the lag almost certainly represents the time taken for auxin to penetrate the coleoptile. When allowance is made for this uptake phase, only about 5 min are available for the molecular events which result in increased extension growth. It can be convincingly argued on kinetic grounds that this time is too short for the synthesis of new RNA species or new proteins [131]. For example, the lag period in induced enzyme synthesis in bacteria growing at 37°C is about 4 min. On this basis, the time taken for synthesis of a new enzyme in plant cells maintained at 25°C would be at least 8–10 min. The data concerning IAA efflux (Fig. 7.2) also support the view that no new enzymes are formed during the lag period: if IAA causes the synthesis of an enzyme controlling extension then the decline in growth rate following the removal of IAA implies that the enzyme has a half-life of less than 5 min. This rate of degradation is 20 times faster than any yet known for plant enzymes. The apparent inability of actinomycin D and cycloheximide to affect the length of the lag period also argues against RNA and protein synthesis being involved. However, the effects of both these inhibitors on the subsequent rate of cell extension indicate that protein and RNA synthesis may be necessary at a later stage.

Thus, if it is supposed that auxin functions by altering enzyme activity, then consideration of the lag period must lead to the assumption that the alteration in activity is mediated by a post-translational control mechanism. There is no concrete evidence concerning the nature of this mechanism, but the effect of KCN (see Fig. 7.3) strongly suggests that the lag period is dependent on the continued operation of oxidative metabolism.

C. Cell wall plasticity

In the classical picture of the vacuolated plant cell, a form of equilibrium is envisaged between the osmotic potential of the cell contents and the equal and opposite wall pressure. In order to achieve cell extension, the osmotic potential must be increased or the wall pressure decreased, or both. The rather limited evidence currently available suggests that osmotic potential is not directly affected by auxin treatment. There is, however, a good deal of evidence, from the experiments of Heyn in 1931 [192] and from many subsequent investigations, that auxin reduces wall pressure by softening the cell wall.

This can be shown in the rather simple type of experiment described diagrammatically in Fig. 7.4. In this experiment, coleoptile sections are incubated for half an hour or less in water or in IAA. The sections are then stretched with a constant force applied at one end until the situation shown in B is reached. When the sections are released from this stretching force the situation in C occurs. From Fig. 7.4 it is obvious that the coleoptile sections have been irreversibly deformed by this treatment and that the deformation is greater in the section treated with IAA. If the sections in C are again stretched to reach B and then released, they revert back to the lengths shown in C. The difference in length between the stretched coleoptile section and the released one is a measure of the *elasticity* of such sections. Comparison of B and C shows that this difference is the same in control and auxin-treated sections. The difference in length between A and C is a measure of the *plasticity* or irreversible deformation of the sections. This is increased by auxin at concentrations which promote growth.

Kinetic analysis of the effects of auxin on the plastic deformation of coleoptiles indicates that the effect may commence within the lag period of auxin action. Further increased plasticity is almost certainly a result of an alteration of the physical nature of the cell wall itself, since sections which have been killed by boiling may be stretched and deformed in the same way as the control sections (i.e. no auxin) in the experiment described above.

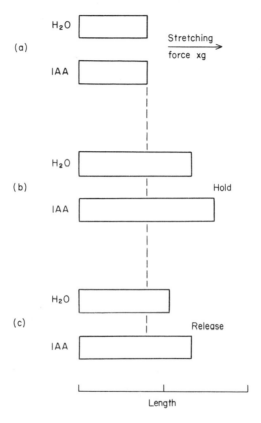

FIG. 7.4. The effect of pre-incubation in IAA solution on the irreversible deformation of oat coleoptile sections. Coleoptile sections were incubated for 0·5 h in water or IAA (A). They were then stretched by applying a standard weight to one end and the lengths measured (B). After releasing the weight, the length of the sections was measured again (C). (Data based on ref. 84.)

D. The molecular basis of cell wall plasticity

When observed with the electron microscope, the plant cell wall is seen to consist of bundles of crystalline or semicrystalline cellulose fibrils embedded in an amorphous matrix. During extension growth, there is some re-orientation of the cellulose fibrils, analogous to the re-orientation of the coils of a spring during stretching. Further, the various layers of fibrils must slip one over the other for extension to occur. It is thus clear that the physical nature of the amorphous matrix, and the extent to which it enmeshes the cellulose fibrils, may be important in determining the slippage and reorientation of the fibrils. The component polysaccharides of the amorphous matrix (i.e. the pectins and hemicelluloses: see Chapter 6 for detailed discussion of these components) are probably present as a gel. It is a property of gels (e.g. agar) that they can maintain a three-dimensional structure without external support. This is merely an outward expression of an internal molecular order amongst the component polymer chains which have joined up to form a network. Two forms of interaction in the cell wall matrix give rise to this internal structure. Firstly, as has recently been demonstrated by Albersheim and his colleagues [234], most of the components of the cell wall are covalently linked (see Fig. 6.3). Xyloglucans are joined to pectic polysaccharides which in turn are linked to the hydroxyproline-rich structural protein ("extensin"). The matrix material is linked to the cellulose fibrils via hydrogen-bonding between the xyloglucans and the cellulose.

The second type of interaction is much weaker, and is best illustrated by studies made on pectin gels. Pectins from a number of plants may be solubilized by boiling in water. On cooling, the extracted pectin sets to a gel which can be melted again by slight warming. The linkages holding the polymer chains together are clearly too weak to be covalent linkages. In order to obtain setting of pectin gels it is often necessary to add calcium ions in order to neutralize the free carboxyl groups, or to add large quantities of sucrose or glycerol in order to reduce the water content. Lowering the water content increases the contacts between the polymer chains, leading not only to intra-chain contraction, but also to clustering of chains, providing the necessary cross-linking to form the gel.

Chemical modification of pectins can inhibit gel formation. Natural pectins frequently have many of their carboxyl groups methylated, but conversion of the methyl group to an ethyl, hydroxy-ethyl or larger group will prevent gel formation. The profound effects of such minor chemical modifications suggests that an intimate fit between adjacent chains is necessary to form the clusters. Thus the extent of chemical modification of the matrix polymer could markedly affect the slippage and re-orientation of the cellulose fibrils. If the cellulose fibrils are trapped in a gel-like matrix network, their ability to re-orientate during extension growth will be hindered. If the degree of cross-linking in the matrix gel is reduced and the gel structure is broken, the resulting viscous liquid will lubricate and thus promote re-orientation of the fibrils.

There are thus a number of ways in which the slippage and re-orientation of the cellulose fibrils may be regulated. Firstly, the hydrogen bonds between the cellulose and the xyloglucan could be broken, releasing the fibrils from linkage with the matrix material. Secondly, the covalent linkages between some of the matrix components could be enzymatically hydrolysed. Recent work has shown that coleoptile cell wall glycosidases have enhanced activity after auxin treatment. There is also evidence that part of the xyloglucan component may be actually released from the cell wall as extension growth commences. Thirdly, the addition of bulky side groups on the polymer chains could interfere with gel formation in the matrix material. One such side group, a branched poly-arabinan, occurs naturally in pectin. Plant cells, tissues and organs with a high potential for growth, such as pollen, seeds and young roots, have an unusually high proportion of arabinose side-chains in the pectic fraction of their cell walls. As a plant cell ages and extension growth diminishes, the proportion of arabinose side chains also diminishes. In this context, it is interesting that the synthetic auxin, 2, 4-D, has been found to promote the incorporation of arabinose into neutral pectins in the cell walls of cultured plant tissue [396]. Another possibility is that small arabinan side groups may be moved from one part of the wall (from "extensin" for example) to another by transglycosylation. Many glycosidases also show glycosyl transferase activity and, as has already been mentioned, auxin increases the activity of these cell wall enzymes.

It is clear that regulation of cell wall hydrolytic enzymes by auxin may be central to understanding the regulation of plasticity and hence of extension. One possible way in which auxin may regulate this activity may now be considered.

E. Acid-induced cell extension and the control of hydrogen ion secretion

The first observations of the effects of low pH on extension growth were made in the 1930s, when it was found that coleoptile sections incubated at pH 3–4 showed a markedly higher rate of extension than coleoptiles incubated at pH 7. This phenomenon has recently been re-examined [378]. The rate of extension of a coleoptile incubated at pH 3–4 in the absence of auxin is only about one-third of that which can be observed in coleoptiles incubated in auxin at optimal concentration and at physiological pH. The cuticle acts as a barrier to the penetration of hydrogen ions into the coleoptile. If the cuticle is removed, then rates of extension comparable to those achieved with auxin at neutral or near neutral pH can be achieved by incubating the coleoptile at pH 5 in the absence of auxin. Acid-induced extension growth is accompanied by the "normal" increase in cell wall plasticity, but there is no lag period. Coleoptiles which have been frozen and thawed in order to break the intracellular membranes still exhibit increased wall plasticity when incubated at acid pH. The frozen and thawed coleoptiles do not exhibit turgor pressure, but if the lost turgor pressure is replaced by a stretching force applied at the ends of the coleoptile, then acid causes an increased rate of extension. The frozen and thawed coleoptiles are unable to synthesize proteins and cell wall components. Neither of these two events can therefore be essential for the regulation of plastic extension.

The following hypothesis has been proposed to explain and develop some of the observations discussed above. It is supposed that coleoptile cells have an anisotropically arranged proton pump located in the plasmalemma. The presence of such pumps in chloroplast and mitochondrial membranes is well

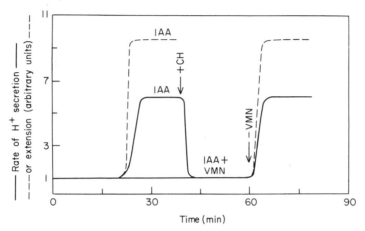

Time (min)

FIG. 7.5. The effects of IAA and various inhibitors on the rate of extension and H⁺ ion secretion from peeled oat coleoptile sections. At zero time, IAA or IAA + valinomycin (VMN) was added to batches of peeled oat coleoptile sections and the length and H⁺ ion secretion (measured as a decline in pH) determined against time and rates of H ion secretion (pH unit/unit time) calculated. At 40 min. cycloheximide (CH) was added to the coleoptile sections incubated just in IAA and the pH measured for a further 15 min. At 60 min. the coleoptiles in IAA and valinomycin (VMN) were transferred to a solution containing IAA alone and the rate of extension and pH decline determined for a further 30 min. (Data calculated from ref. 376.)

established. Auxin combines, perhaps allosterically, with the protein component of the pump, thus activating it. Provided a supply of ATP is available as a source of energy, protons are pumped from the cytoplasm into the wall. Here, cell wall plasticity is increased by one or both of the following processes: acid-labile linkages (hydrogen bonds and/or covalent linkages) may be broken, or a hydrolytic enzyme with an acidic pH optimum may have its catalytic activity increased.

There is in fact a good deal of evidence to support this hypothesis. The allosteric interaction of auxin with a protein is consistent with the shortness of the lag period, with the apparent immunity of the events of the lag period to actinomycin D and cycloheximide, and with the decline in the rate of extension following the efflux of auxin. The requirement for continued oxidative metabolism and hence ATP synthesis during extension growth has been known for many years. Indeed, one of the effects of auxin on coleoptiles is to increase the rate of oxygen uptake. Respiratory inhibitors and anaerobiosis inhibit extension as well as lengthening the lag period. Auxin-induced growth is also severely limited by treatment of coleoptiles with CCCP (carbonylcyanide-M-chlorophenylhydrazone), a respiratory inhibitor which appears to make membranes more permeable to protons. It causes inhibition of extension growth at concentrations that have no detectable effects on respiration, and it may do this by impairing the secretion of hydrogen ions into the wall compartment. Finally, in the presence of auxin, peeled coleoptile sections secrete protons into the incubation medium, as shown by a lowering of the pH of the medium (see Fig. 7.5). The increased hydrogen ion secretion appears to commence at about the same time as the increased extension growth, and provided the presence of auxin is maintained, may continue for several hours, causing the pH of the incubation medium to fall by a unit or more. Both 2, 4-D and IAA enhance proton secretion, but 3, 5-D, an inactive auxin analogue, does not. Abscisic acid, cycloheximide, 2, 4-dinitrophenol, 2, 3, 5-tri-iodobenzoic acid, CCCP and valinomycin all inhibit auxin-induced hydrogen ion secretion from coleoptiles at the same concentrations at which they inhibit extension growth [376]. The effects of valinomycin are illustrated in Fig. 7.5.

There are other observations, however, which must be considered in any critical assessment of this hypothesis. The *initial* rate of extension of a peeled coleoptile at pH 5 in the absence of auxin is comparable to the maximum rate achieved in the presence of auxin at neutral pH, but in acid-incubated coleoptiles, these high rates are not maintained for very long. Thus, after 7 h incubation coleoptiles incubated in auxin are 40% longer than coleoptiles incubated at pH 5 in the absence of auxin. Auxin is clearly necessary for the maintenance of extension growth. A still greater difficulty with the hypothesis lies in the exact timing of hydrogen ion secretion in relation to extension growth. Technical difficulties prevent an exact measurement of the timing of initiation of hydrogen ion secretion after the addition of auxin. In Fig. 7.5, hydrogen ion secretion is shown as commencing at the same time as increased extension growth, but on the basis of the available data, hydrogen ion transport could just as well be stimulated a few minutes later than extension growth. The timing of these events is of course critical if hydrogen ion secretion is proposed as the initial auxin-induced effect. In fact this point has become more critical, since Raven and Smith have indicated that growth in plants is accompanied by a production of an excess of hydrogen ions over hydroxyl ions [373]. These excess ions must be removed either by secretion out of the cell or into the vacuole. It is thus possible that the effect of auxin on hydrogen ion secretion is a consequence of the increased growth rate and not a cause of it.

A further difficulty arises from the effects of cycloheximide on hydrogen ion secretion. In section IVB, it was shown that cycloheximide has no effect on the length of the lag period. However, as is shown clearly in Fig. 7.5, cycloheximide does inhibit hydrogen ion secretion. Whatever the molecular basis of this inhibition, it is an obvious interpretation of these data that the lag period does not involve hydrogen ion secretion. This would support the view that increased hydrogen ion secretion merely accompanies increased extension growth, rather than precedes it. It is to be hoped that further work will help to resolve this problem.

F. Regulation of glucan synthetase activity by auxin

Addition of auxin to coleoptiles or to stem tissue can lead to an increased rate of incorporation of labelled glucose into cell wall material. The increase in incorporation rate generally occurs after the increase in extension growth, and the incorporation into all cell fractions appears to be increased to the same extent. The timing of the response indicates that it is probably a correlative effect. Enzymatic studies suggest that changes in the activities of polysaccharide synthetases are responsible for the increased rate of synthesis of cell wall material [375].

One such enzyme is β1–4 glucan synthetase, the enzyme catalysing the synthesis of cellulose [375]. Excision of stem segments from the plant causes a severe decline in the extractable activity of the enzyme, so that by 8 h after excision, only 10% of the original activity remains. Addition of auxin prevents this, and indeed, under appropriate conditions, 4-fold increases in enzyme activity may be demonstrated in the presence of auxin. The auxin-induced increase in enzyme activity is detectable within 15 min and may be initiated at the same time as increased extension growth. Cycloheximide and actinomycin D, applied at concentrations which severely inhibit protein and RNA synthesis, do not inhibit the auxin-induced increase in enzyme activity. This suggests that the increase in activity is mediated via activation of a pre-existing protein, rather than by *de novo* synthesis. It is only possible at present to speculate on the mechanism of activation; the following observations are relevant to our speculation. Firstly, inhibitors of oxidative metabolism, such as cyanide and azide, prevent the activation, as do high concentrations of mannitol and sucrose. This suggests that continued oxidative metabolism and, perhaps surprisingly, maintenance of full turgor pressure are essential for activation. Secondly, addition of auxin to the membrane fraction with which the enzyme is associated (see Chapter 6) has no effect on enzyme activity. The activation is thus not the result of an allosteric interaction between auxin and enzyme. Thirdly, preliminary evidence indicates that the affinity of the enzyme for the substrate, uridine diphosphate glucose (the glycosyl donor for cellulose synthesis) is enhanced after activation. This suggests that the activation depends on modification of an already active protein, rather than conversion from an inactive state.

V. GIBBERELLIC ACID AND ENZYME SECRETION

A. Production of enzymes in the aleurone

The important biological features of the mature barley seed are the embryo, the starchy endosperm and the aleurone. The aleurone consists of a layer of cells three to five deep, and relatively uniform in size and appearance, which surrounds the endosperm. During germination of the barley seed, gibberellins are synthesized and released by the embryo and diffuse or are transported into the aleurone layer. In the aleurone cells, the intracellular production of a number of hydrolytic enzymes, including α-amylase and protease, is initiated. The enzymes are then actively secreted into the endosperm. Endosperm starch and protein are degraded to give maltose, glucose and amino acids which are then transported via the scutellum to feed the growing embryo. There is little doubt about the main steps in this sequence. Isolated aleurone layers respond to added gibberellins (and not to auxins, cytokinins, abscisins or ethylene) by secreting α-amylase and protease into the incubation medium. Isolated embryos secrete gibberellins. Isolated embryos and aleurones incubated together can bring about the accumulation of α-amylase in the incubation medium. Analysis of intact barley seeds during germination shows that *in vivo* the accumulation of α-amylase occurs in the starchy endosperm. These general features are exhibited by many varieties of barley (*Hordeum vulgare*) and by some varieties of wheat (*Triticum* sp.) and oat (*Avena sativa*). It would, however, be premature to assume that all cereals behave in this manner.

The two enzymes that have received most attention are α-amylase and protease. The time-course for production and secretion of these enzymes is shown in Fig. 7.6. There is a distinct lag period of 6–8 h after gibberellic acid (GA) treatment before increased levels of α-amylase are detectable in the aleurone tissue. Several hours later, increased amounts of α-amylase are detected in the incubation medium, and over a 24 h period, total levels of α-amylase and protease show a 10–20-fold increase over the levels in control samples (i.e. no GA supplied).

Varner and his associates [537] have demonstrated that α-amylase and protease are probably synthesized *de novo* in the aleurone as a result of GA treatment. The aleurone cells contain considerable reserves of protein in the aleurone bodies. These storage proteins are hydrolysed to give amino acids which are then available for protein synthesis in the aleurone. By incubating aleurones in $H_2{}^{18}O$ during the phase of protein hydrolysis, it is possible to

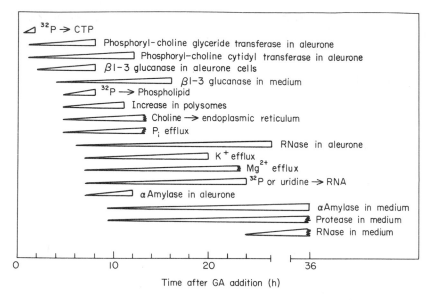

FIG. 7.6. Metabolic changes which occur in barley aleurone cells after gibberellic acid treatment. The left point of the triangle indicates the time at which the metabolic change is initiated and the apex the time at which the metabolic change is at its maximum. Triangles with serrated edges indicate that the time of maximal metabolic change was not determined.

density-label these amino acids. Proteins synthesized from the amino acids acquire the density label and the increase in density may be detected by centrifugation in gradients of caesium chloride (see Chapter 6). Both α-amylase and protease increase in density under these conditions, and the magnitude of the density shift suggests that the whole populations of these two enzymes are synthesized from precursor amino acids.

Although use of the density labelling method clearly distinguishes between activation and synthesis, it does not discriminate between regulation of synthesis and regulation of degradation (see section III and Chapter 6). It is tacitly assumed by many investigators that GA acts to increase the rate of synthesis of α-amylase and protease. Within the context of present knowledge, it could as easily be argued that secretion of the enzymes may result in their stabilization or that intracellular stabilization enables accumulation and secretion to occur (i.e. that GA regulates protein degradation). However, for the purposes of continued discussion, it will be necessary to assume (as do most investigators working in this field) that GA acts on enzyme synthesis.

A number of other hydrolytic enzymes have been reported to be secreted from aleurone cells or half-seeds after GA treatment. These include phosphatase, phytase, several glucosidases, β1–3 glucanase, ribonuclease and β-amylase. There is reason to think that biochemical changes underlying the production of some of these enzymes may be different from those underlying the production of α-amylase and protease. β-amylase, for example, is derived from an inactive precursor in the endosperm starch. By contrast, both β1–3 glucanase and ribonuclease are apparently synthesized *de novo* as judged from the result of density labelling experiments, but unlike α-amylase and protease, both β1–3 glucanase and ribonuclease accumulate in the aleurone cells in the absence of GA. The effect of GA on glucanase and ribonuclease is not a stimulation of their synthesis and accumulation but a stimulation of their secretion. The time-courses of production and release of β1–3 glucanase and ribonuclease after GA treatment are illustrated in Fig. 7.6.

B. Regulation of α-amylase synthesis

The evidence for the view that GA regulates α-amylase synthesis at the transcriptional level is very straightforward. Treatment of aleurones with 5-bromo-uracil, 6-aza-guanine, 5-aza-cytidine, 8-aza-adenine 6-methyl-purine or actino-mycin D (all inhibitors of RNA synthesis) together with GA, inhibits the accumulation of α-amylase in the incubation medium. These data may be taken as indicating the necessity for continued RNA synthesis during α-amylase production and secretion. On the other hand, 5-fluoro-uracil, a selective inhibitor of the synthesis of rRNA and tRNA, only marginally affects the total amount of α-amylase produced. Thus it can be argued that the dependence of α-amylase production on RNA synthesis is a dependence on the synthesis of mRNA. There are two further lines of evidence to support this view. Firstly, as is shown in Fig. 7.7, the inhibitory effects of actinomycin D diminish if the inhibitor is added after the addition of GA. In fact, if actinomycin D is added at the end of the lag period, it has little or no effect. It is thus possible that the lag period represents the time required for the synthesis of the mRNA for α-amylase. Secondly, it has been shown that GA causes a selective increase in the synthesis of a 5–14*s* RNA fraction, the sedimentation coefficient of which is the size that would be expected of the α-amylase mRNA [543].

A number of criticisms of these lines of evidence can be made. Firstly, there is the general criticism that compounds such as actinomycin D and bromo-uracil are relatively non-specific and may inhibit α-amylase production by an effect other than by inhibition of RNA synthesis. Secondly, many of the inhibitors of RNA synthesis only impair α-amylase accumulation from the isolated aleurone but not from half-seeds. The reason for this sensitivity difference is not known but the lack of inhibition of enzyme production in the half seed is difficult to reconcile with a transcriptional control mechanism. The third, and perhaps most significant criticism concerns the time-course for production of the putative mRNA for α-amylase. The increased incorporation

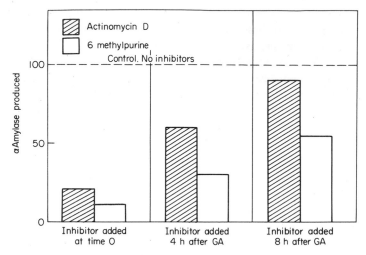

FIG. 7.7. The effect of various times of addition of 6-methylpurine or actinomycin D on total synthesis of α-amylase induced by GA. Barley aleurone layers were incubated in GA in the presence or absence of 6-methylpurine (0·1mM) or actinomycin D (100μg ml⁻¹) added 0, 4 or 8 h after GA. Amylase production in aleurone cells and medium was measured after 24 h. The control equals aleurone layers treated with GA alone for 24 h. (Data calculated from ref. 537.)

of radioactive precursors into the 5–14s RNA fraction is initiated at about 8 h after GA application (i.e. right at the end of the lag period) and then continues for a further 8 h. Actinomycin D added at 8 h has little effect on α-amylase accumulation although it strongly inhibits RNA synthesis. It is therefore very unlikely that the synthesis of the 5–14s RNA is necessary for α-amylase synthesis. Thus, the evidence to support the view that GA regulates α-amylase production at the transcriptional level is not at present very convincing.

The alternative hypothesis that GA regulates α-amylase synthesis at the translational level is more soundly based. The hypothesis embodies the following features. The lag period is the period in which the secretory apparatus, i.e. the rough endoplasmic reticulum, is formed in the aleurone cells. The messenger RNA for α-amylase is already present in the untreated aleurone. The effects of inhibitors of RNA synthesis on α-amylase accumulation are not mediated by an inhibition of synthesis of α-amylase mRNA but of the synthesis of RNA necessary for the formation of rough endoplasmic reticulum. The evidence relating to this hypothesis will now be discussed.

Figure 7.8 shows the effects of actinomycin D (at a lower concentration than used in the experiments illustrated in Fig. 7.7) and 5-fluoro-uracil on the accumulation of α-amylase both in the aleurone and in the medium. As was noted earlier, 5-fluoro-uracil has no effect on the amount of α-amylase produced. The same is true of actinomycin D at the lower concentration used in

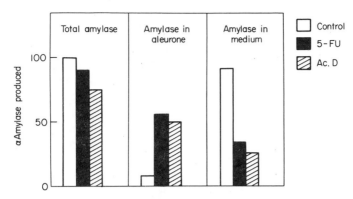

FIG. 7.8. The effect of actinomycin D and 5-fluorouracil on the synthesis and secretion of α-amylase from barley aleurone layers treated with gibberellic acid. Aleurone layers were incubated in gibberellic acid in the presence or absence of actinomycin D (50 μg ml⁻¹) or 5-fluorouracil (2 5mM). After 24 h, the amounts of α-amylase in the medium and in the aleurone cells were separately determined. (Data calculated from refs. 132 and 537.)

these experiments. However, both inhibitors cause a marked reduction in the amount of enzyme that is secreted from the aleurone into the medium. Further, since 5-fluoro-uracil probably does not affect mRNA synthesis, it seems likely that the *secretion* of α-amylase is dependent on the continued synthesis of rRNA and/or tRNA.

The formation of the secretory apparatus has been followed in aleurone cells using electron microscopy [537]. The untreated but imbibed aleurone cell contains aleurone grains with inclusions of phytic acid, numerous small lipid-containing spherosomes, and occasional pieces of endoplasmic reticulum, together with the normal complement of cell organelles. After GA treatment for 12–24 h marked ultrastructural changes are visible [537]. The aleurone grains with associated phytic acid and spherosomes disappear. There is a pronounced accumulation of rough endoplasmic reticulum which is often stacked into layers around the nucleus. The numbers of free polysomes increase and there is usually extensive development of cristae in mitochondria.

Biochemical analysis confirms this general picture and provides greater insight into the molecular events responsible for it. As shown in Fig. 7.6, two of the earliest changes detectable after GA treatment are in the activities of two enzymes important in the lecithin biosynthetic pathway. These are phosphoryl-choline glyceride transferase and phosphoryl-choline cytidyl transferase; both increase about 3-fold in activity. These changes are accompanied or preceded by an increased incorporation of ³²P-orthophosphate into CTP. Several hours after these two events, dramatic changes in lipid metabolism occur. Increased incorporation of labelled choline into microsomal lipid and increased ³²P incorporation into phospholipid are both detectable 4 h after GA treatment and are 4–7-fold higher after a further 4–8 h treatment. Accompanying these changes in lipid metabolism are an approximate doubling of the quantity of polysomes which may be isolated and an increased secretion of inorganic phosphate into the medium. This probably originates from phytic acid.

Electron microscopic examination has also revealed one more interesting fact. Rough endoplasmic reticulum can be formed in the presence of actinomycin D and GA, but large areas of it are fragmented and disorganized. Such an observation supports the view that actinomycin D acts as an inhibitor of α-amylase secretion rather than of synthesis (see Fig. 7.8). This view is further supported by the decline in sensitivity of the aleurone to actinomycin D (shown in Fig. 7.7). The biochemical data suggest that the formation of the rough endoplasmic reticulum is initiated 4 h after gibberellic acid treatment and is completed 4–8 h later, at which time secretion is initiated. The decline in sensitivity to actinomycin D commences at 4 h and is nearly complete 8 h after GA treatment. Taken together, these two pieces of information suggest that once the rough endoplasmic reticulum is formed, its continued function does not require associated RNA synthesis. Although actinomycin D may damage the construction of the rough endoplasmic reticulum by inhibiting RNA synthesis, some recent work carried out on chick fibroblasts suggests that actinomycin D can directly inhibit the synthesis of phospholipids and a similar inhibition could account for its deleterious effects on the secretory apparatus of the aleurone [347].

Direct evidence that the mRNA necessary for α-amylase synthesis is synthesized in the imbibed but untreated aleurone has been obtained using the inhibitor 5-fluoro-uracil (5-FU). If viruses or prokaryotic cells are allowed to synthesize mRNA in the presence of 5-FU the inhibitor is incorporated into the messenger RNA. On subsequent translation, limited misreading is caused and the protein molecules formed still have enzymatic activity but they are more easily denatured and show a reduced thermostability. If aleurone cells are incubated in 5-FU for 24 h and then transferred to a medium containing uracil (to chase out the 5-FU) and GA, the α-amylase molecules formed show a reduced thermostability [73]. Such evidence strongly suggests that messenger RNA for α-amylase is synthesized in the absence of gibberellic acid. Unfortunately, it does not disprove the possibility that the rate of amylase mRNA synthesis is increased after GA treatment.

C. The process of α-amylase secretion

Portions of the rough endoplasmic reticulum round off and form vesicles which move to and concentrate at the periphery of the cell, usually on the endosperm side. During the progress of these vesicles to the plasmalemma they are degranulated (the ribosomes are lost) to form smooth vesicles. Fusion occurs with the plasmalemma and the contents are released with occasional membrane fragments appearing outside the cell. The prior release of $\beta 1-3$ glucanase (see Fig. 7.6) causes extensive damage to the aleurone cell wall and it is supposed that the α-amylase molecules diffuse through these damaged areas. The reason for the temporal spacing of production of these two enzymes now becomes apparent and this is but another illustration of the importance of the kinetic approach to understanding the action of growth substances.

The picture of secretion has been built up largely from studies with the electron microscope and is difficult to substantiate biochemically. But by way of support it has been found that some of the α-amylase in cell-free homogenates is particle-bound. Some study has also been made of the mechanism of release of the enzyme itself from the aleurone cells. These data indicate that 2, 4-dinitrophenol, CCCP or anaerobiosis all inhibit release and thus a continued supply of ATP is necessary. Other substances (GA, cycloheximide or actinomycin D) do not affect α-amylase release directly but only indirectly through control of ER (endoplasmic reticulum) formation and intracellular synthesis of the enzyme.

D. Antagonistic effects of abscisic acid

Although the major emphasis in this section has been placed on GA, some reference must also be made to the effects of abscisic acid (ABA). At low concentrations, ABA brings the synthesis and secretion of α-amylase to a halt within 4–5 h at whatever time it is added to the aleurone. Nearly every biochemical event initiated by GA (Fig. 7.6) can be reversed by the addition of ABA. It seems probable that GA and ABA affect the same initial molecular event, one as an inducer and one as an inhibitor.

E. Summary

The earliest detectable biochemical changes initiated by gibberellic acid are the changes in activity of the two enzymes in the lecithin pathway (Fig. 7.6). These changes begin to occur within 1–2 h after application of GA. Whether these increases in enzyme activity, which are probably concerned with increased membrane synthesis (e.g. for endoplasmic reticulum) can be regarded in any way as basic molecular events is extremely doubtful, since the increases are both inhibited by actinomycin D and cycloheximide. In general terms we can state that the function of GA in the aleurone is probably to control the formation of rough endoplasmic reticulum, although the mechanism involved in this control is by no means clear. Once the rough endoplasmic reticulum is formed, the mRNAs coding for α-amylase and protease, already present in the aleurone, can be translated. This general picture clearly still has many defects, but perhaps these will be eradicated with continued research.

VI. THE ROLE OF ABSCISIC ACID IN EMBRYOGENESIS

If auxins, gibberellins, cytokinins and ethylene are termed effectors, then it would be reasonable to refer to abscisic acid as an inhibitor. The physiological role of ABA seems to be the induction of resting phases in cells, either as a part of a developmental programme, or so that they can survive extreme environmental conditions. In this section an example of the first type will be considered: the prevention of precocious germination in developing seeds. Examples of the second type are to be found in terminal bud formation and stomatal closure.

The system to be described has been extensively researched by Dure and his associates [206, 207] and is concerned with that most neglected area of plant development, embryogenesis. Here they have uncovered a fascinating interplay of developmental factors. A summary of the general features of embryogenesis in cotton (*Gossypium*) is shown in Fig. 7.9. The graph shows the variation with time of the fresh weight of the developing embryo. Two important developmen-

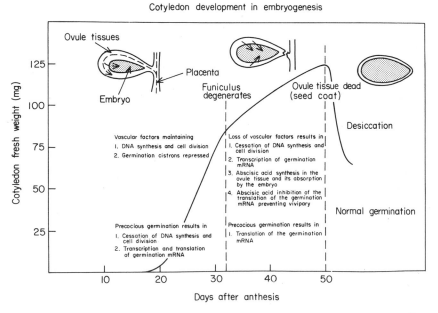

FIG. 7.9. Postulated scheme of developmental events in cotton cotyledon embryogenesis. (from ref. 207.)

tal steps can be delineated. Firstly, at about 85 mg fresh weight, DNA synthesis and cell division cease, probably as a result of a rupture of the connection between the embryo and the parent plant. Secondly, at about 125 mg fresh weight, the whole programme of embryogenesis is finished and the mature seed becomes dehydrated. These two developmental steps are shown as dotted lines on Fig. 7.9. It will be convenient then to refer to three stages of embryonic development, 0–85 mg, stage I; 85–125 mg, stage II and 125 onwards, stage III.

During normal germination, the cotyledons act not only as the first pair of leaves but also as a store of reserve protein and lipid. Both of these materials are hydrolysed during germination to feed the growing plant. Dure and his associates have studied the activity of two cotyledon enzymes important in the breakdown of protein and lipid, a protease and isocitrate lyase (an enzyme in the pathway for conversion of fatty acids to carbohydrates). The molecular regulation of both enzymes appears to be identical and therefore only the protease will be referred to subsequently. The protease has been purified to a single protein component and during germination its activity increases several hundredfold. If germination is allowed to proceed in the presence of cycloheximide, little or no enzyme activity is detectable. If, on the other hand, germination is allowed to proceed in the presence of labelled amino acids, the specific activity of the purified enzyme is some 30-fold higher than that of general cellular protein. Such evidence indicates that the protease accumulates by *de novo* synthesis.

Embryos of all stages down to about 40 mg fresh weight show the rather unusual phenomenon of precocious germination. That is, simply removing the embryo from the ovule and placing it on moist filter paper inhibits further embryogenesis and permits germination to occur. The smaller the embryo, the longer germination may take and below the critical size of 40 mg, the embryo seems to be too immature to germinate at all. During precocious germination, the protease and isocitrate lyase appear in the cotyledons but the more immature the embryo, the longer is the time required for their appearance. For example, stage III embryos produce 16 units of protease in 3 days, stage II embryos (90 mg) produce seven units in 4 days whilst stage I embryos (45 mg) produce only one unit in 6 days. Precocious germination also results in one other change in stage I embryos: cell division, which is normally occurring at this stage, ceases upon removal of the embryo from the ovule.

If stage III embryos are germinated in the presence of actinomycin D, the level of protease increases normally. This is despite the fact that all RNA synthesis is strongly inhibited. Observations like these have been made on a number of seeds and have led to the concept that for certain enzymes in the seed, mRNA has been synthesized and stored in an inactive form during embryogenesis. During germination, this inactive mRNA is then, in some way yet unknown, activated and translated. Using this observation and the phenomenon of prococious germination, Dure has been able to locate the precise time during embryogenesis at which the mRNA for the protease is synthesized. Embryos of 90 mg fresh weight (stage II) produce protease normally when procociously germinated in actinomycin D whereas embryos of 80 mg fresh weight (stage I) produce no protease at all (see Fig. 7.10). This locates the time of formation of the mRNA for the protease at the transition of embryos from stage I to stage II. Thus during prococious germination of stage I embryos, the mRNA for the protease has to be both synthesized and translated, whereas in stage II and III embryos it merely has to be translated. This probably in part accounts for the much slower production of protease exhibited by stage I embryos during prococious germination.

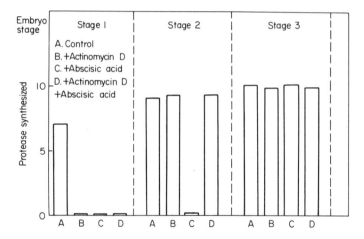

Fig. 7.10. The development of protease activity during precocious germination of stage I, II and III embryos of cotton. Batches of cotton embryos of all three stages were precociously germinated in water (A) actinomycin D (B) abscisic acid (C) or actinomycin D + abscisic acid (D) and protease estimated. Precocious germination times were 6 days for stage I embryos, 5 days for stage II and 4 days for stage III embryos. (Data calculated from ref. 206.)

What stimulus initiates the synthesis of the mRNAs for the enzymes involved in germination? Observations on the developing embryo show that at about 85 mg fresh weight, the vascular connection to the parent plant degenerates and thus the parental source of growth substances and nutrients is lost (Fig. 7.9). While this may act as the necessary stimulus, cell division and DNA synthesis also cease at the same time and it is possible that these events themselves could qualitatively alter genomic activity and in turn mRNA synthesis.

Ungerminated stage II and stage III embryos have no detectable protease; it takes nearly 24 h of germination before protease may be detected. Since the information for the protease has already been synthesized in stage II embryos, some factor must prevent its expression. In stage III embryos, this appears simply to be dehydration but several lines of evidence suggest that this factor for stage II embryos may be ABA. Addition of ABA to stage I and stage II embryos prevents the appearance of the protease during prococious germination. This is shown in Fig. 7.10. Since the information (the mRNA necessary for the protease) has already been transcribed in stage II embryos, ABA must inhibit the expression of this information by some post-transcriptional control. Developing cotton bolls were the tissue in which ABA was first identified and characterized; it becomes detectable in embryos only when their fresh weight is over 90 mg and it may then be extracted from the ovule tissue. Dure and his associates suggest that the degeneration of the funiculus is the signal for the production of ABA by the ovule tissue surrounding the embryo. Diffusion of the ABA into the embryo prevents prococious germination which is apparently a lethal process in cotton.

Two further facts emerge from Fig. 7.10. Stage III embryos can germinate in the presence of ABA. They will not do so normally in the developing boll because of the desiccation which commences with the death of the surrounding ovule tissue. Sensitivity to ABA is obviously lost during the transition from stage II to stage III. An even more surprising result is the inhibition of the ABA effect in stage II embryos by actinomycin D. The literal interpretation of this result is that continued RNA synthesis is necessary for the inhibitory action of ABA to be expressed. Perhaps it is necessary to produce a receptor protein for ABA or perhaps ABA initiates the production of inhibitory RNA molecules. Either way, the possible interpretations suggest further interesting developments in the field.

In this work, much reliance has to be placed on the specificity of action of actinomycin D. There is no doubt that it inhibits RNA synthesis in cotton cotyledons and its inability to inhibit germination in stage II and stage III embryos represents a very useful control. 5-Fluoro-uracil does not duplicate the inhibitory action of actinomycin D in stage I embryos. Furthermore, an extensive analysis of the tRNA species in the developing embryos suggests no significant change which could account for the transition from stage I to stages II and III. Thus, it can be argued that the species of RNA which is different in stage I compared with stage II embryos is messenger RNA.

It can also be argued, however, that all this merely puts a rather simplistic gloss on what is a complex developmental process. In this context, two points should be emphasized. Firstly, the termination of cell division between stage I and stage II has been taken as indicating a termination of DNA synthesis. However, it is clear from studies of developing pea cotyledons that DNA

K

synthesis and cell division are not necessarily directly related (see also Chapter 5). Many cells in pea cotyledons become extensively "polyploid" (an average value of 50C has been recorded) [313]. DNA synthesis thus continues long after the cessation of cell division. If this in fact happens in the cotyledon of stage II cotton embryos, it would somewhat complicate the analysis of precocious germination in such embryos. Secondly, it must be emphasized that post-transcriptional control has only been demonstrated in this system for synthesis of protease and isocitrate lyase. The synthesis of these two enzymes is just one aspect of germination. ABA, however, inhibits the complete process of prococious germination, with all its attendant metabolic changes. It cannot be assumed therefore that the inhibitory effect of ABA on prococious germination is exerted *only* at the level of translation of mRNA.

VII. REGULATION BY AUXIN OF CELLULASE AND THE ORGANIZATION OF NEW MERISTEMS

A. Introduction

One of the major functions of growth substances is the regulation of the meristematic regions in the plant. Excellent examples of auxin control are seen in the regulation of cambial activity, the organization of new root meristems in root and shoot cuttings and (together with cytokinins) in the organization of root or shoot meristems from callus and the control of dormant bud meristems. The biochemical events underlying auxin induction of root meristems has attracted some attention in recent years and it is the purpose of this section to describe this work.

The induction of roots on stem cuttings by application of auxin is unusual in showing no apparent optimum or plateau in the dose–response curve. Usually the higher the concentration of auxin used the more roots will be formed. The two most widely used experimental materials are the 7–9-day-old etiolated pea stem and the 2–3-day-old soya bean hypocotyl. Auxin has been applied to pea stem tissue by removal of the apical hook and painting the cut surface with IAA and to soya bean hypocotyls by spraying with high concentrations of 2, 4-D. Observations with the microscope have indicated the following sequence of events in the apical 1 cm portion of pea stem tissue after IAA application. In the first 24 h, extension growth of the top 1 cm is severely inhibited and lateral swelling occurs. Parenchymatous cells in the cortex become swollen and there is some cell division. After 2 days, there is increased cell division; disintegration of many parenchymatous cells in the cortex commences, leaving lacunae filled with cell and cell wall debris. After 3 days, root primordia become recognizable as developing from the vascular region, probably originating from the cambium. The masses of new cells in these primordia continue to divide, occupying the lacunae left by the disintegration of cortical cells, and 4–5 days after application of auxin, the new roots break through the epidermis of the stem approximately 10 mm below the cut tip. A similar sequence occurs if soya bean hypocotyl tissue is treated with a high concentration of 2, 4-D, but with one marked difference. It is the older tissue, 2–4 cm below the apex, which responds to auxin by cell division and root formation in the hypocotyl.

These observations imply a number of biochemical changes. Cell division must be linked with increases in DNA, RNA, protein and cell wall material. The disintegration of the cortical cells has its origin in the production of cell wall and other hydrolytic enzymes. The organization of root primordia in stem tissue requires the production of new information from the nucleus. This reveals itself in qualitative changes in the synthesis of RNA and protein. The evidence presented in the remainder of this section is concerned with some of these biochemical changes.

The results to be described have been obtained using two different experimental conditions. Firstly, the tissue has been treated with high levels of auxin on the intact plant and secondly, sections of tissue have been treated with lower concentrations of auxin in solution. Many but not all of the biochemical changes which auxin induces in these two experimental states are qualitatively similar but quantitatively lower in the isolated section. The reasons for this difference are not completely understood but seem to be related to the incapacity of the isolated section to maintain an adequate level of protein synthesis. As a consequence, the protein level of the section declines and cell division is unable to occur. It has been suggested that lack of growth substances from the root (possibly cytokinins and gibberellins) may be responsible for this difference. The apparent disadvantage of the isolated section is, however, offset by the ease with which it may be labelled with isotopic precursors, a property not found in the intact plant.

B. Regulation of cellulase and pectinase activity by auxin

It was surmised earlier that the disintegration of cortical cells in the stem prior to primordial formation may be the result, at least in part, of production of cell wall hydrolases. A number of enzymes, cellulase, β1–3 glucanase, pectinase, insoluble and soluble pectinesterase and α-amylase have been assayed in intact stem tissue undergoing auxin-induced root formation. The activity of all of these enzymes, apart from α-amylase, increases but the size of the increase is variable. After 3 days cellulase activity is increased 30–40-fold, pectinase 7–8-fold, pectinesterase and β1–3 glucanase 3–4-fold [99]. It could be concluded from these data that there is a selective alteration in the synthesis of cellulase. As pointed out, however, in section III and elaborated further there, these differences in enzyme activity may arise not from a selective effect on the synthesis of one enzyme but from a general and similar increase in synthesis coupled with differences in the rates of degradation.

There is some evidence to suggest that the increase in cellulase activity may be by *de novo* synthesis. Treatment of stem tissue with actinomycin D, 8-aza-guanine or puromycin, together with IAA, totally abolishes the changes in cellulase activity. This may indicate, at least, that continued RNA and protein synthesis is necessary for cellulase activity to increase. When DNA synthesis, and thus subsequent cell division, is inhibited by 5 fluorodeoxyuridine (FUDR) the increases in cellulase are reduced by about a half. There is thus no obligatory coupling between cell division and increases in cellulase activity.

To substantiate the idea that cellulase is synthesized *de novo*, various subcellular fractions have been analysed for cellulase activity after auxin treatment. On a kinetic basis, the specific activity of cellulase is increased first in the microsomal fraction and these increases then appear later in the supernatant and wall fractions. The cellulase bound to the microsomes cannot be released by detergents but removal of magnesium ions or treatment with RNase result in substantial amounts being solubilized. Incubation of microsomal fractions from treated tissue with the necessary cofactors for *in vitro* protein synthesis results in an increase of about 50% in cellulase activity. This, together with the data obtained from use of inhibitors is strongly suggestive of *de novo* synthesis of cellulase on ribosomes [102].

C. Alterations in nucleic acid metabolism and protein synthesis associated with cellulase formation

Treatment of pea epicotyls with auxin causes the DNA and protein content of the apical 1 cm to increase 2·5-fold over a 3-day period, while the RNA content shows a 4-fold increase. Increases in DNA, RNA and protein have also been observed after 3 days of auxin treatment in the basal region of the soya bean hypocotyl. In this system, DNA and protein show a 4—5-fold increase while RNA increases 10-fold. These results, assuming the DNA content to be some measure of the number of cells, clearly imply that the amount of RNA per cell increases. This may be confirmed in pea by treating with FUDR together with auxin. In the absence of DNA synthesis and cell division, the RNA content of the apical 1 cm increases 2·5-fold and the protein increases to 1·5 times its original level. Actinomycin D and puromycin abolish auxin-induced increases in DNA, RNA and protein [239].

The distribution of RNA amongst subcellular fractions has also been examined. Auxin appears to increase selectively the amount of RNA found in the microsomal fraction as compared to soluble RNA. In the presence of auxin, the ratio of microsomal RNA to soluble RNA was shown to be 6·9 and in control tissue, 4·8.

There are therefore two selective changes: RNA as against DNA and protein, and microsomal RNA as against soluble RNA. The possibility that the synthesis of microsomal RNA, soluble RNA and protein are all increased to the same degree by auxin and that these apparent selective alterations arise from differences in turnover rates must again be considered (see section III, Fig. 7.1). In the only plant for which data are available, *Lemna minor,* the rate of degradation of ribosomal RNA is higher than that of soluble RNA and that of protein [484, 485]. This supports the possibility that differential rates of turnover, coupled with a general stimulation by auxin, are responsible for the observed changes.

The times of initiation of increases in RNA, DNA, protein and cell division are shown in Fig. 7.11. Incubation of pea stem sections or of soya bean

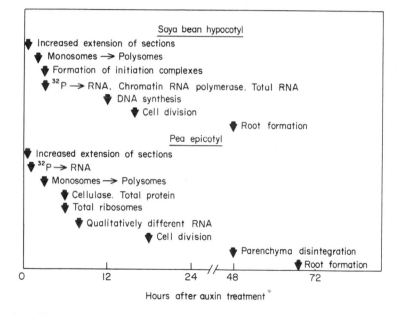

Fig. 7.11. Time of initiation of metabolic changes induced by IAA after addition to soya bean hypocotyl or pea epicotyl.

hypocotyl sections in auxin plus radioactive precursors leads to increased labelling of RNA with a lag period of 1 h in pea and 3 h in soya bean [241, 482]. This labelled RNA has been fractionated on columns of methylated albumin kieselguhr after several hours' treatment and there appears to be a preferential increase in the labelling of ribosomal RNA and transfer RNA, as compared to the AMP-rich RNAs D–RNA and TB–RNA. After 4 h treatment with 2, 4-D the specific activity of ribosomal RNA and transfer RNA shows an increase of about 50%, that of D–RNA about 25% and that of TB–RNA 0%. This difference has already been discussed in section III, Fig. 7.1b and it was pointed out there that the selectivity could arise from the shorter half lives of D–RNA and TB–RNA. The synthesis of all RNA fractions may thus be increased by auxin to the same degree.

This increased synthesis of RNA can be studied *in vitro* by isolating chromatin from treated and untreated intact soya bean hypocotyl tissue and using it as a template for the synthesis of RNA from labelled nucleoside triphosphates. Some 24 h after auxin treatment, the rate of RNA synthesis of the isolated chromatin is 8–9-fold higher than that observed with chromatin from control tissue. An increase in RNA synthesis could arise from more of the genome being unmasked in the presence of excess endogenous RNA polymerase or from an enhanced activity of the polymerase itself. These two possibilities can be distinguished by adding an excess of exogenous RNA polymerase (e.g. from *Escherichia coli*) to the chromatin from control tissue. If auxin causes an unmasking of the genome, then addition of *E. coli* RNA polymerase will have no effect on the rate of RNA synthesis in chromatin from untreated tissue. If, on the other hand, auxin only increases the activity of RNA polymerase, then addition of *E. coli* RNA polymerase to chromatin from untreated tissue will bring its rate of RNA synthesis up to that of chromatin from auxin-treated tissue [339]. When experiments of this type are performed, the results suggest that the major effect of auxin is on the activity of the RNA polymerase but that there is some selective unmasking of the genome as well. More recent work [167] has shown that in soya bean hypocotyl, treatment with 2, 4-D enhances the activity of RNA polymerase I (which synthesizes ribosomal RNA) but not of RNA polymerase II. The increased activity of RNA polymerase I is reflected in an increase in the amount of enzyme protein. It may be surmised that the unmasking of the genome which also occurs may be exploited by RNA polymerase II without the need for an increase in enzyme activity. The base composition of the RNA synthesized by the chromatin isolated from auxin-treated tissue is different from that synthesized by chromatin from control tissue, and use of the technique of DNA–RNA hybridization shows that the RNA made in pea stem tissue is qualitatively different after auxin treatment. The latter change can only just be detected 8 h after auxin treatment of the stem tissue, but increases thereafter. There are thus three pieces of evidence which indicate that the RNA species produced in hypocotyl and epicotyl tissue are qualitatively different after auxin treatment. This view is further supported by the finding that the mRNA which codes for cellulase is detectable after auxin treatment, but not before [497a].

It seems likely that part of this selective change in gene expression is the result of the induction of cell division in the stem tissue. Cell division is a complex biochemical process which requires a unique set of RNA molecules for the proteins concerned with division (see Chapter 5) and since the control tissue in this case is non-dividing, qualitative differences would be expected. In soya bean, DNA polymerase activity of isolated chromatin is measurably higher 12 h after auxin treatment and cell division can be demonstrated as starting just after this time. The timing of cell division in pea tissue has not been so carefully measured but it probably commences at about the same time as in soya bean.

Treatment of the stem tissue with FUDR and auxin eliminates DNA synthesis and cell division. The RNA synthetic capacity of the isolated chromatin in this case is increased only 3–4-fold by auxin (compared to 8–9-fold without FUDR) but experiments with *E. coli* RNA polymerase still indicate some selective alteration in the template [197].

The initiation of cell division is preceded by substantial changes in the ribosome population in both soya bean and pea. Two hours after auxin treatment of pea stems, an increased labelling of polysomes can be detected, and an hour later there is a shift amongst the polysome population itself. The proportion of polysomes containing 10 ribosomes or more is increased 3-fold, largely at the expense of the monomer population [101]. Between 6 and 10 h, the total ribosome population itself doubles and it is probably significant that the increases in total protein and cellulase can first be detected at this time. After 10 h treatment the ratio of polysomes to monosomes is 9 compared to 0·5 in control tissue. This represents a 6-fold increase in the numbers of polysomes.

In soya bean this early auxin-induced shift from monosomes to polysomes also occurs after 3 h auxin treatment. The shift is inhibited by actinomycin D, 6-methyl-purine and cordycepin but not by 5-fluoro-uracil [479]. This again suggests that the synthesis of new ribosomes is not necessary for this early polysome increase but that it is dependent upon the continued synthesis of TB– and D–RNAs (the synthesis of which are not inhibited by 5FU). At the same time the capacity of the individual polysome to carry out *in vitro* protein synthesis is increased. This is the result of a higher rate of formation of initiation complexes. Monomeric ribosomes isolated from auxin-treated tissue show a 3-fold higher level of attached peptidyl-tRNA as measured by incubating the ribosome with radioactive puromycin and counting the peptidyl puromycin formed.

It is not yet clear whether the increased numbers of polysomes can be equated with a higher level of attached messenger RNA. If ribosomes are equally spaced along mRNA in both treated and untreated tissue then the level of polysomes gives a direct indication of the amount of messenger RNA participating in protein synthesis. The evidence available from animal systems suggests that the concept of equal spacing is valid for several metabolic states. If this concept is equally valid for plant systems, auxin increases both the amount of messenger RNA attached to ribosomes and the rate of formation of initiation complexes.

For further discussion of the topic mentioned in this section, the reader's attention is drawn to refs 239 and 483.

D. Summary

The following represents a possible sequence of events during the production of adventitious roots. Auxin affects the template availability of chromatin and increases the activity of the RNA polymerase I. The activity of initiation factors in protein synthesis is increased and the newly synthesized AMP-rich RNAs are combined with monosomes to form polysomes. Several hours after this, new forms of messenger RNA appear as a result of changes in the template availability in chromatin, with a large increase in the total ribosome population (the increased rate of synthesis of ribosomal RNA being related to the increased activity of RNA polymerase I). Cellulase, other cell wall hydrolases and other enzymes are synthesized, increasing the total protein content. Some of the cortical cells are destroyed and their contents degraded, releasing cell division factors such as cytokinins (see Chapter 5). Cambial cells are induced to divide, forming a root primordium which then fills the lacunae left by the degraded cortical cells. Figure 7.11 summarizes this sequence.

This picture is, of course, a very incomplete one. For example there is no substantive evidence to support the idea that the increase in the amount of messenger RNA for cellulase is caused by *de novo* synthesis rather than a decreased degradation rate of the mRNA. The picture has also been constructed using evidence both from sections and from intact plants. Increases in RNA, polysomes and cellulase can be detected in sections but the size of increases is small. Cellulase, for example, only increases some 50% in a section compared to 30–40-fold for the intact plant.

A further complication is the role of ethylene. High concentrations of auxin induce the formation of ethylene and this is probably responsible for the swelling of the parenchymal cells. Ethylene on its own does not initiate cell division but it can cause small increases in cellulase and some cell breakdown. Application of auxin under conditions where ethylene is removed induces cell division but lateral root initiation does not occur. Both growth substances together are essential for the production of lateral roots.

Finally, some mention must be made of the regulation of secretion in this system. To reach its substrate cellulase must be secreted through the plasmalemma. Whether or not the secretory apparatus is available in untreated tissue is unknown although the similarity of many of the biochemical events initiated by gibberellic acid in the aleurone to those initiated by auxin in stem tissue leads to the suggestion that the lag time in cellulase synthesis (about 6 h) could be the time required to form the secretion apparatus, the rough endoplasmic reticulum. Further consideration of this similarity between the effects of gibberellic acid and auxin leads naturally to the conclusion that each initiates an apparently similar set of molecular changes. The marked dissimilarity in the end product is an expression of the different potentials of the two tissues.

VIII. ETHYLENE

A. Control of abscission and cellulase activity

Leaf abscission generally occurs by a break at a defined region at the base of the petiole. This separation or abscission zone is a relatively narrow band of cells which frequently have a minimum of thickening. The abscission zone has been characterized as a zone of weakness in some mature woody plants. Microscopic examination has shown that abscission is preceded by dissolution of the middle lamella and sometimes the primary wall of the cells in the separation zone itself. Such observations indicate a probable role for enzymes hydrolysing the cell wall in the control of abscission.

The physiology of abscission is complex, with many factors affecting the strength of the abscission zone. It is, however, generally conceded that ethylene accelerates and auxin retards the abscission process. The experimental materials used for abscission studies are generally explants of the petiole base region of *Coleus*, *Gossypium* (cotton) or *Phaseolus* (bean). The effect of ethylene upon abscission in these explants is assessed by measurements of the weight required to cause a break at the abscission zone. One unfortunate feature of the material is that control tissue itself shows an increasing weakening of the abscission zone with time and ethylene therefore only accelerates a process that is already occurring.

There are several reasons why it is thought that regulation of cellulase activity may be the molecular means of controlling abscission. There is a good correlation between the break-strength of the abscission zone and its associated cellulase activity. Analysis of the distribution of cellulase through the petiole shows the abscission zone to have a 2–10-fold higher cellulase activity than the surrounding tissue. Ethylene can increase the cellulase activity of the abscission zone up to 4-fold and this increase precedes the change in the break-strength. If abscission is inhibited by treating with auxin together with ethylene, cellulase remains at control values and there is no weakening of the abscission zone.

Two lines of evidence suggest that cellulase is made *de novo* after ethylene treatment. Firstly, cycloheximide can inhibit the weakening of the abscission zone and the increase in cellulase activity [1]. Secondly, if abscission is allowed to proceed in the presence of D_2O, the density of extractable cellulase, determined isopycnically, is increased [278]. There is a lag period of 3–6 h after the addition of ethylene and before increased cellulase levels are detectable. Treatment with actinomycin D in this lag period subsequently impairs cellulase accumulation but later treatment with actinomycin D has little or no effect. Treatment with 5-fluoro-uracil has no effect on cellulase accumulation. General RNA and protein syntheses, as measured by the incorporation of labelled precursors, are also increased in the abscission zone after ethylene treatment, suggesting that cellulase may be only one of a number of enzymes to increase during abscission.

A feature of particular importance in abscission is the necessity for cellulase to be secreted to reach its substrate, the cell wall. Some recent work has indicated that ethylene may control the process of secretion directly. When ethylene is removed from the explants by vigorous flushing with air, further weakening of the abscission zone ceases. Assay of cellulase levels indicate that the activity of the enzyme continues to increase even after the removal of ethylene. Measurements of the distribution of the cellulase between the intracellular and extracellular compartments shows that whereas removal of ethylene prevents further secretion into the extracellular space, intracellular accumulation of cellulase continues unabated [1].

This result raises a rather important point. A number of enzymes (polygalacturonase, pectin *trans*-eliminase, pectinase, arabinase hemicellulase and $\beta 1$–3 glucanase) have been examined in the past to see if there is meaningful variation during the abscission process. Usually, the variations in total level of these enzymes shows no significant change and thus it has been assumed that these enzymes have no role in abscission. The data relating to cellulase clearly imply that what is important is the amount of enzyme which is secreted to the cell wall compartment, and for the enzymes listed above this is unknown. The observation that the middle lamella of cells in the abscission zone is frequently degraded suggests that some of these enzymes must be secreted outwards.

The parallels between secretion of cellulase in the abscission zone cells, auxin regulation of cellulase in pea stem and gibberellin-controlled secretion of α-amylase are too obvious to be missed. The lag period, the requirement for RNA synthesis only in the lag period and the subsequent *de novo* enzyme synthesis and secretion, again imply that the formation of the secretion apparatus and its maintenance represent an important and unifying aspect of the molecular control exerted by these three growth substances.

B. Fruit ripening

The general physiological changes that accompany the ripening of fruits are well known. These include a softening of the cell wall, some dehydration, disappearance of malic acid and starch and the appearance of glucose and sucrose, loss of chlorophyll and the development of anthocyanins. These physiological changes can be related to increases in the activity of cellulase, pectinase, malic decarboxylase, amylase, sucrose synthetase, chlorophyllase and phenylalanine-ammonia-lyase. There is no doubt that ethylene controls this aspect of fruit development and it obviously does so by increasing the activity of all of these enzymes. Not a great deal is known about the mechanism of control. Treatment of unripe fruits with ethylene frequently results in increased incorporation of labelled precursors into RNA and protein and many of the ethylene-induced changes can be inhibited by cycloheximide. There are obviously some parallels with the molecular events in abscission; cellulase and pectinase will have to be secreted, for example, but beyond this little can be said at the present time.

IX. A POSSIBLE UNIFYING HYPOTHESIS

A. Introduction

The preceding five sections have given some idea of the diversity of biochemical responses to growth substances. The physiological effects induced by growth substances show an equally bewildering array of variety. Dormancy, growth rate, flower initiation, sex determination, fruit set, fruit growth, fruit ripening, tuberization, abscission, rooting and senescence, are all known to be affected by every growth substance in one plant or other. The physiological response to any growth substance depends in the first instance on what type of cell is receiving the stimulus. For example, a developing leaf cell in a barley plant will respond to gibberellin by elongating, whereas an aleurone cell from the same plant will respond by producing α-amylase. The specificity of response is built into the cell by its previous developmental programme (see also Chapter 6). The highly specific animal hormones affecting only one type of cell have no counterpart in plants.

Despite the great variety in the physiological effects of growth substances, there is enough evidence to support the following unifying hypothesis for their initial molecular mode of action. According to this hypothesis, the initial site of action of all growth substances is a membrane system in the receptive cell. The physical properties of the membrane and hence the catalytic activities of the membrane-bound enzymes are altered. Modified rates of ion flux through such membranes also result from the initial change in membrane properties. Some of the enzymes in the aqueous phase of the cell are sensitive to the changed ion fluxes, and thus their catalytic activities are also modified.

The arguments for this hypothesis can be broadly divided into the evidence supporting the view that growth substances affect some aspects of membrane physiology and the evidence which indicates how altered ion fluxes could initiate appropriate and rather specific biochemical changes. It is recognized that alteration of the activity of membrane bound enzymes in themselves, located in the Golgi, mitochondrial, microsomal, chloroplast, nuclear, vacuole and plasma membranes may be sufficient to account for some of these specific biochemical effects of growth substances. That a physical change in a membrane system alters associated enzyme activities has been very adequately demonstrated by Lyons and Raison and associated co-workers [289].

B. Evidence relating growth substances to membrane physiology

An examination of the chemical structure and solubility properties of growth substances provides the first clue. Auxins, gibberellins, abscisins and cytokinins are molecules with a hydrophobic group attached to a single charged moiety which is a carboxyl group in the first three and a substituted amino group in cytokinins. Such molecules generally show much greater solubility in organic solvents than in water, particularly in the uncharged form, and it is very likely that in a cellular system they will align themselves on a membrane surface with the hydrophobic grouping buried in the lipid portion of the membrane and the charged group in the aqueous phase. Evidence from two different sources has shown that complexes are formed between lecithin and IAA (detected by solubility of the complex in carbon tetrachloride) and between gibberellin and lecithin (detected by n.m.r. spectroscopy) [507, 524]. Lecithin is, of course, a major constituent of many natural membranes. In both cases, there is almost a 1 : 1 binding of lecithin and growth substance. Although ethylene does not fall into this general chemical classification, it again shows much greater solubility in organic solvents than in water.

The evidence relating auxin to hydrogen ion secretion into the cell wall has already been considered (section IV). To maintain neutrality, other cations may be exchanged for the outgoing protons. The cell wall compartment contains both calcium and potassium ions but cell wall calcium is not mobilized by auxin and thus it seems possible that the level of potassium ions in the protoplast could be increased. This idea is supported by much older evidence which showed that auxin increases the uptake of potassium ions but not of sodium ions from the medium into epicotyl, hypocotyl and root tissue while at the same time inhibiting ammonium ion uptake. The auxin-enhanced uptake of chloride ions into coleoptile and mesocotyl tissue has also been reported. The possible relationship of all these phenomena to growth has already been discussed in section IV.

Other associated phenomena are the effects of auxin on the potential difference across individual coleoptile cells, whole coleoptiles and roots. Auxin hyperpolarizes the plasmalemma of individual coleoptile cells [129] and induces marked changes in the potential differences between the top and bottom of the coleoptile and the bean root when applied at the apex [409]. These changes are fast, occurring within one to a few minutes of the cells being exposed to auxin. In fact by placing a set of electrodes in series down a coleoptile, the rate of auxin movement can be measured as a wave of changing potential differences [327]. Alterations of the potential difference over a whole coleoptile are the cumulative effects of changes in the membrane potential of individual cells which in turn reflect altered ion fluxes.

In elongating stem tissue, one enzyme that responds relatively quickly to auxin is the *membrane-bound* glucan synthetase which synthesizes a cellulose-like product [375 and section IV]. *In vivo* studies on wall synthesis in coleoptile tissue have shown that auxin promotes the synthesis of all cell wall fractions to the same degree. If low levels of calcium salts are added to the incubation medium, cell extension ceases and the promotion of cellulose synthesis is very specifically inhibited. Such data are suggestive of a possible relationship between calcium ions, IAA and glucan synthetase which is worthy of further investigation.

The relationship of gibberellins to membrane formation, enzyme secretion and release of potassium and magnesium ions has already been described (section V). The latter event apparently involves permeability changes in the plasmalemma allowing increased leakage of those ions which are free in the aleurone cells. The enzymes showing the earliest changes in response to gibberellin, phosphoryl-choline cytidyl and phosphoryl-choline glyceride transferases, are both particulate and can be sedimented at relatively low speeds of centrifugation. Thus they must be located *inside* or *on* a membrane system. A further enzyme in the lecithin biosynthetic pathway, choline kinase, which is found in the soluble fraction, shows no activity change in response to added gibberellin.

There is also other evidence to relate gibberellin to changes in membrane permeability. When phospholipids are dispersed in aqueous media they produce self-ordered particles (micelles or liposomes) which display many of the properties of natural membranes. Substances affecting membrane permeability *in vivo*, e.g. drugs or steroids, have been found to have a comparable effect on the permeability of liposomes. If the liposomes are prepared in the presence of an easily detectable substance, e.g. radioactive glucose, then after simple purification the permeability of these membranes to glucose may be assessed by the rate of leakage of radioactive material into the medium. Most lipid bilayers and liposomes exhibit thermal transition points in their rates of leakage which it is believed reflect physical changes in membrane fluidity. It has been found that gibberellin at low concentrations promotes the rate of leakage of a number of small charged or uncharged molecules from liposomes at 25°C and it appears to do this by lowering the thermal transition temperature in the membrane studied from 25 to 15°C [523]. The complex formed between gibberellin and lecithin may explain the permeability and fluidity changes in liposome membranes.

It is well established that there is a relationship between stomatal aperture and abscisic acid [308]. The opening of the guard cells results from a massive influx of potassium ions from the subsidiary cells. The increased osmotic potential causes the guard cells to take up water and the stomata therefore open. The influx of potassium ions is probably regulated by the activity of a hydrogen ion pump situated in the plasmalemma of the guard cells, since hydrogen ions are secreted by guard cells coincidentally with the altered potassium ion flux. Abscisic acid can cause stomatal closure within a few minutes and it can only do this by dissipating the accumulated potassium ions in the guard cells. The probable mechanism is an ABA inhibition of the hydrogen ion pump with a dissipation of the accumulated potassium ions by diffusion. In a completely analogous way, the auxin-induced hydrogen ion secretion from coleoptiles has been shown to be abolished by abscisic acid within minutes, an event preceding the ABA inhibition of extension growth. ABA also increases the hydraulic permeability of carrot cells within 30 min.

Fruit ripening, a developmental sequence controlled by ethylene, has long been associated with changes in membrane permeability. These can be detected as an increased leakage of potassium ions, amino acids and sugars, accompanied by an increase in free space and preceding the structural disorganization associated with the final stages of ripening. Ethylene can also initiate the very rapid senescence of many flower petals. A recent electron microscopic study on *Ipomoea* [301] has indicated that one of the earliest detectable cellular events is a contraction of the vacuole with some dilution of the cytoplasm, accompanied by overall shrinkage of the whole protoplast. This causes a loss of turgor and is obviously the result initially of permeability changes in the tonoplast and the plasmalemma. These turgor losses can be detected visually several hours after ethylene treatment and probably account for the increase in free space which accompanies the senescence of many plant tissues. The direct effects of ethylene on cellulase secretion in the abscission layer have already been commented upon. It remains only to record that abscission can be strongly inhibited by the application of calcium chloride but not by the chlorides of magnesium, potassium, sodium or manganese [364].

Leaf senescence, which can be initiated by leaf detachment and detected as loss of chlorophyll and protein, is a phenomenon that has many parallels with fruit ripening. Large increases in free space and hydraulic permeability can be detected and these may be the result of permeability breakdown in the plasmalemma and tonoplast as described above. Other changes observed microscopically include disorganization of membranes, particularly the endoplasmic reticulum, with accumulation of lipid bodies followed by structural disintegration. Increases in hydrolytic enzymes, notably proteases and RNases, accompany the structural breakdown and the low molecular weight products from the hydrolysis are mobilized into the petiole. All of these processes are inhibited by cytokinins. Detached leaves, for example, can be kept in a green and healthy condition for up to 3 weeks provided the petiole is in a solution of cytokinin. Not only is the transfer of low molecular weight materials from the blade into the petiole inhibited, but cytokinins appear to be able to reverse the process. For example, an area of a detached tobacco leaf that remains green because of localized cytokinin application can attract and cause an accumulation of amino acids, sugars and even inorganic phosphate from untreated portions of the leaf. Even non-protein amino acids such as γ-aminoisobutyric acid are subject to the same accumulation. Cytokinins also promote the uptake of potassium ions, but not of sodium or ammonium ions in detached leaf and cotyledon tissue, and this can be detected several hours after cytokinin application. Incubation of detached leaf tissue in solutions of calcium ions but not of magnesium, potassium or sodium ions, strongly inhibits the loss of chlorophyll and protein and the increase in free space in an analogous way to incubation in cytokinin [365]. The ability of cytokinins to direct the movement of nutrients such as sucrose and ions is now well recognized and it has been suggested that they direct the accumulation of nutrients in fruits and tubers. Such properties suggest a fundamental role of cytokinins in membrane transport and permeability.

The chromoprotein, phytochrome, is not generally considered to be a growth substance although, like growth substances, it does initiate changes in morphogenesis. In the active form, P_{FR}, it affects flower initiation and development, the rephasing of circadian rhythms, control of seed germination and stem and leaf enlargement [434]. The apparent lack of mobility of phytochrome has precluded it from consideration as a growth substance although there seems little reason why it should not move, albeit slowly, through the symplast. Like growth substances, phytochrome has been shown to alter the activities of a number of enzymes including ascorbate oxidase, phenylalanine-ammonia-lyase, glyceraldehyde-3-phosphate dehydrogenase, glycollate oxidase, amylase and lipoxygenase [317]. Again like IAA, phytochrome has been shown to regulate the rate of initiation of protein synthesis on ribosomes [434]. The gross, biochemical, physiological and morphogenetic changes induced by P_{FR} are thus similar in many respects to those of growth substances and there is good justification for discussing its mode of action in this section. There is also excellent evidence to suggest that phytochrome is a photosensitive ion gate for potassium ions [434].

The effects of P_{FR} can be detected extremely quickly after its formation following irradiation with red light. Plastid orientation and the effects on sleep movements of the pulvinus can be detected within 5 min [188]. Again like IAA, phytochrome can alter the potential difference (and thus ion fluxes) between the top and bottom of a coleoptile and this can be detected 15 s after illumination of the apex with red light [328].

Excised leaflets of *Albizzia* close within 30–90 min after brief exposure to red light and subsequent transfer to darkness. If far-red light is given, opening commences after a lag period of 10 min. The pulvinule cells in *Albizzia* contain high levels of potassium ions. During leaf closure, the potassium is lost from the ventral cells and enters the dorsal cells, and on leaf opening the process is reversed [401]. As in stomatal cells, an increased inward flux of potassium ions raises the osmotic potential and thus the turgor pressure and there is a direct correlation between cell turgor and pinna movement. Further evidence relating phytochrome to potassium ion movement has been provided by the demonstration that the circadian movements of the pulvinus of *Phaseolus* can be rephased not only by phytochrome but also by valinomycin, an antibiotic that makes membranes specifically permeable to potassium ions [65]. This relationship has been heightened by the recent proposal [330] that the basic rhythmic process in circadian phenomena is a diurnal variation in membrane permeability. The role of phytochrome in rephasing plant circadian rhythms would in this case again support a membrane function. It has recently been shown that there is a circadian rhythm in potassium ion content in *Gonyaulax* [462].

Further evidence for permeability changes induced by phytochrome is shown by the "Tanada effect" [468]. Root tips of barley will adhere to a glass surface within 30 s if they are irradiated with far-red light and they can be detached just as quickly if red light is subsequently given. Glass is normally positively charged and this Tanada effect is the result of changes in the net charge associated with the root tip. The presence of both calcium ions and IAA is necessary for the effect to occur; the effect can be reversed by very low concentrations of abscisic acid. Recent studies on the intracellular localization of phytochrome point very strongly to location in the plasmalemma in the green alga *Mougeotia* and to a position in one of the cell membranes in higher plants at least for the active form of phytochrome, P_{FR} [434].

Thus there is considerable evidence which suggests that the known regulators of growth and differentiation in plants have their primary mode of action on membrane systems.

C. The possible cellular consequences of regulating ion fluxes

The best known example of ion flux regulation is the action potential seen particularly in nerve cells. This is a consequence of breakdown in selective permeability of the plasmalemma of the nerve cell to potassium and sodium ions. Action potentials have been observed in plants and in the case of *Mimosa* they are used for the transmission of information of a touch stimulus. The rate of movement of the action potential is much slower than in animal systems; the action potential initiates leaflet closure through changes in potassium ion flux and thus turgor. Changes in cellular potassium levels are likely to have more dramatic consequences than mere turgor changes. Many enzymes have a direct and specific requirement for cations for catalytic activity. The cellular processes that have been shown to be dependent upon potassium ion concentration include glycolysis, starch synthesis, oxidative phosphorylation, photosynthetic phosphorylation, TCA cycle, protein and RNA synthesis.

The influence of potassium ions on RNA and protein synthesis requires some further discussion. It is a common observation that the nuclear volume of cells undergoing differentiation may be larger than that of cells which have fully differentiated. Thus differentiation is frequently accompanied by a decline in nuclear volume. Tissue reactivation, e.g. "ageing" of storage tissue discs, germination, or treatment of mature cells with high levels of auxin, causes the nucleus to swell again. This phenomenon is related to chromatin dispersion and although the mechanics of this process are not completely understood, it is characterized by greater ability of molecules such as acridine dyes and actinomycin D to penetrate the chromatin and to bind to the DNA. Dispersed chromatin also exhibits a much higher rate of RNA synthesis than condensed chromatin. Isolated nuclei can be made to swell or contract according to the ion composition of the medium. Nuclei which have been caused to swell by placing them in low concentrations of potassium ions show all the characteristics of dispersed chromatin including higher rates of RNA synthesis. The structure of chromatin is based on an interaction between an acid (DNA) and bases (histones) with a certain degree of modulation by weakly acidic proteins (see Chapter 1). Chromatin of isolated nuclei can be changed from condensed to dispersed by lowering the potassium ion concentration of the ambient medium from 0·2M to 0·01M. In this case, the swelling is caused by repulsion between like charges which were previously neutralized by the potassium ions.

There is considerable specificity in the effects of ions on the expression of particular areas of the genome. Individual bands of some insect salivary gland chromosomes have been shown to respond to specific low concentrations of potassium ions, and others respond to sodium ions, by becoming dispersed and permitting RNA synthesis, a phenomenon referred to as puffing. Ecdysone (an insect hormone initiating the developmental step of moulting) alters the puffing pattern of *Drosophila* chromosomes and changes the ionic balance of potassium and sodium in the nucleus. The cation content of both amphibian and insect nuclei has been shown to change during differentiation and perhaps more intriguingly, ecdysone is able to phase-shift circadian rhythms in insects in an analogous way to that in which phytochrome phase-shifts circadian rhythms in plants.

The ribosome is another organelle composed of nucleic acids and proteins. The effect of low and increasing concentrations of cations on ribosomes is the selective removal of weakly bound proteins necessary for the process of protein synthesis. It would appear then that the ionic content of the cell and in particular the potassium ion content may play an important part in controlling differentiation. Further evidence supports this. The imposition of a potassium ion gradient (and a hydrogen ion or auxin gradient) across a fertilized *Fucus* egg determines the future polarity of development [218]. The rhizoid emerges on the side with the higher potassium concentration. The developing egg also exhibits a potential difference between its two ends and the polarity of development can again be determined by imposing a potential difference across the egg. Germinating *Funaria* and *Equisetum* spores show similar polarity responses to an applied potential difference. In this context, it is of interest that coleoptile cells show a slight potential difference between the top and the bottom of the cell.

Other cations that could be subject to regulation are calcium and magnesium. The function of calcium ions in plants is directly related to membrane integrity [531]. It is believed that calcium interacts with the membrane phospholipids. The removal of calcium from root cells by EDTA causes membrane disorganization and initiates a series of degenerative changes in root cells not dissimilar to those shown by fruit or leaf cells undergoing senescence. Calcium ions are intimately concerned with the mechanism of uptake of other monovalent ions. In the absence of calcium ions, potassium, sodium and chloride ions appear to be in passive diffusion equilibrium across the plasma membrane but upon the addition of calcium this is no longer the case. Membrane hyperpolarization may occur and it is thought that calcium may be required for the operation of one or more electrogenic pumps in the plasma membrane [194].

Other effects of calcium have been reported. Depletion of calcium impairs the polar movement of auxin in coleoptiles and a simple gradient of calcium ions in the stigma provides the directional stimulus for pollen tube growth [155]. Calcium is an essential component of the Tanada effect. Calcium is also an essential component of the aleurone system [537]. It is necessary to add calcium ions together with gibberellin to the isolated aleurone in order to get maximum expression of α-amylase and it was originally thought that calcium merely stabilized the enzyme secreted into solution. While this is certainly so, it has been found that four isozymes of α-amylase are secreted by the aleurone cells. Omission of calcium drastically inhibits the synthesis of only two of these and the indications are that it is events late in the lag period of gibberellin response which are particularly sensitive to calcium deficiency.

A major cellular store of calcium in animals and plants is the mitochondrion, where insoluble calcium phosphate may be accumulated [372]. Some plants also accumulate substantial levels of calcium ions in the chloroplasts. During the formation of calcium phosphate, phosphate is taken up as HPO_4^{2-} together with calcium and a proton is expelled when the insoluble tricalcium phosphate is formed. It is likely that the factors affecting the mobility of this calcium store will thereby affect the operation of the hydrogen ion pump and also affect the permeability of the mitochondrial membrane.

Magnesium is an ion that is required relatively non-specifically by a large number of enzymes. However, it has recently been suggested that the enzymatic activity of ribulose diphosphate carboxylase may be controlled by a flux of magnesium ions into the chloroplast upon exposure to light and by the appropriate flux outwards during darkness [22]. Factors that modify ion fluxes could thus markedly affect the process of photosynthesis.

X. CONCLUDING REMARKS

The evidence outlined in the preceding sections clearly shows that plant growth substances can markedly modify ion fluxes, and further that the modification of ion fluxes could lead to many of the biochemical changes known to be initiated by growth substances. However, we cannot at present assume that all effects of growth substances are mediated via changes in ion fluxes. It remains a possibility that some of the effects of growth substances may be mediated by direct interaction of the growth substances with cell components other than membranes, such as enzymes. Thus, there is clearly scope for a good deal of further research in this area.

SUGGESTIONS FOR FURTHER READING

ABELES, F. B. (1973). "Ethylene in Plant Biology". Academic Press, New York and London.

CLELAND, R. (1974). Mode of action of plant hormones. In "Biochemistry of Hormones" (H. V. Rickenberg, ed.). International review of Science, Vol. 8. Medical and Technical Publishing Co., Lancaster.

IHLE, J. N. and DURE, L. S. (1972). The developmental biochemistry of cotton seed embryogenesis and germination. III. Regulation of the biosynthesis of enzymes utilized in germination. J. biol. Chem. 247, 5048–5055.

KEY, J. L. (1969). Hormones and nucleic acid metabolism. A. Rev. Pl. Physiol., 20, 449–474.

RAVEN, J. A. and SMITH, F. A. (1973). The regulation of intercellular pH as a fundamental biological process. In "Ion Transport in Plants" (W. P. Anderson, ed.). Academic Press, London and New York.

RAY, P. M. (1974). The biochemistry of the action of indoleacetic acid on plant growth. *In* "The Chemistry and Biochemistry of Plant Hormones" (V. C. Runickles, E. Sondheimer and D. C. Walton, eds). Academic Press, New York and London.

SMITH, H. (1975). "Phytochrome and Photomorphogenesis". McGraw-Hill, Maidenhead, Berks.

TREWAVAS, A. J. (1968). Relationship between plant growth hormones and nucleic acid metabolism. *Progress in Phytochemistry,* **1,** 113–161.

YOMO, H. and VARNER, J. E. (1971). Hormonal control of a secretory tissue. *In* "Current Topics in Developmental Biology" (A. A. Moscona, ed.), Vol. 6, pp. 111–144. Academic Press, New York and London.

Bibliography

1. ABELES, F. B. (1973). "Ethylene in Plant Biology". Academic Press, New York and London.
2. ACTON, G. J. and SCHOPFER, P. (1974). Phytochrome-induced synthesis of ribonuclease *de novo* in lupin hypocotyl sections. *Biochem. J.* **142**, 449–455.
3. ALLDRIDGE, N. A. (1964). Anomalous vessel elements in wilty-dwarf tomato. *Bot. Gaz.* **125**, 138–142.
4. ALLENDE, J. E. (1969). Protein biosynthesis in plant systems. *In* "Techniques in Protein Biosynthesis" (P. N. Campbell and J. R. Sargent, eds). Vol. 2. Academic Press, London and New York.
5. AMES, B. N. and HARTMAN, P. E. (1963). The histidine operon. *Cold Spring Harb. Symp. quant. Biol.* **26**, 349–356.
6. ANDERSON, J. W. and FOWDEN, L. (1969). A study of the aminoacyl-sRNA synthetases of *Phaseolus vulgaris* in relation to germination. *Pl. Physiol* **44**, 60–68.
7. ANDERSON, L. E. and LEVIN, D. A. (1970). Chloroplast aldolase is controlled by a nuclear gene. *Pl. Physiol.* **46**, 819–820.
8. ANDERSON, M. B. and CHERRY, J. H. (1969). Differences in leucyl-transfer RNAs and synthetases in soybean seedlings. *Proc. natn. Acad. Sci. U.S.A.* **62**, 202–209.
9. ANKER, P., STROUN, M., GREPPIN, H. and FREDJ, M. (1971). Metabolic DNA in spinach stems in connexion with ageing. *Nature, New Biology* **234**, 184–186.
10. APELBAUM, A., SFAKIOTAKIS, E. and DILLEY, D. R. (1974). Reduction in extractable deoxyribonucleic acid polymerase activity in *Pisum sativum* seedlings by ethylene. *Pl. Physiol.* **54**, 125–128.
11. ASLAM, M., HUFFAKER, R. C and TRAVIS, R. L. (1973). The interaction of respiration and photosynthesis in induction of nitrate reductase activity. *Pl. Physiol.* **52**, 137–141.
12. ATTARDI, G. and AMALDI, F. (1970). Synthesis of ribosomal RNA. *A. Rev. Biochem.* **39**, 183–226.
13. AVADHANI, N. G., LYNCH, M. J. and BUETOW, D. E. (1971). Protein synthesis on polysomes in mitochondria isolated from *Euglena gracilis. Expl Cell Res.* **69**, 226–228.
14. AVERY, O. T., MACLEOD, C. M. and MCCARTY, M. (1944). Studies on the chemical nature of the substance inducing transformation of pneumococcal types. Induction of transformation by a desoxyribonucleic acid fraction isolated from *Pneumococcus* type III. *J. exp. Med.* **79**, 137–158.
15. BAGLIONI, C., BLEIBERG, I. and ZAUDERER, M. (1971). Assembly of membrane-bound polysomes. *Nature, New Biology* **232**, 8–12.
16. BAILEY, C. J., COBB, A. and BOULTER, D. (1970). A cotyledon slice system for the electron autoradiographic study of the synthesis and intracellular transport of seed storage protein of *Vicia faba. Planta* **95**, 103–118.

17. BARD, E., EFRON, D., MARCUS, A. and PERRY, R. P. (1974). Translation capacity of deadenylated messenger RNA. *Cell* **1**, 101–106.

18. BARKER, G. R., BRAY, C. M. and WALTER, T. J. (1974). The development of ribonuclease and acid phosphatase during germination of *Pisum arvense*. *Biochem. J.* **142**, 211–219.

19. BARLOW, P. W. (1973). Mitotic cycles in root meristems. *In* "The Cell Cycle in Development and Differentiation" (M. Balls and F. S. Billett, eds). Cambridge University Press.

20. BARNETT, W. E., PENNINGTON, C. J. and FAIRFIELD, S. A. (1969). Induction of *Euglena* transfer RNA's by light. *Proc. natn. Acad. Sci. U.S.A.* **63**, 1261–1268.

21. BARRETT, T., MARYANKA, D., HAMLYN, P. H. and GOULD, H. J. (1974). Non-histone proteins control gene expression in reconstituted chromatin. *Proc. natn. Acad. Sci. U.S.A.* **71**, 5057–5061.

22. BASSHAM, J. A. (1973). Control of photosynthetic carbon metabolism. *Symp. Soc. exp. Biol.* **27**, 461–484.

23. BAUR, E. (1909). Das Wesen und die Erblishkeitsverhaltnisse der "Varietates albomarginatae hort" von *Pelargonium zonale*. *Z. Vererbungslehre* **1**, 330–351.

24. BELL, P. R. (1970). Are plastids autonomous? *Symp. Soc. exp. Biol.* **24**, 109–128.

25. BELL, P. R. and MÜHLETHALER, K. (1964). The degeneration and reappearance of mitochondria in the egg cells of a plant. *J. Cell Biol.* **20**, 235–248

26. BENDICH, A. (1957). Methods for characterisation of nucleic acids by base composition. *In* "Methods in Enzymology" (S. P. Colowick and N. O. Kaplan, eds), Vol. III. Academic Press, New York and London.

27. BENDICH, A. J. and ANDERSON, R. S. (1974). Novel properties of satellite DNA from musk melon. *Proc. natn. Acad. Sci. U.S.A.* **71**, 1511–1515.

28. BENDICH, A. J. and McCARTHY, B. J. (1970). DNA comparisons among barley, oats, rye and wheat. *Genetics* **65**, 545–565.

29. BENNETT, M. D. (1972). Nuclear DNA content and minimum generation time in herbaceous plants. *Proc. R. Soc. Lond.* B **181**, 109–135.

30. BERLIN, C. M. and SCHIMKE, R. T. (1965). Influence of turnover rates on the responses of enzymes to cortisone. *Molec. Pharmacol.* **1**, 149–156.

31. BERNARDI, G., FAURES, M., PIPERNO, G. and SLONIMISKI, P. P. (1970). Mitochondrial DNA's from respiratory-sufficient and cytoplasmic respiratory-deficient mutant yeast. *J. molec. Biol.* **48**, 23–42.

32. BEVAN, E. A., HERRING, A. J. and MITCHELL, D. (1973). Preliminary characterisation of two species of DS-RNA in yeast and their relationship to the "killer" character. *Nature, Lond.* **245**, 81–86.

33. BICK, M. D. and STREHLER, B. L. (1972). Leucyl-tRNA synthetase activity in old cotyledons: evidence on repressor accumulation. *Mechanisms of Ageing and Development* **1**, 33–42.

34. BIRNSTIEL, M., SPEIRS, J., PURDOM, I., JONES, K. and LOENING, U. E. (1968). Properties and composition of the isolated ribosomal DNA satellite of *Xenopus laevis*. *Nature, Lond.* **219**, 454–463.

35. BISHOP, J. O., MORTON, J. G., ROSBASH, M. and RICHARDSON, M. (1974). Three abundance classes in HeLa cell messenger RNA. *Nature, Lond.* **250**, 199–204.

36. BLAIR, G. E. and ELLIS, R. J. (1973). Protein synthesis in chloroplasts. 1. Light-driven synthesis of the large subunit of fraction 1 protein by isolated pea chloroplasts. *Biochim. biophys. Acta* **319**, 223–234.

37. BLOCH, D. P., MacQUIGG, R. A., BRACK, S. D. and WU, J.-R, (1967). The synthesis of deoxyribonucleic acid and histone in the onion root meristem. *J. Cell Biol.* **33**, 451–468.

38. BOARDMAN, N. K., FRANCKI, R. I. B. and WILDMAN, S. G. (1966). Protein synthesis by cell-free extracts of tobacco leaves. III. Comparison of the physical properties and protein synthesising activities of 70s chloroplast and 80s cytoplasmic ribosomes. *J. molec. Biol.* **17**, 470–489.

39. BOGORAD, L., METS, L. J., MULLINIX, K. P., SMITH, H. J. and STRAIN, G. C. (1974). Possibilities for intracellular integration: the ribonucleic acid polymerases of chloroplasts and nuclei, and genes specifying chloroplast ribosomal proteins. *Biochem. Soc. Symp.* **38**, 17–41.

40. BONNER, J. (1965). "The Molecular Biology of Development". Clarendon Press, Oxford.

41. BONNER, T. I., BRENNER, D. J., NEUFELD, B. R. and BRITTEN, R. J. (1973). Reduction in the rate of DNA reassociation by sequence divergence. *J. molec. Biol.* **81**, 123–135.

42. BÖRNER, T., KNOTH, R., HERRMANN, F. and HAGEMANN, R. (1972). Struktur und Funktion der genetischen Information in den Plastiden. V. Das Fehlen von ribosomaler RNS in den Plastiden der Plastommutante "Mrs. Parker" von *Pelargonium zonale* Ait. *Theoret. appl Genetics* **42**, 3–11.

43. BOULTER, D., ELLIS, R. J. and YARWOOD, A. (1972). Biochemistry of protein synthesis in plants. *Biol. Rev.* **47**, 113–175.

44. BOURQUE, D. P. and WILDMAN, S. G. (1973). Evidence that nuclear genes code for several chloroplast ribosomal proteins. *Biochem. biophys. Res. Commun.* **50**, 532–537.

45. BRADBEER, J. W. (1973). The synthesis of chloroplast enzymes. *In* "Biosynthesis and its Control in Plants" (B. V. Milborrow, ed.). Academic Press, London and New York.

46. BRADBURY, E. M., INGLIS, R. J. and MATTHEWS, H. R. (1974). Control of cell division by very lysine rich histone (f1) phosphorylation. *Nature, Lond.* **247**, 257–261.

47. BRADSHAW, M. J. and EDELMAN, J. (1969). Enzyme formation in higher plants. The production of a gibberellin preceding invertase synthesis in aged tissue. *J. exp. Bot.* **20**, 87–93.

48. BRADY, T. (1973). Feulgen cytophotometric determination of the DNA content of the embryo proper and suspensor cells of *Phaseolus coccineus*. *Cell Differentiation* **2**, 65–75.

49. BRADY, T and CLUTTER, M. E. (1972). Cytolocalisation of ribosomal cistrons in plant polytene chromosomes. *J. Cell Biol.* **53**, 827–832.

50. BRANDLE, E. and ZETSCHE, K. (1973). Localisation of the α-amanitin sensitive RNA polymerase in nuclei of *Acetabularia*. *Planta* **111**, 209–217.

51. BRAWERMAN, G. (1974). Eukaryotic messenger RNA. *A. Rev. Biochem.* **43**, 621–642.

52. BRETSCHER, M. S. (1968). Polypeptide chain termination: an active process. *J. molec. Biol.* **34**, 131–136.

53. BREWIN, N. J. and NORTHCOTE, D. H. (1973). Variations in the amounts of 3′,5′-cyclic AMP in plant tissues. *J. exp. Bot.* **24**, 881–888.

302					Bibliography

54. BRIARTY, L. G., COULT, D. A. and BOULTER, D. (1969). Protein bodies of developing seeds of *Vicia faba. J. exp. Bot.* **20**, 358–372.
55. BRIGHTWELL, M. D., LEECH, C. E., O'FARRELL, M. K., WHISH, W. J. D. and SHALL, S. (1975). Poly(adenosine diphosphate ribose) polymerase in *Physarum polycephalum.* Biochem. J. **147**, 119–129.
56. BRITTEN, R. J. and KOHNE, D. E. (1968). Repeated sequences in DNA. *Science, N.Y.* **161**, 529–540.
57. BROCHERT, R. and McCHESNEY, J. D. (1973). Time course and localisation of DNA synthesis during wound healing of potato tuber tissue. *Devl Biol.* **35**, 293–301.
58. BROOKS, R. R. and MANS, R. J. (1973). Accentuated type I reversibility of maize DNA denaturation: thermal stability conferred by guanine-cytosine-rich tracts. *Biochim. biophys. Acta* **312**, 14–32.
59. BROWN, R. and DYER, A. F. (1972). Cell division in higher plants. *In* "Plant Physiology, a Treatise" (F. C. Steward, ed.), Vol. VIc. Academic Press, New York and London.
60. BRYANT, J. A. and AP REES, T. (1971). Nucleic acid synthesis and induced respiration by disks of carrot storage tissue. *Phytochemistry* **10**, 1191–1197.
61. BRYANT, J. A., STEVENS, C. and WEST, G. A. Previously unpublished data.
62. BRYANT, J. A. and WILDON, D. C. (1971). Criticism of the use of methylated-albumin-kieselguhr chromatography for the isolation of metabolically labile DNA. *Biochim. biophys. Acta* **232**, 624–629.
63. BRYANT, J. A., WILDON, D. C. and WONG, D. (1974). Metabolically labile DNA in aseptically grown seedlings of *Pisum sativum. Planta* **118**, 17–24.
64. BRYSK, M. M. and CHRISPEELS, M. J. (1972). Isolation and partial characterization of a hydroxy-proline-rich cell wall glycoprotein and its cytoplasmic precursor. *Biochim. biophys. Acta* **257**, 421–432.
65. BUNNING, E. and MOSER, I. (1972), Influence of valinomycin on circadian leaf movements of *Phaseolus. Proc. natn. Acad. Sci. U.S.A.* **69**, 2732–2733.
66. BURGESS, R. R. (1971). RNA polymerase. *A. Rev. Biochem.* **40**, 711–740
67. BURKARD, G., GUILLEMAUT, P., STEINMETZ, A. and WEILL, J. H. (1974). Transfer ribonucleic acid and transfer ribonucleic acid-recognizing enzymes in bean cytoplasm, chloroplasts, etioplasts and mitochondria. *Biochem. Soc. Symp.* **38**, 43–56.
68. BURKARD, G. and KELLER, E. B. (1974). Poly (A) polymerase and poly (G) polymerase in wheat chloroplasts. *Proc. natn. Acad. Sci. U.S.A.* **71**, 389–393.
69. BURZIO, L. and KOIDE, S. S. (1973). Activation of the template activity of isolated rat liver nuclei for DNA synthesis and its inhibition by NAD. *Biochem. biophys. Res. Commun.* **53**, 572–579.
70. CALLAN, H. G. (1967). On the organisation of genetic units in chromosomes. *J. Cell Sci.* **2**, 1–7.
71. CALLAN, H. G. (1973). Replication of DNA in eukaryotic chromosomes. *Br. med. Bull.* **29**, 192–195.
72. CALVAYRAC, R., BUTOW, R. A. and LEFFORT-TRAN, M. (1972). Cyclic replication of DNA and changes in mitochondrial morphology during the cell cycle of *Euglena gracilis* (Z). *Expl Cell Res.* **71**, 422–432.
73. CARLSON, P. S. (1972). Notes on the mechanism of action of gibberellic acid. *Nature, New Biology* **237**, 39–41.
74. CHAMBRON, P., WEILL, J. D., DOLY, J., STROSSER, M. T. and MANDEL, P. (1966). On the formation of a novel adenylic compound by enzymatic extracts of liver nuclei. *Biochem. biophys. Res. Commun.* **25**, 638–645.

75. CHAN, P. H. and WILDMAN, S. G. (1972). Chloroplast DNA codes for the primary structure of the large subunit of Fraction I protein. *Biochim. biophys. Acta* **277**, 677–680.
76. CHANG, L. M. S. and BOLLUM, F. J. (1972). Variation of deoxyribonucleic acid polymerase activities during rat liver regeneration. *J. biol. Chem.* **247**, 7948–7950.
77. CHANG, L. M. S., BROWN, M. and BOLLUM, F. J. (1973). Induction of DNA polymerase in mouse L cells. *J. molec. Biol.* **74**, 1–8.
78. CHARGAFF, E. (1950). Chemical specificity of nucleic acids and mechanism of their enzymatic degradation. *Experientia* **6**, 201–209
79. CHEN, D., SCHULTZ, G. and KATACHALSKI, E. (1971). Early ribosomal RNA transcription and appearance of cytoplasmic ribosomes during germination of the wheat embryo. *Nature, New Biology* **231**, 69–72.
80. CHEN, J. L. and WILDMAN, S. G. (1970). "Free" and membrane-bound ribosomes and nature of products formed by isolated chloroplasts incubated for protein synthesis. *Biochim. biophys. Acta* **209**, 207–219.
81. CHIANG, K.-S. and SUEOKA, N. (1967). Replication of chloroplast DNA in *Chlamydomonas reinhardi* during vegetative cell cycle: its mode and regula tion. *Proc. natn. Acad. Sci. U.S.A.* **57**, 1506–1513.
82. CHRISPEELS, M. J. (1970). Synthesis and secretion of hydroxyproline-containing macromolecules in carrots. II. *In vivo* conversion of peptidyl proline to peptidyl hydroxyproline. *Pl. Physiol.* **45**, 223–227.
83. CHUN, E. H. L., VAUGHAN, M. H. and RICH, A. (1963). The isolation and characterisation of DNA associated with chloroplast preparations. *J. molec. Biol.* **7**, 130–141.
84. CLELAND, R. (1967). Auxin and the mechanical properties of the cell wall. *Ann. Acad. Sci. N. Y.* **144**, 3–18.
85. CLOWES, F. A. L. and JUNIPER, B. E. (1968). "Plant Cells". Blackwell Scientific Publications, Oxford.
86. COEN, D., DEUTSCH, J., NETTER, P., PETROCHILO, E. and SLONIMISKI, P. P. (1970). Mitochondrial genetics. I. Methodology and phenomenology. *Symp. Soc. exp. Biol.* **24**, 449–496.
87. COOK, J. R. (1966). The synthesis of cytoplasmic DNA in synchronised *Euglena*. *J. Cell Biol.* **29**, 369–373.
88. CORRENS, C. (1909). Vererbungsuersuche mit blass (gelb) grunen and buntblattrigen Sippen bei *Mirabilis jalapa, Urtica pilulifera* und *Lunaria annua*. *Z. Vererbungslehre* **1**, 291–329.
89. COVEY, S. N. and GRIERSON, D. (1976). Subcellular distribution and properties of poly(A)-containing RNA from cultured plant cells. *Eur. J. Biochem.*, **63**, 599–606.
90. COX, B. J. and TURNOCK, G. (1973). Synthesis and processing of ribosomal RNA in cultured plant cells. *Eur. J. Biochem.* **37**, 367–376.
91. CRAFTS, A. S. and CURRIER, H. B. (1963). On sieve tube formation. *Protoplasma* **57**, 188–202.
92. CRICK, F. H. C. (1957). The structure of nucleic acids and their role in protein synthesis. *Biochem. Soc. Symp.* **14**, 25–26.
93. CRICK, F. H. C. (1966). Codon-anticodon pairing: the wobble hypothesis. *J. molec. Biol.* **19**, 548–555.
94. CRICK, F. (1971). General model for the chromosomes of higher organisms. *Nature, Lond.* **234**, 25–27.

95. CRONSHAW, J. and ESAU, K. (1967). Tubular and fibrillar components of mature and differentiating sieve elements. *J. Cell Biol.* **34**, 801–815.
96. CULLIS, C. A. (1973). DNA differences between flax genotrophs. *Nature, Lond.* **243**, 515–516.
97. D'AMATO, F. (1952). Polyploidy in the differentiation and function of tissues and cells in plants. *Caryologia* **4**, 311–358.
98. DAS, N. K. (1963). Chromosomal and nucleolar RNA synthesis in root tips during mitosis. *Science, N. Y.* **140**, 1231–1233.
99. DATKO, A. H. and MCLACHLAN, G. A. (1968). Indoleacetic acid and the synthesis of glucanases and pectic enzymes. *Pl. Physiol.* **43**, 735–742.
100. DAVIDSON, J. N. (1972). "The Biochemistry of the Nucleic Acids" (7th edition). Chapman and Hall, London.
101. DAVIES, E. and LARKINS, B. A. (1973). Polyribosomes from peas. *Pl. Physiol.* **52**, 339–345.
102. DAVIES, E. and MCLACHLAN, G. A. (1969). Generation of cellulase activity during protein synthesis by pea microsomes *in vitro*. *Archs Biochem. Biophys.* **129**, 581–587.
103. DELANGE, R. J., FAMBOROUGH, D. M., SMITH, E. L. and BONNER, J. (1969). Calf and pea histone IV. III. Complete amino acid sequence of pea seedling histone IV. Comparison with the homologous calf thymus histone. *J. biol. Chem.* **244**, 5669–5679.
104. DETCHON, P and POSSINGHAM, J. V. (1973). Chloroplast ribosomal ribonucleic acid synthesis in cultured spinach leaf tissue. *Biochem. J.* **136**, 829–836.
105. DIENER, T. O. (1975). Viroids. *In* "Modifications of the Information Content of Plant Cells" (Proceedings of 2nd John Innes Symposium, R. Markham, D. R. Davies, D. A. Hopwood and R. W. Horne, eds) North-Holland, Amsterdam.
106. DJORDJEVIC, B. and SZYBALSKI, W. (1960). Genetics of human cell lines. III. Incorporation of 5-bromo- and 5-iododeoxyuridine into the deoxy-ribonucleic acid of human cells and its effect on radiation sensitivity. *J. exp. Med.* **112**, 509–513.
107. DOOLITTLE, W. F. (1972). Ribosomal ribonucleic acid synthesis and maturation in the blue-green alga *Anacystis nidulans*. *J. Bacteriol.* **111**, 316–324.
108. DUDOCK, B. J., KATZ, G., TAYLOR, E. A. and HOLLEY, R. W. (1969). Primary structure of wheat germ phenylalanine transfer RNA. *Proc. natn. Acad. Sci. U.S.A.* **62**, 941–945.
109. DUNHAM, V. L. and CHERRY, J. H. (1973). Multiple DNA polymerase activity solubilized from higher plant chromatin. *Biochem. biophys. Res. Commun.* **54**, 403–410.
110. DUNN, J. J. and STUDIER, W. (1973). T_7 early RNAs and *E. coli* ribosomal RNAs are cut from large precursor RNAs *in vivo* by ribonuclease III. *Proc. natn. Acad. Sci. U.S.A.* **70**, 3296–3300.
111. DURRANT, A. and JONES, T. W. A. (1971). Reversion of induced changes in amount of nuclear DNA in *Linum*. *Heredity* **27**, 431–439.
112. EAGLESHAM, A. R. J. and ELLIS, R. J. (1974). Protein synthesis in chloroplasts. II. Light-driven synthesis of membrane proteins by isolated pea chloroplasts. *Biochim. biophys. Acta.* **335**, 396–407.
113. ECKSTEIN, H., PADUCH, V. and HITZ, H. (1967). Synchronized yeast cells. 3. DNA synthesis and DNA polymerase after inhibition of cell division by X-rays. *Eur. J. Biochem.* **3**, 224–231.
114. EDELMAN, J. and HANSON, A. D. (1971a). Sucrose suppression of chlorophyll synthesis in carrot callus cultures *Planta* **98**, 150–156.

115. EDELMAN, J. and HANSON, A. D. (1971b). Sucrose suppression of chlorophyll synthesis in carrot tissue cultures: the role of invertase. *Planta* **101**, 122–132.

116. EDELMAN, M., EPSTEIN, H. T. and SCHIFF, J. A. (1966). Isolation and characterization of DNA from the mitochondrial fraction of *Euglena. J. molec. Biol.* **17**, 463–469.

117. EDELMAN, M., SCHIFF, J. A. and EPSTEIN, H. T. (1965). Studies of chloroplast development in *Euglena* XII. Two types of satellite DNA. *J. molec. Biol.* **11**, 769–774.

118. ELLIS, R. J., BLAIR, G. E. and HARTLEY, M. R. (1974).The nature and function of chloroplast protein synthesis. *Biochem. Soc. Symp.* **38**, 137–162.

119. ELLIS, R. J. and HARTLEY, M. R. (1971). Sites of synthesis of chloroplast proteins. *Nature, New Biology* **233**, 193–196.

120. ELLIS, R. J. and HARTLEY, M. R. (1974). Nucleic acids of chloroplasts. *In* "Nucleic Acids" (K. Burton, ed.). International Review of Science, Biochemistry Series, Vol. 6. Medical and Technical Publishing Co. Lancaster.

121. ENGLEMAN, E. M. (1965). Sieve elements of *Impatiens sultanii*. 2. Developmental aspects. *Ann. Bot.* **29**, 103–118.

122. EPHRUSSI, B. (1953). "Nucleo-cytoplasmic Relations in Micro-organisms". Clarendon Press, Oxford.

123. ERHAN, S., REISHER, S., FRANKO, E. A., KAMATH, S. A. and RUTMAN, R. J. (1970). Evidence for "the wedge", initiator of DNA replication. *Nature, Lond.* **225**, 340–342.

124. ERIKSON, R. O. (1964). Synchronous cell and nuclear division in tissues of the higher plants. *In* "Synchrony in cell division and growth". (E. Zeuthen, ed.). Interscience Publishers, New York.

125. ERIKSSON, T. (1966). Partial synchronization of cell division in suspension cultures of *Haplopappus gracilis*. *Physiol. Plantarum* **19**, 900–910.

126. ESAU, K. (1964). Aspects of ultrastructure of phloem. *In* "The Formation of Wood in Forest Trees" (M. H. Zimmerman, ed.). Academic Press, New York and London.

127. ESAU, K. and CHEADLE, V. I. (1965). Cytologic studies on phloem. *Univ. Calif. Publ. Bot.* **36**, 253–344.

128. ESAU, K. and GILL, R. H. (1965). Observations on cytokinesis. *Planta* **67**, 168–181.

129. ETHERTON, B. (1970). Effect of 3 indoleacetic acid on membrane potentials of oat coleoptile cells. *Pl. Physiol.* **45**, 527–528.

130. EVANS, M. L. and HOKANSON, R. (1969). Timing of the response of coleoptiles to the application and withdrawal of various auxins. *Planta* **85**, 85–95.

131. EVANS, M. L. and RAY, P. M. (1969). Timing of the auxin response in coleoptiles and its implications regarding auxin action. *J. gen. Physiol.* **53**, 1–20.

132. EVINS, M. H. and VARNER, J. E. (1972). Hormonal control of polyribosome formation in barley aleurone layers. *Pl. Physiol.* **49**, 348–352.

133. FAIRFIELD, S. A. and BARNETT, W. E. (1971). On the similarity between the tRNA's of organelles and prokaryotes. *Proc. natn. Acad. Sci. U.S.A.* **68**, 2972–2976.

134. FALK, H. (1969). Rough thylakoids: polysomes attached to chloroplast membranes. *J. Cell Biol.* **42**, 582–587.

135. FAUMAN, M., RABIWITZ, M. and GETZ, G. M. (1969). Base composition and sedimentation properties of mitochondrial RNA of *Saccharomyces cerevisiae*. *Biochim. biophys. Acta.* **182**, 355–360.

135a. FINCH, J. T., NOLL, M. and KORNBERG, R. D. (1975). Electron microscopy of defined lengths of chromatin. *Proc. natn. Acad. Sci. U.S.A.* **72**, 3320–3322.
136. FIRTEL, R. A., BAXTER, L. and LODISH, H. (1973). Actinomycin D and the regulation of enzyme biosynthesis during development of *Dictyostelium discoideum. J. molec. Biol.* **79**, 315–327.
137. FLAMM, W. G. (1972). Highly repetitive sequences of DNA in chromosomes. *Int. Rev. Cytol.* **32**, 2–51.
138. FLAMM, W. G., BERNHEIM, N. J. and BRUBACKER, P. E. (1971). Density gradient analysis of newly replicated DNA from synchronized mouse lymphoma cells. *Expl Cell Res.* **64**, 97–104.
139. FLAMM, W. G., WALKER, P. M. B. and McCALLUM, M. (1969). Some properties of the single strands isolated from the DNA of the nuclear satellite of the mouse (*Mus musculus*). *J. molec. Biol.* **40**, 423–443.
140. FLOWERS, H. M., BATRA, K. K., KEMP, J. and HASSID, W. Z. (1968). Biosynthesis of insoluble glucans from uridine-diphosphate-D-glucose with enzyme preparations from *Phaseolus aureus* and *Lupinus albus. Pl. Physiol.* **43**, 1703–1709.
141. FRANKLIN, R. E. and GOSLING, R. G. (1953). Molecular configuration in sodium thymonucleate. *Nature, Lond.* **171**, 740–741.
142. FRASER, R. S. S. (1969). The effects of two TMV strains on the synthesis and stability of chloroplast rRNA in tobacco leaves. *Molec. gen. Genetics* **106**, 73–79.
143. FRASER, R. S. S. (1973). The synthesis of tobacco mosaic virus RNA and ribosomal RNA in tobacco leaves. *J. gen. Virol.* **18**, 267–279.
144. FRENCH, C. J. and SMITH, H. (1975). An inactivator of phenylalanine-ammonia lyase from gherkin hypocotyls. *Phytochemistry* **14**, 963–966.
145. FREUDENBERG, K. (1959). Biosynthesis and constitution of lignin. *Nature, Lond.* **183**, 1152–1155.
146. FURIYAMA, Y. and LUCK, D. J. L. (1973). Ribosomal RNA synthesis in mitochondria of *Neurospora crassa. J. molec. Biol.* **73**, 425–437.
147. GAHAN, P. B. and MAPLE, A. J. (1966). The behaviour of lysosome-like particles during cell differentiation. *J. exp. Bot.* **17**, 151–155.
148. GARDINER, M. and CHRISPEELS, M. J. (1975). Involvement of the Golgi apparatus in the synthesis and secretion of hydroxyproline-rich cell wall glycoproteins. *Pl. Physiol.* **55**, 536–541.
149. GEFTER, M. L. and RUSSELL, R. L. (1969). Role of modifications in tyrosine transfer RNA: a modified base affecting ribosome binding. *J. molec. Biol.* **39**, 145–157.
150. GIANNATTASIO, M., SICA, G. and MACCHIA, V. (1974). Cyclic AMP phosphodiesterase from dormant tubers of Jerusalem artichoke. *Phytochemistry* **13**. 2729–2733.
151. GILL, B. S. and KIMBER, G. (1974). The Giemsa C-banded karyotype of rye. *Proc. natn. Acad. Sci. U.S.A.* **71**, 1247–1249.
152. GLEASON, F. K. and HOGENKAMP, H. P. C. (1970). Ribonucleotide reductase from *Euglena gracilis*, a deoxyadenosylcobalamin-dependent enzyme. *J. biol Chem.* **245**, 4894–4899.
153. GOLINSKA, B. and LEGOCKI, A. B. (1973). Purification and some properties of elongation factor-1 from wheat germ. *Biochim. biophys. Acta* **324**, 156–170.
154. GOULD, A. R., BAYLISS, M. W. and STREET, H. E. (1974). Studies on the growth in culture of plant cells. XVII. Analysis of the cell cycle of asynchronously

dividing *Acer pseudoplatanus* L. cells in suspension culture. *J. exp. Bot.* **25**, 468–478.

155. GREEN, P. B. (1969). Cell morphogenesis. *A. Rev. Pl. Physiol.* **20**, 365–395.

156. GRESSEL, J., ROSNER, A. and COHEN, N. (1975). Temperature of acrylamide polymerization and electrophoretic mobility of nucleic acids *Analyt. Biochem.* **69**, 84–91.

157. GRIENINGER, G. E. and ZETSCHE, K. (1972). Die Activität von Phosphoglucose-isomerase und Phosphoglucomutase während der Morphogenese kernhaltiger und kernlose Acetabularien. *Planta* **104**, 329–351.

158. GRIERSON, D. (1974). Characterisation of ribonucleic acid components from leaves of *Phaseolus aureus*. *Eur. J. Biochem.* **44**, 509–515.

159. GRIERSON, D. (1975). The hybridisation of ^{3}H-poly(uridylic acid) to DNA from *Phaseolus aureus*. *Planta* **127**, 87–92.

160. GRIERSON, D., COVEY, S. N. and GILES, A. W. (1975). The effect of light on RNA synthesis in developing leaves. *In* "Light and Plant Development". Proceedings of 22nd Nottingham Easter School in Agricultural Science (H. Smith, ed.), Butterworths, London.

161. GRIERSON, D., COVEY, S. N. and GILES, A. W. Previously unpublished data.

162. GRIERSON, D. and LOENING, U. E. (1974). Ribosomal RNA precursors and the synthesis of chloroplast and cytoplasmic ribosomal ribonucleic acid in leaves of *Phaseolus aureus*. *Eur. J. Biochem.* **44**, 501–507.

163. GRIERSON, D., ROGERS, M. E., SARTIRANA, M. L. and LOENING, U. E. (1970). The synthesis of ribosomal RNA in different organisms: structure and evolution of the rRNA precursors. *Cold Spring Harb. Symp. quant. Biol.* **35**, 589–598.

164. GRIERSON, D. and SMITH, H. (1973). The synthesis and stability of ribosomal RNA in blue-green algae. *Eur. J. Biochem.* **36**, 280–285.

165. GROOT, G. S. P., FLAVELL, R. A., VAN OMMEN, G. J. B. and GRIVELL, L. A. (1974). Yeast mitochondrial RNA does not contain poly (A). *Nature, Lond.* **252**, 167–168.

166. GUDERIAN, R. H., PULLIAM, R. L. and GORDON, M. P. (1972). Characterization and fractionation of tobacco leaf transfer RNA. *Biochim. biophys. Acta* **262**, 50–65.

167. GUILFOYLE, T. J., LIN, C. Y., CHEN, Y. M., NAGAO, R. T. and KEY, J. L. (1975). Enhancement of soybean RNA polymerase I by auxin. *Proc. natn. Acad. Sci. U.S.A.* **72**, 69–72.

168. GUILLEMAUT, P., BURKARD, G., STEINMETZ, A. and WEILL, J. H. (1973). Comparative studies on the tRNAs^{met} from the cytoplasm, chloroplasts and mitochondria of *Phaseolus vulgaris*. *Plant. Sci. Lett.* **1**, 141–149.

169. GUILLEMAUT, P., BURKARD, G and WEILL, J. H. (1972). Characterization of *n*-formyl-methionyl-tRNA in bean mitochondria and etioplasts. *Phytochemistry* **11**, 2217–2219.

170. GURDON, J. B. and WOODLAND, H. R. (1968). The cytoplasmic control of nuclear activity in animal development. *Biol. Rev.* **43**, 233–267.

171. HADDON, L. E. and NORTHCOTE, D. H. (1975). Quantitative measurement of the course of bean callus differentiation. *J. Cell Sci.* **17**, 11–26.

172. HAGER, A., HENZEL, H. and KRAUSS, A. (1971). Versuche und Hypothese zur Primarwirkung des Auxins beim Strechungswachstum. *Planta* **100**, 47–75.

173. HALVORSON, H. O., CARTER, B. L. A. and TAURO, P. (1971). Synthesis of enzymes during the cell cycle. *In* "Advances in Microbial Physiology" (A. H.

Rose and J. F. Wilkinson, eds), Vol. 6, pp. 47–106. Academic Press, London and New York.

174. HÄMMERLING, J. (1963). Nucleo-cytoplasmic interactions in *Acetabularia* and other cells. *A. Rev. Pl. Physiol.* **14**, 65–92.

175. HARLAND, J., JACKSON, J. F. and YEOMAN, M. M. (1973). Changes in some enzymes involved in DNA biosynthesis following induction of division in cultured plant cells. *J. Cell Sci.* **13**, 121–138.

176. HARRIS, H. (1974). "Nucleus and Cytoplasm" (3rd edition). Clarendon Press, Oxford.

177. HARRIS, P. J. and NORTHCOTE, D. H. (1970). Patterns of polysaccharide biosynthesis in differentiating cells of maize root tips. *Biochem. J.* **120**, 479–491.

178. HARRIS, P. J. and NORTHCOTE, D. H. (1971). Polysaccharide formation in plant Golgi bodies. *Biochim. biophys. Acta* **237**, 56–64.

179. HARTLEY, M. R. Previously unpublished results.

180. HARTLEY, M. R., WHEELER, A. and ELLIS, R. J. (1975). Protein synthesis in chloroplasts. V. Translation of messenger RNA for the large sub-unit of fraction I protein in a heterologous cell-free system. *J. molec. Biol.* **91**, 67–77.

181. HARTLEY, M. R. and ELLIS, R. J. (1973). Ribonucleic acid synthesis in chloroplasts. *Biochem. J.* **134**, 249–262.

182. HASELKORN, R. and ROTHMAN-DENES, L. B., (1973). Protein synthesis. *A. Rev. Biochem.* **42**, 397–438.

183. HECHT, N. B. and DAVIDSON, D. (1973). The presence of a common active subunit in low and high molecular weight murine DNA polymerases. *Biochem. biophys. Res. Commun.* **51**, 299–305.

184. HECHT, S. M., FAULKNER, R. D. and HAWRELAK, S. D. (1974). Competitive inhibition of beef heart cyclic AMP phosphodiesterase by cytokinins and related compounds. *Proc. natn. Acad. Sci. U.S.A.* **71**, 4670–4674.

185. HECKER, L. I., EGAN, J., REYNOLDS, R. J., NIX, C. E., SCHIFF, J. A. and BARNETT, W. E. (1974). The sites of transcription and translation for *Euglena* chloroplastic aminoacyl-tRNA synthetases. *Proc. natn. Acad. Sci. U.S.A.* **71**, 1910–1914.

186. HEIMER, Y. M. and FILNER, P. (1971). Regulation of the nitrate assimilation pathway in cultured tobacco cells. III. The nitrate uptake system. *Biochim. biophys. Acta* **230**, 362–372.

187. HEMBERG, T. (1947). Studies of auxins and growth-inhibiting substances in the potato tuber and their significance with regard to its rest period. *Acta Horti Bergiani* **14**, 133–220.

188. HENDRICKS, C. B. and BORTHWICK, H. A. (1967). The function of phytochrome in regulation of plant growth. *Proc. natn. Acad. Sci. U.S.A.* **58**, 2125–2130.

189. HEREFORD, L. M. and HARTWELL, L. H. (1973). Role of protein synthesis in the replication of yeast DNA. *Nature, New Biology* **244**, 129–131.

190. HERMANN, E. C. and SCHMIDT, R. R. (1965). Synthesis of phosphorus-containing macromolecules during synchronous growth of *Chlorella pyrenoidosa*. *Biochim. biophys. Acta* **95**, 63–75.

191. HERMANN, R. G. (1969). Are chloroplasts polyploid? *Expl Cell Res.* **55**, 414–416.

192. HEYN, A. N. J. (1931). Der Mechanismus der Zellstreckung. *Rec. Trav. Bot. Neerl.* **28**, 1–113.

193. HIGGINS, T. J. V., MERCER, J. F. B. and GOODWIN, P. B. (1973). Poly-(A) sequences in plant polysomal RNA. *Nature, New Biology* **246,** 68–70.

194. HIGINBOTHAM, N. (1973). Electropotentials of plant cells, *A. Rev. Pl. Physiol.* **24,** 25–46.

195. HNILICA, L. S. (1972). "The Structure and Biological Function of Histones", Chemical Rubber Company Press, Cleveland, Ohio.

196. HOCK, B. and BEEVERS, H. (1966). Development of the glyoxylate-cycle enzymes in water melon seedlings (*Citrullus vulgaris* Schrad.) *Z. Pflanzenphysiol.* **55,** 405–414.

197. HOLM, R. E. and KEY, J. L. (1971). Inhibition of auxin-induced DNA synthesis and chromatin activity by 5-FUDR in soybean hypocotyls, *Pl. Physiol.* **47,** 606–608.

198. HOOBER, J. K. and BLOBEL, G. (1969). Characterisation of the chloroplastic and cytoplasmic ribosomes of *Chlamydomonas reinhardi. J. molec. Biol.* **41,** 121–138.

199. HOPKINS, H. A., SITZ, T. O. and SCHMIDT, R. R. (1970). Selection of synchronous *Chlorella* cells by centrifugation to equilibrium in Ficoll. *J. Cell Physiol.* **70,** 231–234.

200. HOTTA, Y. and STERN, H. (1963a). Molecular facets of mitotic regulation. I. Synthesis of thymidine kinase. *Proc. natn. Acad. Sci. U.S.A.* **49,** 649–654.

201. HOTTA, Y. and STERN, H. (1963b). Molecular facets of mitotic regulation. II. Factors underlying the removal of thymidine kinase. *Proc. natn. Acad. Sci. U.S.A.* **49,** 861–865.

202. HOWARD, A. and PELC, S. R. (1953). Synthesis of desoxyribonucleic acid in normal and irradiated cells and its relation to chromosome breakage. *Heredity* (suppl.) **6,** 261–273.

203. HOWLAND, G. P. (1975). Dark-repair of ultraviolet-induced pyrimidine dimers in the DNA of wild carrot protoplasts. *Nature, Lond.* **254,** 160–161.

204. HUGUET, T. and JOUANIN, L. (1972). The heterogeneity of wheat nuclear DNA. *Biochim. biophys. Acta* **262,** 431–440.

205. HUSSEY, G. (1975). Totipotency in tissue explants and callus of some members of the Liliaceae, Iridaceae and Amaryllidaceae, *J. exp. Bot.* **26,** 253–262.

206. IHLE, J. N. and DURE, L. S. (1970). Hormonal regulation of translation inhibition requiring RNA synthesis. *Biochem. biophys. Res. Commun.* **38,** 995–1001.

207. IHLE, J. N. and DURE, L. S. (1972). The developmental biochemistry of cotton seed embryogenesis and germination. III. Regulation of the biosynthesis of enzymes utilised in germination. *J. biol. Chem.* **247,** 5048–5055.

208. INGLE, J., KEY, J. L. and HOLM, R. E. (1965). Demonstration and characterisation of a DNA-like RNA in excised plant tissue. *J. molec. Biol.* **11,** 730–746.

209. INGLE, J., PEARSON, G. C. and SINCLAIR, J. (1973). Species distribution and properties of nuclear satellite DNA in higher plants. *Nature, New Biology* **242,** 193–197.

210. INGLE, J., POSSINGHAM, J. V., WELLS, R., LEAVER, C. J. and LOENING, U. E. (1970). The properties of chloroplast ribosomal-RNA. *Symp. Soc. exp. Biol.* **24,** 303–325.

211. INGLE, J. and SINCLAIR, J. (1972). Ribosomal RNA genes and plant development. *Nature, Lond.* **235,** 30–32.

212. ITZHAKI, R. F. (1971). The arrangement of proteins on the deoxyribonucleic acid in chromatin. *Biochem. J.* **125,** 221–224.

213. JACKSON, M. and INGLE, J. (1973). The nature of polydisperse ribonucleic acid in plants. *Biochem. J.* **131,** 523–533.
214. JACOB, F. and MONOD, J. (1961). Genetic regulatory mechanisms in the synthesis of proteins. *J. molec. Biol.* **3,** 318–356.
215. JACOBS-LORENA, M. and BAGLIONI, C. (1970). A study of ribosomal subunits during cell-free protein synthesis. *Biochim. biophys. Acta* **224,** 165–173.
216. JACOBSON, A., FIRTEL, R. A. and LODISH, H. F. (1974). Synthesis of messenger RNA and ribosomal RNA precursors in isolated nuclei of the cellular slime mould *Dictyostelium discoideum. J. molec. Biol.* **82,** 213–230.
217. JACOBSON, A. B., SWIFT, H. and BOGORAD, L. (1963). Cytochemical studies concerning the occurrence and distribution of RNA in plastids of *Zea mays. J. Cell Biol.* **17,** 557–570.
218. JAFFE, L. F. (1969). On the centripetal course of development, the *Fucus* egg, and self-electrophoresis. *Symp. Soc. devl Biol.* **28,** 83–108.
219. JAKOB, K. M. and BOVEY, F. (1969). Early nucleic acid and protein synthesis and mitoses in the primary root tips of germinating *Vicia faba. Expl Cell Res.* **54,** 118–126.
220. JARVIS, B. C., FRANKLAND, B. and CHERRY, J. H. (1968). Increased DNA template and RNA polymerase associated with the breaking of seed dormancy. *Pl. Physiol.* **43,** 1734–1736.
221. JEFFS, R. A. and NORTHCOTE, D. H. (1966). Experimental induction of vascular tissue in an undifferentiated plant callus. *Biochem. J.* **101,** 146–152.
222. JEFFS, R. A. and NORTHCOTE, D. H. (1967). The influence of indol-3-yl acetic acid and sugar on the pattern of induced differentiation in plant tissue culture. *J. Cell Sci.* **2,** 77–88.
223. JOHN, P. C. L., McCULLOUGH, W., ATKINSON, A. W., FORDE, B. G. and GUNNING, B. E. (1973). The cell cycle in *Chlorella. In* "The Cell Cycle in Development and Differentiation" (M. Balls and F. S. Billett, eds). Cambridge University Press.
224. JOHNSON, C. B., HOLLOWAY, B. R., SMITH, H. and GRIERSON, D. (1973). Isozymes of acid phosphatase in germinating peas. *Planta* **115,** 1–10.
225. JOHRI, M. M. and VARNER, J. E. (1968). Enhancement of RNA synthesis in isolated pea nuclei by gibberellic acid. *Proc. natn. Acad. Sci. U.S.A.* **59,** 269–276.
226. JONES, B. L., NAGABHUSHAN, N. and ZALIK, S. (1972). Isolation and partial characterization of chloroplast and cytoplasmic ribosomes and ribosomal subunits from wheat. *Symp. Biol. hung.* **13,** 249–262.
227. JONES, K. W. (1970). Chromosomal and nuclear location of mouse satellite DNA in individual cells. *Nature, Lond.* **225,** 912–915.
228. JOUANNEAU, J. P. (1971). Controle par les cytokinines de la synchronisation des mitoses dans les cellules de tabac. *Expl Cell Res.* **67,** 329–337.
229. KAHL, G. (1973). Genetic and metabolic regulation in differentiating plant storage tissue cells. *Bot. Rev.* **39,** 274–299.
230. KALOUSEK, F. and MORRIS, N. R. (1969). Deoxyribonucleic acid methylase activity in pea seedlings. *Science, N.Y.* **164,** 721–722.
231. KANABUS, J. and CHERRY, J. H. (1971). Isolation of an organ-specific leucyl-tRNA synthetase from soybean seedlings. *Proc. natn. Acad. Sci. U.S.A.* **68,** 873–876.
232. KATTERMAN, F. R. H. and ENDRIZZI, J. E. (1973). Studies on the 70-*S* ribosomal

content of a plastid mutant in *Gossypium hirsutum. Pl. Physiol.* **51,** 1138–1139.

233. KAWASHIMA, N. and WILDMAN, S. G. (1972). Studies on fraction I protein. IV. Mode of inheritance of primary structure in relation to whether chloroplast or nuclear DNA contains the code for a chloroplast protein. *Biochim. biophys. Acta* **262,** 42–49.

234. KEEGSTRA, K., TALMADGE, K. W., BAUR, W. D. and ALBERSHEIM, P. (1973). The structure of plant cell walls. III. A model of the walls of cultured sycamore cells based on the interconnections of the macromolecular components. *Pl. Physiol.* **51,** 188–196.

235. KEIR, H. M. and CRAIG, R. K. (1973). Regulation of deoxyribonucleic acid synthesis. *Biochem. Soc. Trans.* **1,** 1073–1077.

236. KELLER, S. J., BIEDENBACH, S. A. and MEYER, R. R. (1973). Partial purification of a chloroplast DNA polymerase from *Euglena gracilis. Biochem. biophys. Res. Commun.* **50,** 620–628.

237. KENDE, H. (1971). The cytokinins. *Int. Rev. Cytol.* **31,** 301–338.

238. KESSLER, B. (1971). Isolation, characterization and distribution of a DNA ligase from higher plants. *Biochem. biophys. Acta* **240,** 496–505.

239. KEY, J. L. (1969). Hormones and nucleic acid metabolism. *A. Rev. Pl. Physiol.* **20,** 449–474.

240. KEY, J. L. and INGLE, J. (1964). Requirement for the synthesis of DNA-like RNA for the growth of excised plant tissue. *Proc. natn. Acad. Sci. U.S.A.* **52,** 1382–1388.

241. KEY, J. L. and INGLE, J. (1968). RNA metabolism in response to auxin. *In* "Biochemistry and Physiology of Growth Substances" (G. Setterfield and F. Wightman, eds). Runge Press, Ottawa.

242. KING, P. J., COX, B. J., FOWLER, M. W. and STREET, H. E. (1974). Metabolic events in synchronized cell cultures of *Acer pseudoplatanus* L. *Planta* **117,** 109–122.

243. KING, P. J. and STREET, H. E. (1973). Growth patterns in cell cultures. *In* "Plant Tissue and Cell Culture" (H. E. Street, ed.). Blackwell Scientific Publications, Oxford.

244. KIRK, J. T. O. (1964). DNA-dependent RNA synthesis in chloroplast preparations. *Biochem. biophys. Res. Commun.* **14,** 393–397.

245. KIRK, J. T. O. (1967). Effect of methylation of cytosine residues on the buoyant density of DNA in caesium chloride solution. *J. molec. Biol.* **28,** 171–172.

246. KIRK, J. T. O. (1971). Will the real chloroplast DNA please stand up? *In* "Autonomy and Biogenesis of Mitochondria and Chloroplasts" (N. K. Boardman, A. W. Linnane and R. M. Smillie, eds). North-Holland, Amsterdam.

247. KIRKHAM, M. B., GARDNER, W. R. and GERLOFF, G. C. (1972). Regulation of cell division and cell enlargement by turgor pressure. *Pl. Physiol.* **49,** 961–962.

248. KISLEV, N. and EISENSTADT, J. M. (1972). Protein synthesis in mitochondria of *Euglena gracilis. Eur. J. Biochem.* **31,** 226–229.

249. KISLEV, N., SELSKY, M. I., NORTON, C. and EISENSTADT, J. M. (1972). tRNA and tRNA aminoacyl synthetases of chloroplasts, mitochondria and cytoplasm from *Euglena gracilis. Biochim. biophys. Acta* **287,** 256–269.

250. KISLEV, N., SWIFT, H. and BOGORAD, L. (1965). Nucleic acids of chloroplasts and mitochondria in Swiss chard. *J. Cell Biol.* **25,** 327–344.

251. KLEINSCHMIDT, A. K., LANG, D., JACHERTS, D. and ZAHN, R. K. (1962).

Darstellung und Langenmessungen des gesamten Desoxyribonucleinsäure-
inhaltes von T2-Bakteriophagen. *Biochim. biophys. Acta* **61**, 857–864.

252. KNUTSEN, G. (1968). Repressed and derepressed synthesis of phosphatases
during synchronous growth of *Chlorella pyrenoidosa. Biochim. biophys. Acta*
161, 205–214.

253. KOLODNER, R. and TEWARI, K. K. (1972a). Molecular size and conformation of
chloroplast deoxyribonucleic acid from pea leaves. *J. biol. Chem.* **247**, 6355–
6364.

254. KOLODNER, R. and TEWARI, K. K. (1972b). Physicochemical characterization of
mitochondrial DNA from pea leaves. *Proc. natn. Acad. Sci. U.S.A.* **69**,
1830–1834.

255. KORNBERG, A. (1969). The active centre of DNA polymerase. *Science, N.Y.*
163, 1410–1418.

255a. KORNBERG, R. D. (1974). Chromatin structure: a repeating unit of histones and
DNA. *Science, N.Y.* **184**, 868–871.

256. KOVACS, C. J. and VAN'T HOF, J. (1970). Synchronization of a proliferative
population in a cultured plant tissue. Kinetic evidence for a Gl/S population. *J.
Cell Biol.* **47**, 536–539.

257. KUMAR, K. V. and FRIEDMAN, D. L. (1972). Initiation of DNA synthesis in
HeLa cell-free system. *Nature, New Biology* **239**, 74–76.

258. KUNG, S. D., THORNBER, J. P. and WILDMAN, S. G. (1972). Nuclear DNA codes
for the photosystem II chlorophyll-protein of chloroplast membranes. *FEBS
Lett.* **24**, 185–188.

258a. LACY, E. and AXEL, R. (1975). Analysis of DNA of isolated chromatin
subunits. *Proc. natn. Acad. Sci. U.S.A.* **72**, 3978–3982.

259. LAMPORT, D. T. A. and NORTHCOTE, D. H. (1960). Hydroxyproline in primary
cell walls of higher plants. *Nature, Lond.* **188**, 665–666.

260. LANGE, H. and ROSENSTOCK, G. (1963). Physiologisch-anatomische Studien
zum Problem der Wundheilung. II. Kausalanalytische Untersuchungen zur
Theorie des Wundreizes. *Beitr. Biol. Pfl.* **39**, 383–488.

261. LANZANI, G. A., BOLLINI, R. and SOFFIENTINI, A. N. (1974). The heterogeneity
of the elongation factor EF-1 from wheat embryos. *Biochim. biophys. Acta*
335, 275–283.

262. LARSSON, A. (1969). Ribonucleotide reductase from regenerating rat liver. *Eur.
J. Biochem.* **11**, 113–121.

263. LARSSON, A. (1973). Ribonucleotide reductase from regenerating rat liver. II.
Substrate phosphorylation level and effect of deoxyadenosine triphosphate.
Biochim. biophys. Acta **324**, 447–451.

264. LEAVER, C. J. (1966). "The Correlation between Nucleic Acid Synthesis and
Induced Enzyme Activity in Plant Tissue Slices". Ph.D. thesis, University of
London.

265. LEAVER, C. J. (1974). The biogenesis of plant mitochondria. *In* "The Chemistry
and Biochemistry of Plant Proteins" (J. B. Harborne and C. F. van Sumere,
eds), pp. 137–165. Academic Press, London and New York.

266. LEAVER, C. J. and HARMEY, M. R. (1974). Plant mitochondrial nucleic acids.
Biochem. Soc. Symp. **38**, 175–194.

267. LEAVER, C. J. and INGLE, J. (1971). The molecular integrity of chloroplast
ribosomal ribonucleic acid. *Biochem. J.* **123**, 235–244.

268. LEAVER, C. J. and KEY, J. L. (1970). rRNA synthesis in plants. *J. molec. Biol.*
49, 671–680.

269. LEFFLER, H. R., O'BRIEN, T. J., GLOVER, D. V. and CHERRY, J. H. (1971). Enhanced DNA polymerase activity of chromatin from soybean hypocotyl treated with 2, 4-D. *Pl. Physiol.* **48,** 43–45.

270. LEGOCKI, A. B. and MARCUS, A. (1970). Polypeptide synthesis in extracts of wheat germ. Resolution and partial purification of the soluble transfer factors. *J. biol. Chem.* **245,** 2814–2818.

271. LEGON, S., DARNBROUGH, C. H., HUNT, T. and JACKSON, R. J. (1973). Initiation of eukaryotic protein synthesis: native 40-*S* subunits bind initiator tRNA before associating with mRNA. *Biochem. Soc. Trans.* **1,** 553–557.

272. LEIS, J. P. and KELLER, E. B. (1970). Protein chain-initiating methionine tRNAs in chloroplasts and cytoplasm of wheat leaves. *Proc. natn. Acad. Sci. U.S.A.* **67,** 1593–1599.

273. LEIS, J. P. and KELLER, E. B. (1971). *N*-formyl-methionyl-tRNA$_f$ of wheat chloroplasts. Its synthesis by a wheat transformylase. *Biochemistry* **10,** 889–894.

274. LEVINE, R. P. (1969). The analysis of photosynthesis using mutant strains of algae and higher plants. *A. Rev. Pl. Physiol.* **20,** 523–540.

275. LEVINE, R. P. (1970). Interactions between nuclear and organelle genetic systems. *Brookhaven Symp. Biol.* **23,** 503–533.

276. LEVINE, R. P. and GOODENOUGH, U. W. (1970). The genetics of photosynthesis and of the chloroplast in *Chlamydomonas reinhardi*. *A. Rev. Genetics* **4,** 397–408.

277. LEVY, C. C., SCHMUKLER, M., FRANK, J. J., KARPETSKY, T. P., JEWETT, P. B., HIETER, P. A., LE GENDRE, S. M. and DORR, R. G. (1975). Possible role for poly(A) as an inhibitor of endonuclease activity in eukaryotic cells. *Nature, Lond.* **256,** 340–341.

278. LEWIS, L. N. and VARNER, J. E. (1970). Synthesis of cellulase during abcission of *Phaseolus vulgaris* leaf explants. *Pl. Physiol.* **46,** 194–199.

279. LIMA-DE-FARIA, A. (1959). Differential uptake of tritiated thymidine into hetero- and euchromatin in *Melanoplus* and *Secale*. *J. biophys. biochem. Cytol.* **6,** 457–466.

280. LIN, C. Y., KEY, J. L. and BRACKER, C. E. (1966). Association of D-RNA with polyribosomes in the soybean root. *Pl. Physiol.* **41,** 976–982.

281. LODISH, H. (1971). α- and β-globin mRNA. Different amounts and rates of initiation of translation. *J. biol. Chem.* **246,** 7131–7138.

282. LOENING, U. E. (1967). The fractionation of higher-molecular-weight ribonucleic acid by polyacrylamide-gel electrophoresis. *Biochem. J.* **102,** 251–257.

283. LOENING, U. E. (1968). Molecular weights of ribosomal RNA in relation to evolution. *J. molec. Biol.* **38,** 355–365.

284. LOENING, U. E. (1969). The determination of the molecular weight of ribonucleic acid by polyacrylamide-gel electrophoresis. The effects of changes in conformation. *Biochem. J.* **113,** 131–138.

285. LOENING, U. E. (1970). The mechanism of synthesis of ribosomal RNA. *Symp. Soc. gen. Microbiol.* **20,** 77–106.

286. LOENING, U. E. and INGLE, J. (1967). Diversity of RNA components in green plant tissues. *Nature, Lond.* **215,** 363–367.

287. LOFTFIELD, R. B. (1972). Mechanism of aminoacylation of tRNA. *Prog. Nucleic Acid Res. molec. Biol.* **12,** 87–128.

288. LONGO, C. P. (1968). Evidence for *de novo* synthesis of isocitratase and malate synthetase in germinating peanut cotyledons. *Pl. Physiol.* **43,** 660–664.

289. LYONS, J. M. (1973). Chilling injury in plants. *A. Rev. Pl. Physiol.* **24,** 445–466.
290. LYTTLETON, J. W. (1962). Isolation of ribosomes from spinach chloroplasts. *Expl Cell. Res.* **26,** 312–317.
291. MACLEOD, R. D. (1973). ³H-thymidine labelled DNA in the primary root of *Vicia faba* L. *Z. Pflanzenphysiol.* **68,** 379–381.
292. MAHER, E. P. and FOX, D. P. (1973). Multiplicity of ribosomal RNA genes in *Vicia* species with different nuclear DNA content. *Nature, New Biology* **245,** 170–172.
293. MANNING, J. E. and RICHARDS, O. C. (1972). Synthesis and turnover of *Euglena gracilis* nuclear and chloroplast deoxyribonucleic acid. *Biochemistry* **11,** 2036–2043.
294. MANNING, J. E., WOLSTENHOLME, D. R. and RICHARDS, O. C. (1972). Circular DNA molecules associated with chloroplasts of spinach, *Spinacia oleracea. J. Cell Biol.* **53,** 594–601.
295. MANNING, J. E., WOLSTENHOLME, D. R., RYAN, R. S., HUNTER, J. A. and RICHARDS, O. C. (1971). Circular chloroplast DNA from *Euglena gracilis. Proc. natn. Acad. Sci. U.S.A.* **68,** 1169–1173.
296. MARCUS, A., WEEKS, D. P., LEIS, J. P. and KELLER, E. B. (1970). Protein chain initiation by methionyl-tRNA in wheat embryo. *Proc. natn. Acad. Sci. U.S.A.* **67,** 1681–1687.
297. MARKS, D. B., PAIK, W. K. and BORUN, T. W. (1973). The relationship of histone phosphorylation to deoxyribonucleic acid replication and mitosis during the HeLa S-3 cell cycle. *J. biol. Chem.* **248,** 5660–5667.
298. MARX-FIGINI, M. (1966). Comparison of the biosynthesis of cellulose *in vitro* and *in vivo* in cotton bolls. *Nature, Lond.* **210,** 754–755.
299. MASTERS, M. and PARDEE, A. B. (1965). Sequence of enzyme synthesis and gene replication during the cell cycle of *Bacillus subtilis. Proc. natn. Acad. Sci. U.S.A.* **54,** 64–70.
300. MATTHEWS, M. B. (1973). Mammalian messenger RNA. In "Essays in Biochemistry (P. N. Campbell and F. Dickens, eds), Vol. 9, pp 59–102. Academic Press, London and New York.
301. MATHILE, P. and WINKENBACH, F. (1971). Function of lysosomes and lysosomal enzymes in the senescing corolla of the Morning Glory. *J. exp. Bot.* **22,** 759–771.
302. MATSUMOTO, H., GREGOR, D. and REINERT, J. (1975). Changes in chromatin of *Daucus carota* cells during embryogenesis. *Phytochemistry* **14,** 41–47.
303. MATTHEWS, R. E. F. (1970). "Principles of Plant Virology". Academic Press, New York and London.
304. McLENNAN, A. G. and KEIR, H. M. (1973). Deoxyribonucleic acid polymerases from *Euglena gracilis. Biochem. Soc. Trans.* **1,** 866–868.
305. MENG, R. L. and VANDERHOEF, L. N. (1972). Mitochondrial tyrosyl transfer ribonucleic acid in soybean seedlings. *Pl. Physiol.* **50,** 298–302.
306. MESELSON, M. and STAHL, F. W. (1958). The replication of DNA in *Escherichia coli. Proc. natn. Acad. Sci. U.S.A.* **44,** 671–682.
307. MICHAELIS, P. (1954). Cytoplasmic inheritance in *Epilobium* and its theoretical significance. *In* "Advances in Genetics" (M. Demerel, ed.) Vol. 6, pp. 287–401. Academic Press, New York and London.
308. MILBORROW, B. V. (1974). Chemistry and physiology of abscisic acid. *A. Rev. Pl. Physiol.* **25,** 259–307.

309. MILKAREK, C., PRICE, R. and PENMAN, S. (1974). The metabolism of a poly-(A)-minus mRNA fraction in HeLa cells. *Cell* **3**, 1–10.
310. MILLER, C. O. (1975). Cell-division factors from *Vinca rosea* L. crown gall tumour tissue. *Proc. natn. Acad. Sci. U.S.A.* **72**, 1883–1886.
311. MILLER, O. L. (1964). Fine structure of lampbrush chromosomes. *J. Cell Biol.* **23**, 109a.
312. MILLER, O. L. and HAMKALO, B. L. (1973). Electron microscopy of active genes. *Symp. Fed. Eur. biochem. Socs* **23**, 367–378.
313. MILLERD, A. and SPENCER, D. (1975). Changes in RNA-synthesising activity and template activity in nuclei from cotyledons of developing pea seeds. *Aust. J. Pl. Physiol.* **1**, 331–341.
314. MIRSKY, A. E. and RIS, H. (1949). Variable and constant components of chromosomes. *Nature, Lond.* **163**, 666–667.
315. MIRSKY, A. E. and RIS, H. (1951). The desoxy-ribonucleic acid content of animal cells and its evolutionary significance. *J. gen. Physiol.* **34**, 451–462.
316. MITCHISON, J. M. (1971). "The Biology of the Cell Cycle". Cambridge University Press, London.
317. MOHR, H. (1972). "Lectures on Photomorphogenesis". Springer-Verlag, Berlin.
318. MOLLOY, G. R. and SCHMIDT, R. R. (1970). Studies on the regulation of ribulose-1, 5-diphosphate carboxylase synthesis during the cell cycle of the eucaryote *Chlorella*. *Biochem. biophys. Res. Commun.* **40**, 1125–1133.
319. MONDAL, H., GANGULY, A., DAS, A., MANDEL, R. K. and BISWAS, B. B. (1972). RNA polymerase from eukaryotic cells. Effects of factors and rifampicin on the activity of RNA polymerase from chromatin of coconut nuclei. *Eur. J. Biochem.* **28**, 143–150.
320. MOORE, A. L., BORCK, K. M. and BAXTER, R. (1971). The incorporation of amino acids into the protein of isolated soya bean mitochondria. *Planta* **97**, 299 309.
321. MORY, Y. Y., CHEN, D. and SARID, S. (1974). Deoxyribonucleic acid polymerase from wheat embryos. *Pl. Physiol.* **53**, 377–381.
322. NAGL, W. (1972). Evidence of DNA amplification in the orchid *Cymbidium in vitro*. *Cytobios* **5**, 145–154.
323. NAGL, W. (1974). Role of heterochromatin in the control of cell cycle duration. *Nature, Lond.* **249**, 53–54.
324. NASS, M. M. K., NASS, S. and AFZELIUS, B. A. (1965). The general occurrence of mitochondrial DNA. *Expl Cell Res.* **37**, 516–539.
325. NEILSON-JONES, W. (1925). Polarity phenomena in seakale roots. *Ann. Bot.* **39**, 359–372.
326. NELSON, O. E. and BURR, B. (1973). Biochemical genetics of higher plants. *A. Rev. Pl. Physiol.* **24**, 493–518.
327. NEWMAN, I. A. (1963). Electric potentials and auxin translocation in *Avena*. *Aust. J. biol. Sci.* **16**, 629–646.
328. NEWMAN, I. A. and BRIGGS, W. R. (1972). Phytochrome-mediated electric potential changes in seedlings. *Pl. Physiol.* **50**, 687–693.
329. NILES, R. M. and MOUNT, M. S. (1974). Cyclic nucleotide phosphodiesterase from carrot. *Phytochemistry* **13**, 2735–2740.
330. NJUS, D., SULZMAN, F. M. and HASTINGS, J. W. (1974). Membrane model for the circadian clock. *Nature, Lond.* **248**, 116–120.
331. NOMURA, M. (1970). Bacterial ribosomes. *Bact. Rev.* **34**, 228–277.

332. NORRIS, R. D., LEA, P. J. and FOWDEN, L. (1973). Aminoacyl-tRNA synthe-
     tases in *Triticum aestivum* L. during seed development and germination. *J. exp.
     Bot.* **24,** 615–625.
333. NORTHCOTE, D. H. (1963). Changes in the cell walls of plants during differentia-
     tion. *Symp. Soc. exp. Biol.* **17,** 157–174.
334. NORTHCOTE, D. H. (1969a). Fine structure of cytoplasm in relation to synthesis
     and secretion in plant cells. *Proc. R. Soc. Lond.* B **173,** 21–30.
335. NORTHCOTE, D. H. (1969b). The synthesis and metabolic control of polysac-
     charides and lignin during the differentiation of plant cells. *In* "Essays in
     Biochemistry" (P. N. Campbell and G. D. Greville, eds), Vol 5, pp 89–137.
     Academic Press, London and New York.
336. NORTHCOTE, D. H. (1971). Organisation of structure, synthesis and transport
     within the plant during cell division and growth. *Symp. Soc. exp. Biol.* **25,** 51–
     69.
337. NORTHCOTE, D. H. and PICKETT-HEAPS, J. D. (1966). A function of the Golgi
     apparatus in polysaccharide synthesis and transport in the root-cap cells of
     wheat. *Biochem. J.* **98,** 159–167.
338. OBENDORF, R. L. and MARCUS, A. (1974). Rapid increase in adenosine 5′-
     triphosphate during early wheat embryo germination. *Pl. Physiol.* **53,** 779–
     781.
339. O'BRIEN, T. J., JARVIS, B. C., CHERRY, J. H. and HANSON, J. B. (1968).
     Enhancement by 2,4-D of chromatin RNA polymerase in soybean hypocotyl
     tissue. *Biochim. biophys. Acta* **169,** 35–43.
340. IKAMURA, S., MIYASAKA, K. and NISHI, A. (1973). Synchronisation of carrot cell
     culture by starvation and cold treatment. *Expl Cell Res.* **78,** 467–470.
341. PADILLA, G. M. and COOK, J. R. (1964). The development of techniques for
     synchronizing flagellates. *In* "Synchrony in Cell Division and Growth" (E.
     Zeuthen, ed.). Interscience Publishers, New York.
342. PARDUE, M. L. and GALL, J. G. (1970). Chromosomal location of mouse satellite
     DNA. *Science, N.Y.* **168,** 1356–1358.
343. PARENTI, R., GUILLE, E., GRISVARD, J., DURANTE, M., GIORGI, L. and BUIATTI,
     M. (1973). Transient DNA satellite in dedifferentiating pith tissue. *Nature,
     New Biology* **246,** 237–239.
344. PARISH, J. H. (1972). "Principles and Practice of Experiments with Nucleic
     Acids". Longman, London.
345. PARTHIER, B. (1972). Sites of synthesis of chloroplast proteins. *Symp. biol. hung.*
     13, 235–248.
346. PARTHIER, B., KRAUSPE, R. and SAMTLEBEU, S. (1972). Light-stimulated syn-
     thesis of aminoacyl-tRNA synthetases in greening *Euglena gracilis. Biochim.
     biophys. Acta* **277,** 335–341.
347. PASTAN, I. and FRIEDMAN, R. M. (1968). Actinomycin D inhibition of phospho-
     lipid synthesis in chick embryo cells. *Science, N.Y.,* **160,** 316–317.
348. PASTAN, I. and PERLMAN, R. (1970). Cyclic adenosine monophosphate in
     bacteria. *Science, N.Y.* **169,** 339–344.
349. PAUL, J. and MORE, I. R. (1972). Properties of reconstituted chromatin and
     nucleohistone complexes. *Nature, New Biology* **239,** 134–135.
350. PAYNE, P. I., CORRY, M. J. and DYER, T. A. (1973). Nucleotide sequence
     analysis of cytoplasmic 5-S ribosomal ribonucleic acid from five species of
     flowering plants. *Biochem. J.* **135,** 845–851.

351. PAYNE, P. I. and DYER, T. A. (1971). Characterization of cytoplasmic and chloroplast 5-S ribosomal ribonucleic acid from broad-bean leaves. *Biochem J.* **124**, 83–89.

352. PAYNE, P. I. and DYER, T. A. (1972). Plant $5 \cdot 8s$ RNA is a component of 80s but not 70s ribosomes. *Nature, New Biology* **235**, 145–147.

353. PAYNE, P. I. and LOENING, U. E. (1970). RNA breakdown accompanying the isolation of pea root microsomes. An analysis by polyacrylamide gel electrophoresis. *Biochim. biophys. Acta* **224**, 128–135.

354. PEARSON, C. J. and INGLE, J. (1972). The origin of stress-induced satellite DNA in plant tissues. *Cell Differentiation* **1**, 43–51.

355. PEDERSON, T. (1972). Chromatin structure and the cell cycle. *Proc. natn. Acad. Sci. U.S.A.* **69**, 2224–2228.

356. PELC, S. R. (1968). Turnover of DNA and function. *Nature, Lond.* **219**, 162–163.

357. PELC, S. R. and LA COUR, L. F. (1959). The incorporation of H^3-thymidine in newly differentiated nuclei of roots of *Vicia faba. Experientia* **15**, 131–133.

358. PERRY, R., GREENBURG, J. R. and TARTOF, K. D. (1970). Transcription of ribosomal, heterogeneous nuclear and messenger RNA in eukaryotes. *Cold Spring Harb. Symp. quant. Biol.* **35**, 577–587.

359. PESTKA, S. (1971). Inhibitors of ribosome functions. *A. Rev. Biochem.* **40**, 697–710.

360. PETERSEN, A. J. and ANDERSON, E. C. (1964). Quantity production of synchronized mammalian cells in suspension culture. *Nature, Lond.* **203**, 642–643.

361. PETES, T. D., BYERS, B. and FANGMAN, W. L. (1973). Size and structure of yeast chromosomal DNA. *Proc. natn. Acad. Sci. U.S.A.* **70**, 3072–3076.

362. PETES, T. D. and FANGMAN, W. L. (1972). Sedimentation properties of yeast chromosomal DNA. *Proc. natn. Acad. Sci. U.S.A.* **69**, 1188–1191.

363. POLLARD, C. J. (1964). The specificity of ribosomal ribonucleic acids of plants. *Biochem. biophys. Res. Commun.* **17**, 171–176.

364. POOVAIAH, B. W. and LEOPOLD, A. C. (1973a). Inhibition of abcission by calcium. *Pl. Physiol.* **51**, 848–851.

365. POOVAIAH, B. W. and LEOPOLD, A. C. (1973b). Deferral of leaf senescence with calcium. *Pl. Physiol.* **52**, 236–239.

366. PRESCOTT, D. M. and BENDER, M. A. (1962). Synthesis of RNA and protein during mitosis in mammalian tissue culture cells. *Expl Cell Res.* **26**, 260–268.

367. PRESSEY, R. and SHAW, R. (1966). Effect of temperature on invertase, invertase inhibitor and sugars in potato tubers. *Pl. Physiol.* **41**, 1657–1661.

368. QUASTLER, H. and SHERMAN, F. G. (1959). Cell population kinetics in the intestinal epithelium of the mouse. *Expl Cell Res.* **17**, 420–438.

369. RAMACHANDRAN, G. N., ed. (1967). "Treatise on Collagen". Academic Press, London and New York.

370. RAO, P. N. and JOHNSON, R. T. (1970). Mammalian cell fusion: studies on the regulation of DNA synthesis and mitosis. *Nature, Lond.* **225**, 159–164.

371. RASCH, E., SWIFT, H. and KLEIN, R. M. (1959). Nucleoprotein changes in plant tumour growth. *J. biophys. biochem. Cytol.* **6**, 11–34.

372. RASMUSSEN, H., GOODMAN, D. B. P. and TENENHOUSE, A. (1972). The role of cyclic AMP and calcium in cell activation. *Critical Reviews in Biochemistry.* **1**, 75–148.

373. RAVEN, J. A. and SMITH, F. A. (1973). The regulation of intracellular pH as a

fundamental biological process. *In* "Ion Transport in Plants" (W. P. Anderson, ed), pp. 271–278. Academic Press, London and New York.

374. RAWSON, J. R., CROUSE, E. J. and STUTZ, E. (1971). The integrity of the 25-*S* ribosomal RNA from *Euglena gracilis* 87-*S* ribosomes. *Biochim. biophys. Acta* **246**, 507–516.

375. RAY, P. M. (1973). Regulation of β-glucan synthetase activity by auxin in pea stem tissue. *Pl. Physiol.* **51**, 601–614.

376. RAYLE, D. L. (1973). Auxin induced hydrogen ion secretion in *Avena* coleoptiles and its implications. *Planta* **114**, 63–73.

377. RAYLE, D. L. and CLELAND, R. (1972). The *in-vitro* acid-growth response: relation to *in-vivo* growth responses and auxin action. *Planta* **104**, 282–296.

378. RAYLE, D. L., HAUGHTON, P. M. and CLELAND, R. E. (1970). An *in vitro* system that simulates plant cell extension growth. *Proc. natn. Acad. Sci. U.S.A.* **67**, 1814–1817.

379. REDDI, K. K. (1969). Studies on the formation of tobacco mosaic virus ribonucleic acid. VIII. Fate of parental viral RNA in the host cell. *Proc. natn. Acad. Sci. U.S.A.* **62**, 604–611.

380. REES, H. and JONES, R. N. (1972). The origin of wide species variation in nuclear DNA content. *Int. Rev. Cytol.* **32**, 53–92.

381. REIJNDERS, L., SLOOF, P., SIVAL, J. and BORST, P. (1973). Gel electrophoresis of RNA under denaturing conditions. *Biochim. biophys. Acta* **324**, 320–333.

382. RENNER, O. (1936). Zur Kenntnis der nichtmendelnden Buntheit der Laubblatter. *Flora, Jena,* **30**, 218–290.

383. RHOADES, M. M. (1955). Interaction of genic and non-genic hereditary units and the physiology of non-genic inheritance. *In* "Encyclopedia of Plant Physiology" (W. Ruhland, ed.), Vol 1. Springer-Verlag, Berlin.

384. RICHARDS, O. C. and RYAN, R. S. (1973). Synthesis and turnover of *Euglena gracilis* mitochondrial DNA. *J. molec. Biol.* **82**, 57–75.

385. RIDLEY, S. M. and LEECH, R. M. (1970). Division of chloroplasts in an artificial environment. *Nature, Lond.* **227**, 463–465.

386. RIS, H. and KUBAI, D. F. (1970). Chromosome structure. *A. Rev. Genet.* **4**, 263–290.

387. RIS, H. and PLAUT, W. (1962). Ultrastructure of DNA-containing areas in the chloroplast of *Chlamydomonas. J. Cell Biol.* **13**, 383–391.

388. ROBERTS, B. E. and PATERSON, B. M. (1973). Efficient translation of tobacco mosaic virus RNA and rabbit globin 9-*S* RNA in a cell-free system from commercial wheat germ. *Proc. natn. Acad. Sci. U.S.A.* **70**, 2330–2334.

389. ROBERTS, K. and NORTHCOTE, D. H. (1970). The structure of sycamore callus cells during division in a partially synchronised suspension culture. *J. Cell Sci.* **6**, 299–321.

390. ROBINSON, N. E. and BRYANT, J. A. (1975). Development of chromatin-bound and soluble DNA polymerase activities during germination of *Pisum sativum* L. *Planta,* **127**, 69–75.

391. ROGERS, M. E., LOENING, U. E. and FRASER, R. S. S. (1970). rRNA precursors in plants. *J. molec. Biol.* **49**, 681–692.

392. ROSNER, A., EDELMAN, M. and GRESSEL, J. (1974). Thermal denaturation of nucleic acids in polyacrylamide gels. *Analyt. Biochem.* **58**, 602–608.

393. ROURKE, A. W. and HEYWOOD, S. M. (1972). Myosin synthesis and specificity of eukaryotic initiation factors. *Biochemistry* **11**, 2061–2066.

394. RUBERY, P. H. and FOSKET, D. E. (1969). Changes in phenylalanine-ammonia-lyase activity during differentiation in *Coleus* and soybean. *Planta* **87**, 54–62.

395. RUBERY, P. H. and NORTHCOTE, D. H. (1968). Site of phenylalanine-ammonia-lyase activity and synthesis of lignin during xylem differentiation. *Nature, Lond.* **219**, 1230–1234.

396. RUBERY, P. H. and NORTHCOTE, D. H. (1970). The effect of auxin on the synthesis of cell wall polysaccharides in cultured sycamore cells. *Biochim. biophys. Acta* **222**, 95–108.

397. SAGER, R. (1972). "Cytoplasmic Genes and Organelles". Academic Press, New York and London.

398. SAGHER, D., EDELMAN, M. and JAKOB, K. M. (1974). Poly-(A)-associated RNA in plants. *Biochim. biophys. Acta* **349**, 32–38.

399. SAKABE, K. and OKAZAKI, R. (1966). A unique property of the replicating region of chromosomal DNA. *Biochim. biophys. Acta* **129**, 651–654.

400. SARROUY-BALAT, H., DELSENEY, M. and JULIAN, R. (1973). Plant hormones and DNA synthesis: evidence for a bacterial origin of rapidly labelled heavy satellite DNA. *Plant Sci. Lett.* **1**, 287–292.

401. SATTER, R. L. and GALSTON, A. W. (1971). Phytochrome-controlled nyctinasty in *Albizzia*. *Pl. Physiol.* **48**, 740–746.

402. SCHERRER, K. (1973). Messenger RNA in eukaryotic cells: the life history of duck globin messenger RNA. *Karolinska Symposia on Research Methods in Reproductive Endocrinology* **8**, 95–128.

403. SCHIFF, J. A. (1973). The development, inheritance and origin of the plastid in *Euglena*. *In* "Advances in Morphogenesis" (M. Abercrombie and J. Brachet, eds), Volume 10, pp. 265–312. Academic Press, New York and London.

404. SCHMIDT, R. R. (1969). Control of enzyme synthesis during the cell cycle of *Chlorella*. *In* "The Cell Cycle: Gene–Enzyme Interactions". (G. M. Padilla, G. L. Whitson and I. L. Cameron, eds). Academic Press, New York and London

405. SCHÖNHERR, O. T., WANKA, F. and KUYPER, C. M. A. (1970). Periodic change of deoxyribonuclease activity in synchronous cultures of *Chlorella*. *Biochim. biophys. Acta* **224**, 74–79.

406. SCHOTZ, F. (1954). Uber Plastidenkonkurrenz bei *Oenothera*. *Planta* **43**, 182–240.

407. SCHREIER, M. H. and STAEHELIN, T. (1973). Initiation of eukaryotic protein synthesis: met-tRNA$_f$40-*S* ribosome initiation complex catalysed by purified initiation factors in the absence of mRNA. *Nature, New Biology* **242**, 35–38.

408. SCHWARTZ, J. H., MEYER, R., EISENSTADT, J. M. and BRAWERMAN, G. (1967). N-formyl-methionine in initiation of protein synthesis in cell-free extracts of *Euglena gracilis*. *J. molec. Biol.* **25**, 571–574.

409. SCOTT, B. I. H. (1967). Electric fields in plants. *A. Rev. Pl. Physiol.* **18**, 409–418.

410. SCOTT, N. S. (1973). Ribosomal RNA cistrons in *Euglena gracilis*. *J. molec. Biol.* **81**, 327–336.

411. SCOTT, N. S. and INGLE, J. (1973). The genes for cytoplasmic ribosomal ribonucleic acid in higher plants. *Pl. Physiol.* **51**, 677–684.

412. SCOTT, N. S., MUNS, R. and SMILLIE, R. M. (1970). Chloroplast and cytoplasmic ribosomes in *Euglena gracilis*. *FEBS Lett.* **10**, 149–152.

413. SCOTT, N. S., SHAH, V. C. and SMILLIE, R. M. (1968). Synthesis of chloroplast DNA in isolated chloroplasts. *J. Cell Biol.* **38**, 151–157.

413a. SCRAGG, A. H., JOHN, P. C. L. and THURSTON, C. F. (1975). Post-transcription control of isocitrate lyase induction in the eukaryotic alga *Chlorella fusca*. *Nature, Lond.* **257,** 498–501.

414. SEITZ, U. and SEITZ, U. (1973). Biosynthese der ribosomalen RNS bei der blaugrünen Alge *Anacystis nidulans*. *Arch. Mikrobiol.* **90,** 213–222.

415. SEN, S., PAYNE, P. I. and OSBORNE, D. J. (1975). Early RNA synthesis during the germination of rye (*Secale cereale*) embryos and the relationship to early protein synthesis. *Biochem. J.* **148,** 381–387.

416. SETTERFIELD, G. (1963). Growth regulation in excised slices of Jerusalem artichoke tuber tissue. *Symp. Soc. exp. Biol.* **17,** 98–126.

417. SHAIN, Y. and MAYER, A. M. (1968). Activation of enzymes during germination: amylopectin 1,6 glucosidase in peas. *Physiologia. Pl.* **21,** 765–776.

418. SHELDRAKE, A. R. and NORTHCOTE, D. H. (1968). Some constituents of xylem sap and their possible relationship to xylem differentiation. *J. exp. Bot.* **19,** 681–689.

419. SHEN, S. R-C., and SCHMIDT, R. R. (1966). Enzymic control of nucleic acid synthesis during synchronous growth of *Chlorella pyrenoidosa*. *Archs Biochem. Biophys.* **115,** 13–20.

420. SHENKIN, A. and BURDON, R. H. (1974). Deoxyadenylate-rich and deoxyguanylate-rich regions in mammalian DNA. *J. molec. Biol.* **85,** 19–39.

421. SHIH, D. S. and KAESBURG, P. (1973). Translation of brome mosiac virus ribonucleic acid in a cell-free system derived from wheat embryo. *Proc. natn. Acad. Sci. U.S.A.* **70,** 1799–1803.

422. SHIMADA, H. and TERAYAMA, H. (1972). DNA synthesis in isolated nuclei from the brains of rats at different post-partal stages and the infant rat brain cytosol factor stimulating the DNA synthesis in infant rat brain nuclei. *Biochim. biophys. Acta* **287,** 415–426.

423. SHIPLEY, R. A. and CLARKE, R. E. (1972). "Tracer Methods for *in vivo* Kinetics". Academic Press, New York and London.

424. SHIPP, W. S., KIERAS, F. J. and HASELKORN, R. (1965). DNA associated with tobacco chloroplasts. *Proc. natn. Acad. Sci. U.S.A.* **54,** 207–212.

425. SHODELL, M. (1972). Environmental stimuli in the progression of BHK/21 cells through the cell cycle. *Proc. natn. Acad. Sci. U.S.A.* **69,** 1455–1459.

426. SIMARD, A. and ERHAN, S. (1972). Incorporation of tritiated thymidine in tobacco pith tissues induced by a substance isolated from Erlich ascites fluid. *Can. J. Bot.* **50,** 719–722.

427. SIMPSON, R. T. (1973). Structure and function of chromatin. *Adv. Enzymol.* **38,** 41–108.

428. SINCLAIR, J., WELLS, R., DEUMLING, B. and INGLE, J. (1975). The complexity of satellite deoxyribonucleic acid in a higher plant. *Biochem. J.* **149,** 31–38.

429. SINGH, S. and WILDMAN, S. G. (1973). Chloroplast DNA codes for the ribulose diphosphate carboxylase catalytic site on fraction I proteins of *Nicotiana* species. *Molec. gen. Genetics.* **124,** 187–196.

430. SINNOTT, W. (1960). "Plant Morphogenesis". McGraw-Hill, New York.

431. SINNOTT, E. W. and BLOCH, R. (1940). Cytoplasmic behaviour during the division of vacuolate plant cells. *Proc. natn. Acad. Sci. U.S.A.* **26,** 223–227.

432. SINNOTT, E. W. and BLOCH, R. (1945). The cytoplasmic basis of intercellular patterns in vascular differentiation. *Am. J. Bot.* **32,** 151–156.

433. SKOOG, F. and ARMSTRONG, D. J. (1970). Cytokinins. *A. Rev. Pl. Physiol.* **21,** 359–384.

434. SMITH, H. (1975). "Phytochrome and Photomorphogenesis". McGraw-Hill, Maidenhead, Berks.
435. SMITH-JOHANNSEN, H. and GIBBS, S. P. (1972). Effects of chloramphenicol on chloroplast and mitochondrial ultrastructure in *Ochromonas danica*. *J. Cell Biol.* **52**, 598–614.
436. SOBOTA, A. E., LEAVER, C. J. and KEY, J. L. (1968). A detailed evaluation of the possible contribution of bacteria to radioactive precursor incorporation into nucleic acids of plant tissues. *Pl. Physiol.* **43**, 907–913.
437. SOLAO, P. B. and SHALL, S. (1971). Control of DNA replication in *Physarum polycephalum* I. Specific activity of NAD pyrophosphorylase in isolated nuclei during the cell cycle. *Expl Cell Res.* **69**, 295–300.
438. SOUTHERN, E. M. (1970). Base sequence and evolution of guinea-pig α-satellite DNA. *Nature, Lond.* **227**, 794–798.
439. SPENCER, D. and WHITFELD, P. R. (1966). The nature of the ribonucleic acid of isolated chloroplasts. *Archs Biochem. Biophys.* **117**, 337–346.
440. SPENCER, D. and WHITFELD, P. R. (1967a). Ribonucleic acid synthesizing activity of spinach chloroplasts and nuclei. *Archs Biochem. Biophys.* **121**, 336–345.
441. SPENCER, D. and WHITFELD, P. R. (1967b). DNA synthesis in isolated chloroplasts. *Biochem. biophys. Res. Commun.* **28**, 538–542.
442. SPENCER, D. and WHITFELD, P. R. (1969). The characteristics of spinach chloroplast DNA polymerase. *Archs Biochem. Biophys.* **132**, 477–488.
443. SPIRIN, A. S. (1969). Structure of the ribosome. *Prog. Biophys. molec. Biol.* **19**, 133–174.
444. SRIVASTAVA, B. I. S. (1968). Acceleration of senescence and of the increase of chromatin-associated nucleases in excised barley leaves by abscisin II and its reversal by kinetin. *Biochim. biophys. Acta* **169**, 534–536.
445. STAEHELIN, T. (1974). Cited in *Nature, Lond.* **249**, 613.
446. STERN, H. and HOTTA, Y. (1973). Biochemical controls of meiosis. *A. Rev. Genetics* **7**, 37–66.
447. STEVENS, C., JENNS, S. M. and BRYANT, J. A. (1975). Deoxyribonucleic acid polymerases and deoxyribonucleases in *Pisum sativum*. *Biochem. Soc. Trans.*, **3**, 1126–1128.
448. STEWARD, F. C., MAPES, M. O., KENT, A. E. and HOLSTEN, R. D. (1964). Growth and development of cultured plant cells. *Science, N.Y.* **143**, 20 27.
449. STEWART, G. R. and SMITH, H. (1972). Effects of abscisic acid on nucleic acid synthesis and induction of nitrate reductase in *Lemna polyrhiza*. *J. exp. Bot.* **23**, 875–885.
450. STOUT, E. R. and ARENS, M. Q. (1970). DNA polymerase from maize seedlings. *Biochim. biophys. Acta* **213**, 90–100.
451. STRATTON, B. R. and TREWAVAS, A. J. Previously unpublished results.
452. STREET, H. E. (1973). Plant cell cultures: their potential for metabolic studies. *In* "Biosynthesis and its Control in Plants" (B. V. Milborrow, ed), pp. 93–125. Academic Press, London and New York.
453. STUTZ, E. (1970). The kinetic complexity of *Euglena gracilis* chloroplast DNA. *FEBS Lett.* **8**, 25–28.
454. STUTZ, E. and NOLL, H. (1967). Characterization of cytoplasmic and chloroplast polysomes in plants: evidence for three classes of ribosomal RNA in Nature. *Proc. natn. Acad. Sci. U.S.A.* **57**, 774–781.

455. STUTZ, E. and VANDREY, J. P. (1971). Ribosomal DNA satellite of *Euglena gracilis* chloroplast DNA. *FEBS Lett.* **17,** 277–280.
456. SUEOKA, N. (1960). Mitotic replication of deoxyribonucleic acid in *Chlamydomonas reinhardi. Proc. natn. Acad. Sci. U.S.A.* **46,** 83–91.
457. SURZYCKI, S. J., GOODENOUGH, U. W., LEVINE, R. P. and ARMSTRONG, J. J. (1970). Nuclear and chloroplast control of chloroplast structure and function in *Chlamydomonas reinhardi. Symp. Soc. exp. Biol.* **24,** 13–37.
458. SUSSMAN, M. and SUSSMAN, R. P. (1969). Patterns of RNA synthesis and of enzyme accumulation and disappearance during cellular slime mould cytodifferentiation. *Symp. Soc. gen. Microbiol.* **19,** 403–435.
459. SUYAMA, Y. and BONNER, W. D. (1966). DNA from plant mitochondria. *Pl. Physiol.* **41,** 383–388.
460. SVETAILO, E. N., PHILLIPPOVICH, I. I. and SISSAKIAN, N. M. (1967). Differences in sedimentation properties of chloroplast and cytoplasmic ribosomes from pea seedlings. *J. molec. Biol.* **24,** 405–415.
461. SWARTZ, M. N., TRAUTNER, T. A. and KORNBERG, A. (1962). Enzymatic synthesis of deoxyribonucleic acid. XI. Further studies on nearest neighbour base sequences in deoxyribonucleic acids. *J. biol. Chem.* **327,** 1961–1967.
462. SWEENEY, B. M. (1974). The potassium content of *Gonyaulax polyedra* and phase changes in circadian rhythm of stimulated bioluminescence by short exposure to ethanol and valinomycin. *Pl. Physiol.* **53,** 337–342.
463. SWIFT, H. H. (1950). The desoxyribose nucleic acid content of animal nucleic. *Physiol. Zool.* **23,** 169–198.
464. SZALAY, A., MUNSCHE, D., WOLLGIEHN, R. and PARTHIER, B. (1972). Ribosomal RNA and ribosomal precursor RNA in *Anacystis nidulans. Biochem. J.* **129,** 135–140.
465. TAKEBE, I., AOKI, S. and SAKAI, F. (1975). Replication and expression of tobacco mosiac virus genome in isolated tobacco leaf protoplasts. *In* "Modification of the Information Content of Plant Cells" (Proceedings of 2nd John Innes Symposium, R. Markham, D. R. Davies, D. A. Hopwood and R. W. Horne, eds). North-Holland, Amsterdam.
466. TAKEHISA, S. and UTSUMI, S. (1973). Heterochromatin and Giemsa banding of metaphase chromosomes of *Trillium kamtschaticum* Pallas. *Nature, New Biology* **243,** 286–287.
467. TAMIYA, H. (1966). Synchronous cultures of algae. *A. Rev. Pl. Physiol.* **17,** 1–26.
468. TANADA, T. (1968). A rapid photoreversible response of barley root tips in the presence of 3-indole acetic acid. *Proc. natn. Acad. Sci. U.S.A.* **59,** 376–380.
469. TAYLOR, J. H. (1960). Nucleic acid synthesis in relation to the cell division cycle. *Ann. Acad. Sci. N.Y.* **90,** 409–421.
470. TAYLOR, J. H., WOODS, P. S. and HUGHES, W. L. (1957). The organization of chromosomes as revealed by autoradiographic studies using tritium-labeled thymidine. *Proc. natn. Acad. Sci. U.S.A.* **43,** 122–128.
471. TERRY, O. W. and EDMUNDS, L. N. (1970). Phasing of cell division by temperature cycles in *Euglena* cultured autotrophically under continuous illumination. *Planta* **93,** 106–127.
472. TEWARI, K. K. and WILDMAN, S. G. (1967). DNA polymerase in isolated tobacco chloroplasts and nature of polymerized product. *Proc. natn. Acad. Sci. U.S.A.* **58,** 689–696.
473. TEWARI, K. K. and WILDMAN, S. G. (1970). Information content in the chloroplast DNA. *Symp. Soc. exp. Biol.* **24,** 147–179.

474. THOMAS, H. (1973). Gel electrophoresis of ribosomal components from seeds of *Pisum sativum* L. *Expl Cell Res.* **77**, 298–302.

474a. THOMAS, J. O. and KORNBERG, R. D. (1975). An octomer of histones in chromatin and free in solution. *Proc. natn. Acad. Sci. U.S.A.* **72**, 2626–2630.

475. THOMPSON, L. R. and MCCARTHY, B. J. (1971). The effects of cytoplasmic extracts on DNA synthesis *in vitro*. *Fedn. Proc. Fedn. Am. Socs. exp. Biol.* **30**, 1177.

476. THORNBURG, W. and SIEGEL, A. (1973). Characterisation of the rapidly reassociating deoxyribonucleic acid of *Cucurbita pepo L.* and the sequences complementary to ribosomal and transfer ribonucleic acid. *Biochemistry* **12**, 2759–2765.

477. TIMMIS, J. N. and INGLE, J. (1973). Environmentally induced changes in rRNA gene redundancy. *Nature, New Biology* **244**, 235–236.

478. TIMMIS, J. N. and INGLE, J. (1974). The nature of the variable DNA associated with environmental induction in flax. *Heredity* **33**, 339–346.

479. TRAVIS, R. L., ANDERSON, J. M. and KEY, J. L. (1973). Influence of auxin and incubation on the relative level of polysomes in excised soybean hypocotyl. *Pl. Physiol.* **52**, 608–612.

480. TRAVIS, R. L., JORDAN, W. R. and HUFFAKER, R. C. (1969). Evidence for an inactivating system of nitrate reductase in *Hordeum vulgare* L. during darkness that requires protein synthesis. *Pl. Physiol.* **44**, 1150–1156.

481. TRENDELENBURG, M. F., SPRING, H., SCHEER, U. and FRANKE, W. W. (1974). Morphology of nucleolar cistrons in a plant cell. *Proc. natn. Acad. Sci. U.S.A.* **71**, 3626–3630.

482. TREWAVAS, A. J. (1968a). Effect of IAA on RNA and protein synthesis. *Archs Biochem. Biophys.* **123**, 324–335.

483. TREWAVAS, A. J. (1968b). Relationship between plant growth hormones and nucleic acid metabolism. *Prog. Phytochem.* **1**, 113–161.

484. TREWAVAS, A. (1970). The turnover of nucleic acids in *Lemna minor*. *Pl. Physiol.* **45**, 742–751.

485. TREWAVAS, A. (1972). Control of protein turnover rates in *Lemna minor*. *Pl. Physiol.* **49**, 47–51.

486. TRIPLETT, E. L., STEENS-LIEVENS, A. and BALTUS, E. (1965). Rates of synthesis of acid phosphatases in nucleate and enucleate *Acetabularia* fragments. *Expl Cell Res.* **38**, 366–378.

487. TWARDOWSDKI, T. and LEGOCKI, A. B. (1973), Purification and some properties of elongation factor-2 from wheat germ. *Biochim. biophys. Acta* **324**, 171–183.

488. URBANCZYK, J. and STUDZINSKI, G. P. (1974). Chromatin-associated DNA endonuclease activities in HeLa cells. *Biochem. biophys. Res. Commun.* **59**, 616–622.

489. VANDERHOEF, L. N. and KEY, J. L. (1970). The fractionation of transfer ribonucleic acid from roots of pea seedlings. *Pl. Physiol.* **46**, 294–298.

490. VAN'T HOF, J. (1963). DNA, RNA and protein synthesis in the mitotic cycle of pea root meristem cells. *Cytologia* **28**, 30–35.

491. VAN'T HOF, J. (1968). Experimental procedures for measuring cell population kinetics parameters in plant root meristems. *In* "Methods in Cell Physiology" (D. M. Prescott, ed). Academic Press, New York and London.

492. VAN'T HOF, J. and KOVACS, C. J. (1972). Mitotic cycle regulation in the meristems of cultured roots: the principal control point hypothesis. *Adv. exp. Med. Biol.* **18**, 15–30.

493. VARNER, J. E. and RAM CHANDRA, G. (1964). Hormonal control of enzyme synthesis in barley endosperm. *Proc. natn. Acad. Sci. U.S.A.* **52,** 100–106.
494. VARSHAVSKY, A. J., ILYIN, YU. V. and GEORGIEV, G. P. (1974). Very long stretches of free DNA in chromatin. *Nature, Lond.* **250,** 602–606.
495. VAUGHAN, D. and MACDONALD, I. R. (1967). The effect of inhibitors on the increase in invertase activity and RNA content of beet discs during ageing. *J. exp. Bot.* **18,** 587–593.
496. VENDRELY, R. and VENDRELY, C. (1948). La teneur du noyau cellulaire en acide désoxyribonuclèique à travers les organes, les individus et les espèces animales. *Experientia* **4,** 434–436.
497. VENIS, M. A. (1973). Hormone receptor proteins. *Commentaries in Plant Science* **3,** 21–28.
497a. VERMA, D. P. S., MACLACHLAN· G. A., BYRNE, H. and EWINGS, D. (1975). Regulation and *in vitro* translation of messenger ribonucleic acid for cellulase from auxin-treated pea epicotyls. *J. biol. Chem.* **250,** 1019–1926.
498. VERMA, D. P. S., NASH, D. T. and SCHULMAN, H. (1974). Isolation and *in vitro* translation of soybean leghaemoglobin mRNA. *Nature, Lond.* **251,** 74–77.
499. WALKER, P. M. B. (1971). "Repetitive" DNA in higher organisms. *Prog. Biophys. molec. Biol.* **23,** 147–190.
500. WALKER, P. M. B. and YATES, H. B. (1952). Nuclear components of dividing cells. *Proc. R. Soc. Lond.* B **140,** 274–299.
501. WANKA, F. and MOORS, J. (1970). Selective inhibition by cycloheximide of nuclear DNA synthesis in synchronous cultures of *Chlorella. Biochem. biophys. Res. Commun.* **41,** 85–90.
502. WANKA, F. MOORS, J. and KRIJZER, F. N. C. Dissociation of nuclear DNA replication from concomitant protein synthesis in synchronous cultures of *Chlorella. Biochem. biophys. Res. Commun.* **269,** 153–161.
503. WATSON, J. D. and CRICK, F. H. C. (1953a). A structure for deoxyribose nucleic acid. *Nature, Lond.* **171,** 737–738.
504. WATSON, J. D. and CRICK, F. H. C. (1953b.). Genetical implications of the structure of deoxyribonucleic acid. *Nature, Lond.* **171,** 964–967.
505. WEEKS, D. P. and MARCUS, A. (1971). Preformed messenger of quiescent wheat embryos. *Biochim. biophys. Acta* **232,** 671–684.
506. WEEKS, D. P., VERMA, D. P. S., SEAL, S. N. and MARCUS, A. (1972). Role of ribosomal subunits in eukaryotic protein chain initiation. *Nature, Lond.* **236,** 167–168.
507. WEIGL, J. (1969). Wechsewirkung pflanzlicher Wachstumhormone mit Membranen. *Z. Naturf.* **24b.,** 1046–1052.
508. WEINTRAUB, H. (1972). A possible role for histone in the synthesis of DNA. *Nature, Lond.* **240,** 449–453.
509. WEISBLUM, B. and DAVIES, J. (1968). Antibiotic inhibitors of the bacterial ribosome. *Bact. Rev.* **32,** 493–528.
510. WELLAUER, P. K., REEDER, R. H., CARROLL, D., BROWN, D. D., DEUTCH, A., HIGASHINAKAGAWA, T. and DAWID, I. B. (1974). Amplified ribosomal DNA from *Xenopus laevis* has heterogeneous spacer lengths. *Proc. natn. Acad. Sci. U.S.A.* **71,** 2823–2827.
511. WELLS, R. and BIRNSTIEL, M. (1969). Kinetic complexity of chloroplastal deoxyribonucleic acid and mitochondrial deoxyribonucleic acid from higher plants. *Biochem. J.* **112,** 777–786.
512. WELLS, R. and INGLE, J. (1970). The constancy of the buoyant density of

chloroplast and mitochondrial deoxyribonucleic acids in a range of higher plants. *Pl. Physiol.* **46**, 178–179.

513. WELLS, R. and SAGER, R. (1971). Denaturation and renaturation kinetics of chloroplast DNA from *Chlamydomonas reinhardi*. *J. molec. Biol.* **58**, 611–622.

514. WETMORE, R. H. and RIER, J. P. (1963). Experimental induction of vascular tissues in callus of angiosperms. *Am. J. Bot.* **50**, 418–430.

514a. WHEELER, A. M. and HARTLEY, M. R. (1975). Major mRNA species from spinach chloroplasts do not contain poly(A). *Nature, Lond.* **257**, 66–67.

515. WHITEHOUSE, H. L. K. (1973). "Towards an Understanding of the Mechanism of Heredity" (3rd edition). Edward Arnold, London.

516. WILKINS, M. H. F., STOKES, A. R. and WILSON, H. R. (1953). Molecular structure of deoxypentose nucleic acids. *Nature, Lond.* **171**, 738–740.

517. WILKINS, M. H. F., ZUBAY, G. and WILSON, H. R. (1959). X-ray diffraction studies of the molecular structure of nucleohistone and chromosomes. *J. molec. Biol.* **1**, 179–185.

518. WILLIAMSON, D. H. and FENNELL, D. J. (1974). Apparent dispersive replication of yeast mitochondrial DNA as revealed by density labelling experiments. *Molec. gen. Genet.* **131**, 193–207.

519. WILSON, R. H., HANSON, J. B. and MOLLENHAUER, H. H. (1968). Ribosome particles in corn mitochondria. *Pl. Physiol.* **43**, 1847–1877.

520. WIMBER, D. E. (1966). Duration of the nuclear cycle in *Tradescantia* root tips at three temperatures as measured with H3-thymidine. *Am. J. Bot.* **53**, 21–24.

521. WOLEDGE, J., CORY, M. J. and PAYNE, P. I. (1974). Ribosomal RNA homologies in flowering plants: comparison of the nucleotide sequences in 5·8-*S* rRNA from broad bean, dwarf bean, tomato, sunflower and rye. *Biochim. biophys. Acta* **349**, 339–350.

522. WOLLGIEHN, R. (1972). RNA synthesis in isolated chloroplasts. *Symp. biol. hung.* **13**, 201–211.

523. WOOD, A. and PALEG, L. C. (1974). Alteration of liposomal membrane fluidity by gibberellic acid. *Aust. J. Pl. Physiol.* **1**, 31–40.

524. WOOD, A., PALEG, L. C. and SPOTSWOOD, T. M. (1974). Hormone-phospholipid interaction. *Aust. J. Pl. Physiol.* **1**, 167–169.

525. WOOD, H. N. and BRAUN, A. C. (1973). 8-bromo-adenosine-3′:5′-cyclic monophosphate as a promoter of cell division in excised tobacco pith parenchyma tissue. *Proc. natn. Acad. Sci. U.S.A.* **70**, 447–450.

526. WOOD, H. N., LIN, M. C. and BRAUN, A. C. (1972). The inhibition of plant and animal adenosine-3′:5′-cyclic monophosphate phosphodiesterases by a cell-division-promoting substance from tissues of higher plant species. *Proc. natn. Acad. Sci. U.S.A.* **69**, 403–406.

527. WOODARD, J., RASCH, E. and SWIFT, H. (1961). Nucleic acid and protein metabolism during the mitotic cycle in *Vicia faba*. *J. biophys. biochem. Cytol.* **9**, 445–462.

528. WOODING, F. B. P. (1968). Radioautographic and chemical studies of incorporation into sycamore vascular tissue walls. *J. Cell Sci.* **3**, 71–80.

529. WOODING, F. B. P. and NORTHCOTE, D. H. (1964). The fine structure and development of the companion cell of the phloem of *Acer pseudoplatanus*. *J. Cell Biol.* **24**, 117–128.

530. WYATT, G. R. (1951). The purine and pyrimidine composition of deoxypentose nucleic acids. *Biochem. J.* **48**, 584–590.

531. Wyn-Jones, R. G. and Lunt, O. R. (1967). The function of calcium in plants. *Bot. Rev.* **33,** 407–426.

532. Yamada, M., Miwa, M. and Sugimura, T. (1971). Studies on poly-(adenosine diphosphate-ribose). X. Properties of a partially purified poly-(adenosine diphosphate-ribose) polymerase. *Archs Biochem. Biophys.* **146,** 579–586.

533. Yarwood, A., Boulter, D. and Yarwood, J. N. (1971). Methionyl-tRNAs and initiation of protein synthesis in *Vicia faba* (L.). *Biochem. biophys. Res. Commun.* **44,** 353–361.

534. Yarwood, A., Payne, E. S., Yarwood, J. N. and Boulter, D. (1971). Aminoacyl-tRNA binding and peptide chain elongation on 80s plant ribosomes. *Phytochemistry* **10,** 2305–2311.

535. Yatsu, L. Y. and Jacks, T. J. (1968). Association of lysosomal activity with aleurone grains in plant seeds. *Archs Biochem. Biophys.* **124,** 466–471.

536. Yeoman, M. M. and Aitchison, P. A. (1973). Growth patterns in tissue (callus) cultures. *In* "Plant Tissue and Cell Culture" (H. E. Street, ed.). Blackwell Scientific Publications, Oxford.

537. Yomo, H. and Varner, J. E. (1971). Hormonal control of a secretory tissue. *In* "Current Topics in Developmental Biology" (A. A. Moscona and A. Monroy, eds), Vol. 6, pp. 111–144. Academic Press, New York and London.

538. Zachau, H. G., Dütting, D. and Feldman, H. (1966). The structures of two serine transfer ribonucleic acids. *Hoppe Seyler's Z. physiol. Chem.* **347,** 212–235.

539. Zalik, S. and Jones, B. L. (1973). Protein biosynthesis. *A. Rev. Pl. Physiol.* **24,** 47–68.

540. Zetsche, K. (1966). Regulation der UDP-Glucose 4-Epimerase Synthese in kernhaltigen und kernlosen Acetabularien. *Biochim. biophys. Acta* **124,** 332–338.

541. Zetterberg, A. and Killander, D. (1965). Quantitative cytophotometric and autoradiographic studies on the rate of protein synthesis during interphase in mouse fibroblasts *in vitro*. *Expl Cell Res.* **40,** 1–11.

542. Zielke, H. R. and Filner, P. (1971). Synthesis and turnover of nitrate reductase induced by nitrate in cultured tobacco cells. *J. biol. Chem.* **246,** 1772–1779.

543. Zwar, J. A. and Jacobsen, J. V. (1972). A correlation between a ribonucleic acid fraction selectively labeled in the presence of gibberellic acid and amylase synthesis in barley aleurone layers. *Pl. Physiol.* **49,** 1000–1006.

Index

327

M

In the search for solutions to the world food shortage, the study of the biochemistry of plant genes and their function is rapidly assuming great importance. Scientists are becoming aware that the genetic content of plant cells and its expression may be artificially modified so as to stimulate the production of more plentiful and higher quality crop yields. It is clearly imperative that those who will be engaged in trying to improve yields have an understanding of the biochemistry of plant genes and especially of their expression.

This is the first book to deal specifically with gene expression in plants; as a result, advanced undergraduates, research workers including postgraduates, and also lecturers working in the fields of botany, plant physiology, molecular biology, biochemistry and genetics should find this volume particularly welcome. Not only will it be appreciated for its value as a work of reference on the subject but also for its synthesis of currently available data. The inclusion of some of the authors' own ideas should prove a stimulating addition.

The first three chapters describe the physico-chemico properties of DNA and RNA and show how these properties relate to the occurrence and role of these molecules in living plants. They also deal with the process of gene expression and its control at the biochemical level. The fourth chapter discusses the genomes and protein synthesising systems of the organelles involved in energy metabolism. The final chapters relate gene expression to plant growth, development and differentiation and link the biochemical controls of gene expression to the growth and physiology of the plant.

Illustration

Field trials with different varieties of rice (photograph courtesy of the International Rice Research Institute, Philippines) and a microdensitometer trace illustrating the fractionation by poly-acrylamide gel electro-phoresis of the nucleic acids extracted from dark-grown pea seedlings (trace courtesy of Miss Sally Greenway).